Earthquake Geotechnical Engineering Design

GEOTECHNICAL, GEOLOGICAL AND EARTHQUAKE ENGINEERING

Volume 28

For further volumes:
http://www.springer.com/series/6011

Michele Maugeri • Claudio Soccodato
Editors

Earthquake Geotechnical Engineering Design

Springer

AGI
Associazione Geotecnica Italiana

Editors
Michele Maugeri
Department of Civil and Environmental Engineering
University of Catania
Catania, Italy

Claudio Soccodato
General Secretary, Italian Geotechnical Society
Rome, Italy

ISSN 1573-6059 ISSN 1872-4671 (electronic)
ISBN 978-3-319-03181-1 ISBN 978-3-319-03182-8 (eBook)
DOI 10.1007/978-3-319-03182-8
Springer Cham Heidelberg New York Dordrecht London

Library of Congress Control Number: 2014931392

Printed on acid-free paper

Springer is part of Springer Science+Business Media (www.springer.com)

Foreword

Earthquake Geotechnical Engineering has developed greatly in the last 30 years since the First International Conference on Recent Advances in Geotechnical Earthquake Engineering and Soil Dynamics that was held in St. Louis in the year 1981. During the opening address of the conference given by D. E. Hudson, he said: "I began to wonder if Geotechnical Earthquake Engineering is in fact any different from Earthquake Engineering. Even for the detailed problems of steel and concrete structural design, the importance of soil-structure interaction may be a critical matter. Geotechnical Engineering is indeed the foundation on which the whole subject is built."

The importance of the soil characterisation in the local site seismic response was outlined some years before in 1974 by H. B. Seed, who grouped 104 earthquake recordings in four soil categories – rock, hard soil, alluvial sandy soil and soft clay – showing the significant amplification of the records on the rock going from hard soil to soft clay.

Following the first conference in St. Louis in 1981, some other International Conferences devoted to Earthquake Geotechnical Engineering and Soil Dynamics were held in many parts of the world. Among these, recently, three International Conferences have been held; two in Italy, one of the UE countries more prone to seismic risk and one in Japan. In 2008, an International Conference commemorating the 1908 Messina and Reggio Calabria Earthquake was held in Reggio Calabria by the Italian Geotechnical Society (AGI). During the conference, K. Ishihara in its inaugural lecture showed that the peak ground design acceleration has been increased from 0.2 to 0.3 g in the 1970s up to 0.6–0.8 g at the present time. Because of that there is a need to shift from pseudo-static approach to performance based design. This topic has been debuted in the International Conference on Performance Based Design in Earthquake Geotechnical Engineering held in Tokyo in 2009. Following this conference, the Second International Conference on Performance Based Design in Earthquake Geotechnical Engineering was held in Taormina (Italy) last year, organised by the TC-203 and ETC-12 of the ISSMGE and by AGI.

Earthquake geotechnical engineering design is progressing very well in the last 5 years. This design is made in several steps. The first one is the evaluation of the site amplification, which is reported in Part I of this book. The second one is related to the evaluation of the soil foundation stability against natural slope failure and liquefaction (Part II). Following these two steps, the volume is devoted to the design of geotechnical works under earthquake loading. Part III is devoted to the design of levees and dams (including natural slopes). Part IV is devoted to the design of foundations and to soil structure interaction analysis; Part V is devoted to underground structures and Part VI is devoted to the new topics, i.e. to the design of reinforced earth retaining wall and landfills.

Part I is subdivided in two chapters and is related to the evaluation of site characterisation and site amplification.

Chapter 1 deals with the site characterisation, based on in situ measurements of shear wave velocity. A new, reliable shear wave velocity model is presented as a prerequisite for the assessment of seismic site response at regional and local scale. The model is based on surface wave test because of their cost effectiveness and efficiency. Three case histories, with pseudo 2D/3D shear waves velocity models, in different geological settings and with different characteristics of the experimental dataset are reported and discussed.

Chapter 2 reports a global dataset of more than 3,000 ground motion records from 536 sites from Greece, Italy, Turkey, USA and Japan, used to propose a new site classification system and soil amplification factors for elastic acceleration response spectra. The new classification system incorporates parameters such as the thickness of soil deposits, the average shear wave velocity to the seismic bedrock and the fundamental period of the site. The dataset is also used to derive soil amplification factors and to compare it with the soil classes amplification of Eurocode 8 (EC8).

Part II is subdivided in three chapters, all devoted to the liquefaction that occurred extensively during the recent earthquakes of 2011 Tohoku-Pacific Ocean earthquake in Japan.

Chapter 3 reports on the liquefaction that occurred in the reclaimed land in Kanto area, located more than 200 km away from the earthquake fault, in soil containing more 50 % of non-plastic fines. Also almost all sand deposits along the Tokyo bay area reclaimed in 1960s or later liquefied, while in good contrast, those older than that did not. To investigate the ageing effect on liquefaction strength of sands containing fines, a series of basic laboratory tests combining innovative miniature cone penetration and subsequent cyclic undrained loading were carried out in a modified triaxial apparatus on sand specimens containing fines.

Chapter 4 reports also on liquefaction in Tokyo Bay and Kanto Region. The Tone River along the Tokyo Bay suffered the liquefaction-associated damage, despite being located at a distance of about 450–500 km from the epicenter of the earthquake. As a measure to estimate its destructiveness, the ground settlements resulting from liquefaction were calculated based on volume decrease characteristics of sandy soils and their outcome was compared with the settlements actually observed on the ground surface.

Chapter 5 is devoted to settlement as inclination of buildings caused by the earthquake. In Tokyo Bay area, the very long duration of the main shock and an aftershock 29 min later probably caused serious settlement and inclination of houses. The maximum inclination in Urayasu City was about 6 %; it is difficult to live in these buildings after the earthquake though no damage to walls and windows were observed. Tilting of buildings are due to differential settlements. Several factors affected the non-uniform settlement. Among differential factors affecting the differential settlement, the effect of adjacent houses was dominant.

Part III is subdivided in two chapters, devoted to river levees, dams and related artificial slopes.

Chapter 6 analyses the damage caused to river levees by the 2011 Tohoku-Pacific Ocean earthquake in Japan. To mitigate the internal liquefaction inside the levees, not only the field investigation technique but also numerical analysis for performance prediction has to be newly developed.

Chapter 7 is devoted to earthquake performance design of dams using destructive potential factors. The seismic performance design of dams is based on good estimates of sliding displacements and crest settlements. Theoretical result has shown a good correlation between sliding displacement of slope and destructiveness potential factor PD. Crest dam vertical settlement recorded from seismic performance in real Chile earthquakes by destructive potential evaluation confirms the good forecast of the theoretical values obtained in terms of PD compared with PGA.

Part IV is subdivided in three chapters, devoted to foundations and soil-structure interaction.

Chapter 8 deals with an interesting performance-based design, related to the macro-element concept which seems to be very promising. The macro-element theory in its different versions (elasto-perfectly plastic, elasto-strain-hardening plastic, bounding surface plastic and hypo-plastic) is first introduced and the mechanical response of shallow foundations under monotonic/cyclic loading, as it results from experimental tests, is outlined.

Chapter 9 is devoted to large-scale modelling of ground and soil-structure earthquake response. A strategy to incorporate ground response and soil-structure interaction (SSI) is implemented based on the domain reduction method (DRM) for three-dimensional earthquake simulation. Vibration properties and seismic response behaviour for the connector and the soil domain are examined. Different scenarios of bridge response are considered and compared, including fixed-base uniform excitation, and multiple-support excitation with and without the full ground/foundation soil domain.

Chapter 10 outlines the relevance of soil structure interaction in seismic displacement based design of structures. A brief summary of the seismic design method known as "direct displacement–based design" (DDBD) is presented, with some discussion of the appropriate seismic design input to be used, on the applicability to retaining structures and on the relevance of including nonlinear soil-structure interaction in the DDBD and the tools to account for it, with reference to shallow foundations.

Part V is devoted to the performance and seismic design of underground structures. Chapter 11 analyses the seismic performance during past earthquakes of underground structures, tunnels, subways, metro stations and parking lots. A general presentation of the methods used for the seismic design of underground structures is presented and discussed. The main issues discussed herein cover the following topics: (i) force based design against displacement based design, (ii) deformation modes of rectangular underground structures under seismic excitation, (iii) seismic earth pressures on underground structures, (iv) seismic shear stresses distribution on the perimeter of the structure, (v) appropriateness of the presently used impedance functions to model the inertial and the kinematic soil-structure interaction effects, (vi) design seismic input motion, taking into account the incoherence effects and the spatial variation of the motion and (vii) effect of the build environment (i.e. city-effects) on the seismic response of underground structures.

Part VI is subdivided in two chapters and it is related to the new topics reinforced earth wall and landfills.

Chapter 12 is devoted to the behaviour of reinforced soil walls during the 2012 Tohoku earthquake in Japan. Approximately 1,600 case histories on the seismic performance of such walls during the 2011 Great East Japan Earthquake were collected and analyzed. Statistical data on the seismic damage revealed that all types of reinforced soil walls performed well during the 2011 earthquake. A case study closely investigated was discussed in terms of economical solution and in terms also of performance required.

Chapter 13 deals with the performance based seismic design of geosynthetic barriers for waste containment. A performance-based methodology for seismic analysis and design of the geosynthetic elements of waste containment systems, including landfills and heap leach pads, has been developed. The methodology offers a rational alternative to the current state of practice for seismic design of geosynthetic containment system elements in which a decoupled Newmark-type displacement analysis is used to calculate a permanent seismic displacement. This calculated displacement is generally considered to be an index of the performance of the containment system during an earthquake. However the Newmark-type design methodology does not gives explicit evaluation of the stresses and strains in the geosynthetic elements of the containment system. So, a finite difference model of waste-liner system interaction has been developed using the computer code FLACTM. This analysis provides a basis for direct performance based seismic design of geosynthetic elements not only in waste containment systems but in a variety of other civil structures that employ geosynthetic elements wherein earthquake ground motions cause relative displacement between the geosynthetic element and the surrounding soil.

The contents of the chapters have been presented as state of art (SOA) papers or keynote (KN) papers at the Second International Conference on Performance Based Design in Earthquake Geotechnical Engineering in Taormina. To the topic of performance based design in earthquake engineering will be also devoted a special number of *Bulletin of Earthquake Engineering* with some others SOA and KN delivered during the Taormina Conference.

The guest editors are indebted to the Italian Geotechnical Society (AGI) for being the main sponsor for the publication of this volume.

Catania, Italy Michele Maugeri
Rome, Italy Claudio Soccodato

Contents

Part I
Site Characterisation and Site Amplification

Chapter 1
Spatially Constrained Inversion of Surface Wave Data to Build Shear Wave Velocity Models

Sebastiano Foti and Laura Valentina Socco

Abstract A reliable shear wave velocity model is a prerequisite for the assessment of seismic site response at regional and local scale. 2D or 3D velocity models can be obtained by interpolation of tests at different locations. Surface wave tests are often used in this context because of their cost effectiveness and efficiency. As any characterization method based on the solution of an inverse problem, surface wave analysis suffers from solution non-uniqueness. Laterally constrained inversion provides a robust framework for the interpretation of surface wave data in the perspective of building pseudo-2D/3D shear wave velocity models. In the chapter three case histories in different geological settings and with different characteristics of the experimental dataset are reported and discussed.

1.1 Introduction

Accurate predictions of seismic site response and soil-structure interaction are particularly important for performance base design. Modern computational tools allow advanced numerical simulations to be performed. Indeed, efficient codes have been established for the simulations of wave propagation at the basin scale. At a different scale, a direct evaluation of the soil-structure interaction under seismic loads is required for assessing the performance of a single construction. In both cases, the reliability of the result is strongly dependent on the availability of an accurate shear wave velocity model, which has to be obtained with geophysical tests.

S. Foti (✉)
Department of Structural Geotechnical and Building Engineering,
Politecnico di Torino, Torino, Italy
e-mail: sebastiano.foti@polito.it

L.V. Socco
Department of Environment, Land and Infrastructure Engineering,
Politecnico di Torino, Torino, Italy

M. Maugeri and C. Soccodato (eds.), *Earthquake Geotechnical Engineering Design*, Geotechnical, Geological and Earthquake Engineering 28,
DOI 10.1007/978-3-319-03182-8_1,

Surface wave methods are appealing for the characterization. Indeed they are very cost effective and flexible as they do not require drilling of boreholes and they can be performed with light and portable equipment. Moreover they allow a significant volume of the subsoil to be tested, which is a prerequisite for the construction of a representative model. On the other side it has to be considered that the interpretation of surface wave data is quite complex as it requires advanced signal processing tools and the solution of an inverse problem. The latter is inherently ill-posed from a mathematical point of view. The main consequence is the non-uniqueness of the solution as different profiles may honor equally well the available experimental information, i.e. they are equivalent solutions of the inverse problem. The consequences of solution non-uniqueness for seismic site response in the simple case of horizontally layered media have been studied by Foti et al. (2009). To reduce the uncertainties associated to solution non-uniqueness, it may be useful to introduce additional constraints, such as a-priori information from borehole logs or from other geophysical methods (Socco et al. 2010a; Foti 2013).

The interpretation of surface wave data is typically based on a horizontally stratified model with linear (visco-)elastic homogeneous layers. The obtained solution is a 1D profile of shear wave velocity versus depth. Most attempts to obtain a 2D or 3D shear wave velocity model are based on the fusion of 1D profiles from multiple realizations of the test at adjacent locations. This is easily implemented with moving receiver arrays.

Considering the uncertainties associated to solution non-uniqueness, the possibility of linking adjacent profiles with a constrained inversion can lead to a more robust and reliable model. This strategy, initially proposed for vertical electrical soundings by Auken and Christiansen (2004), has proven to be very effective on surface wave data (Socco et al. 2009). Moreover the Laterally Constrained Inversion (LCI) can also integrate available a-priori information and/or data from other geophysical tests (Wisèn and Christiansen 2005; Garofalo et al. 2012).

More advanced approaches for getting 2D models, as for example full wave waveform inversion, are very promising, but still under development (e.g. Tran and Hiltunen 2012). Issues to be resolved are in particular related to the very high computational requirements and to the need for adequate strategies for the regularization of the solution.

After a discussion on the use of surface wave data for the construction of pseudo 2D-3D shear wave velocity models and an outline of the Laterally Constrained Inversion approach, this chapter reports three applications to site characterization for seismic projects.

1.2 Surface Waves Methods and Lateral Variations

The analysis of the propagation of surface waves has been initially exploited for the characterization of the Earth's crust in seismological studies (Aki and Richards 2002; Romanowicz 2002). First applications in engineering date back to late 1950s

(Jones 1958), but the diffusion of the technique started with the introduction of the SASW (Spectral Analysis of Surface Waves) method (Nazarian and Stokoe 1984). Nowadays surface wave methods have gained a wide popularity, especially in the fields of applied seismology and near surface geophysics. A detailed discussion of the fields of application and of current trends is reported by Socco et al. (2010b).

Active-source methods are typically preferred for shallow applications. Passive methods based on microtremors allow the medium to be characterized at depth without the need for heavy and costly sources. An overview on the various methods for surface wave analysis is provided by Foti et al. (2011).

The propagation of surface waves is influenced by the physico-mechanical properties of a zone of limited thickness close to the ground surface. Monitoring of particle motion on the ground surface can be used for the solution of an inverse problem aimed at the characterization of the subsoil. Although a full-waveform inversion is in principle feasible (e.g. Tran and Hiltunen 2012), most implementations are based on the evaluation of an experimental dispersion curve, i.e. the relationship phase velocity of surface waves versus frequency. The inverse problem is then solved assuming a stack of linear elastic homogeneous layers as reference model. This assumption has strong implications in presence of lateral variations. Indeed in such conditions there is a discrepancy between the model and the actual medium. The latter is eventually represented as a layered medium with fictitious parameters.

In general, the results of surface wave tests are representative of the subsoil below the whole length of the testing array, for a depth which is linked to the maximum available wavelength (hence proportional to the inverse of the lowest usable frequency in the dataset). This aspect has also to be taken into consideration when comparing the results of surface wave methods with those of intrusive methods such has seismic cross-hole or down hole tests. Indeed the latter are local measurements which are representative only of soil behaviour in the limited space between the source and the receiver (or first to second receiver).

A set of surface wave tests for different positions of the array can be used to obtain a sequence of 1D profiles, which can be considered a pseudo-2D/3D model. In practice the subsoil is modeled locally as an horizontally layered medium (Fig. 1.1). Such a procedure can be implemented with different strategies:

- a series of independent acquisitions with conventional arrays at nearby locations;
- a moving array (streamer) towed by a vehicle on land or by a vessel for marine applications;
- subsets of data extracted with a moving spatial window from long seismic surveys.

Land streamers are widely used because they allow data to be collected in a fast and effective manner. A string of geophones is mounted on a rough support that allows it to be towed by a vehicle on land (van der Veen et al. 2001; Vangkilde-Pedersen et al. 2006). The string is moved at different positions and the test is repeated for each position. Typically a certain overlap is used between different

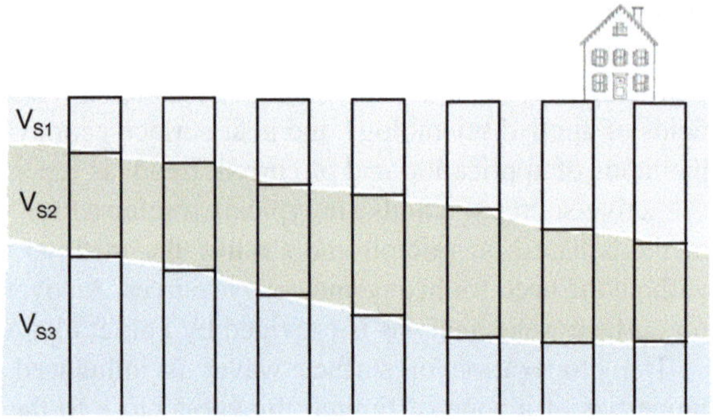

Fig. 1.1 Local approximation of a medium with lateral variations through a series of 1D models obtained with surface wave methods (pseudo-2D model)

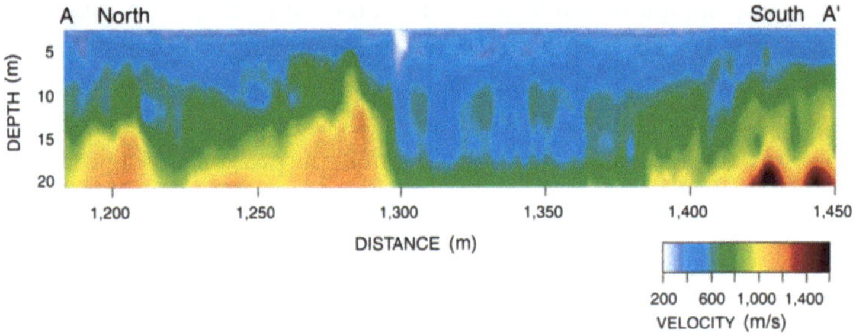

Fig. 1.2 Example of graphical representation of a pseudo-2D surface wave survey as a shear wave velocity map, obtained from the interpolation of the set of 1D profiles. A fault is identified as the low velocity zone in the middle of the profile (Ivanov et al. 2006)

positions to obtain a "continuous" reconstruction. The concept of land streamer has been borrowed from marine applications, in which a string of hydrophones is towed by a vessel for the acquisition of seismic data.

Results from adjacent and overlapped 1D tests are often represented in terms of velocity maps (Fig. 1.2), obtained via interpolation. Although effective from a graphical point of view, this representation may give the false perception that a true 2D reconstruction is obtained.

A major source of uncertainty in pseudo-2D velocity models is solution non-uniqueness, which affects the results of each 1D inversion. Figure 1.3 reports an example of equivalent profiles obtained with a Monte Carlo inversion of surface wave data. As shown in Fig. 1.3b, the theoretical dispersion curves for the shear wave velocity profiles of Fig. 1.3a are very close to the experimental dispersion curve. Accounting also for the unavoidable uncertainties in the experimental measurements and their processing, the best fitting profile has to be considered as a possible solution rather than the only solution of the inverse problem. This aspect is often masked by the use of local search methods (e.g. the least square algorithm) for the solution of the inverse problem. With local search methods the solution may be trapped in a local minimum and it is not possible to evaluate the occurrence of equivalent solutions that may fit equally well the experimental data.

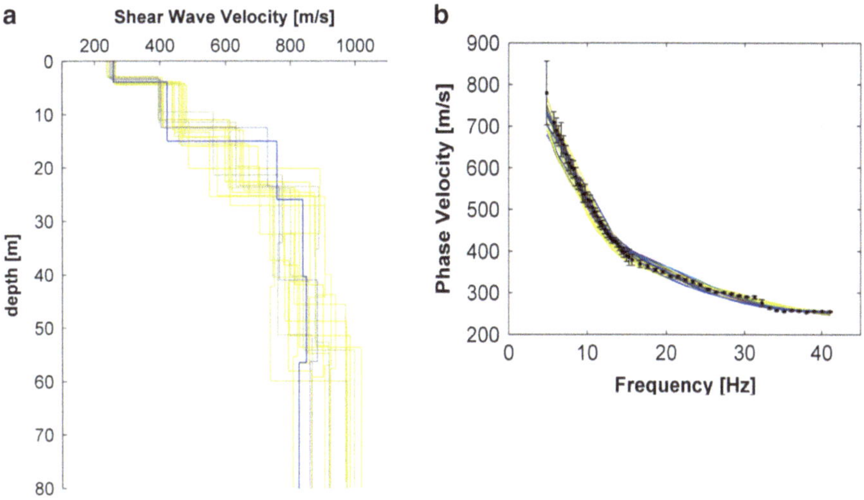

Fig. 1.3 Example of uncertainty due to solution non-uniqueness in surface wave analysis: (**a**) set of statistically equivalent shear wave velocity profiles; (**b**) theoretical dispersion curves for the equivalent profiles compared to the experimental dispersion curve (*black dots*) and its related uncertainty (*black error bars*). A color scale with darker color corresponding to better fitting is adopted for both the equivalent profiles and their associated theoretical dispersion curves (Foti et al. 2009)

Uncertainties from solution non-uniqueness are influenced by several factors, for one the availability of experimental information over a wide enough frequency range. With active-source tests, the resolution is high and the uncertainty is low close to the ground surface because it is easy to collect high frequency data. Additional information, e.g. from other geophysical tests or from conventional borehole logs, may be helpful to reduce the uncertainty by introducing further constraints in the inversion (e.g. Socco et al. 2010a; Foti 2013). In particular, the joint inversion of different datasets is a promising strategy for increasing the reliability of the solution (Hering et al. 1995; Hayashi et al. 2005; Arai and Tokimatsu 2005; Piatti et al. 2012). For example, joint inversion of surface wave dispersion curves and H/V spectra for single stations may lead to significant improvements in the identification of the seismic bedrock (Scherbaum et al. 2003).

In the case of adjacent 1D surveys, constraints may be introduced between neighboring models by linking model parameters. This strategy has been originally proposed by Auken and Christiansen (2004) for the inversion of vertical electrical soundings. Applications to surface wave testing (Fig. 1.4) have been subsequently proposed by Wisèn and Christiansen (2005) and Socco et al. (2009). The inversion is performed by minimizing a single misfit function, which contains:

- The norm of the distance between experimental and theoretical dispersion curves;
- The constraints between adjacent models;
- Available a-priori information.

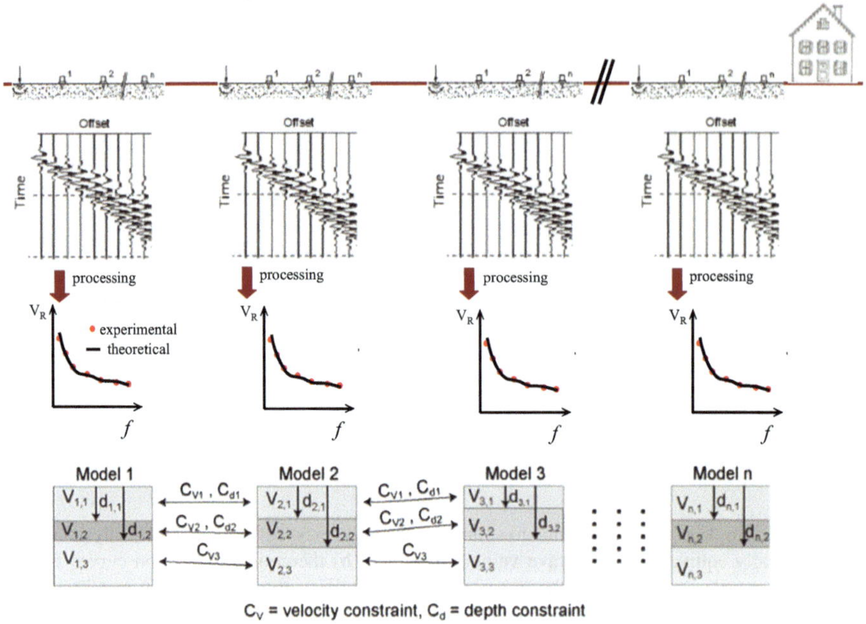

Fig. 1.4 Schematic outline of laterally constrained inversion of surface waves

The final result is a group of 1D velocity models which accounts for all the above information. Hence in principle it is more robust than single inversions which only account for the information on surface wave propagation. The crucial point is the strength of the constraints between adjacent models. With very weak constraints, the solution is very close to the conventional solution obtained with independent inversion for each profile. On the other end, very strong constraints lead to a solution which is close to the assumption of a single horizontally layered medium for the whole investigated volume. The choice of constraint strength should take into account also geological information about expected lateral variations at the site. An usual strategy relies on multiple preliminary inversions with different levels of constraint. The optimal strength is then selected as the maximum value that does not affect significantly the misfit on the dispersion curve for the single profiles. Indeed the best fitting solution for each individual dataset is necessarily obtained without lateral constraints. Increasing the constraint strength, the misfit evaluated for each single dispersion curve is expected to increase.

Imposing lateral constraints may not be adequate in situations in which sharp boundaries are present in the subsoil, e.g. when in presence of a buried fault. In such conditions, lateral constraints have to be imposed on the sides, allowing for the discontinuity. Several approaches for recognizing the presence of a sharp discontinuity in long seismic surveys have been tested by Bergamo (2012). A possible strategy is based on the evaluation of phase difference versus offset (Strobbia and Foti 2006) which in presence of a sharp boundary should be characterized by two different trends.

Validation of the laterally constrained inversion of surface waves with synthetic data is reported by Boiero and Socco (2010) showing that it is possible to reconstruct reliably lateral variations also in presence of topographic variations. Validation on a physical model at laboratory scale has been performed by Bergamo (2012).

1.3 Case Histories

The three case histories are related to projects in zones of medium to high seismic hazard, for which the evaluation of seismic site response is relevant. Three different geological settings are considered: an alluvial fan, which is typical of alpine regions; a flat depositional environment, with marked stratigraphic discontinuities; an alluvial basin of limited extension, which represents a condition in which 3D stratigraphic effects may a play very important role, justifying the necessity of advanced numerical simulations.

The three examples are also different for the type of seismic dataset. In the first case, the surface wave data have been extracted from a dataset collected for a seismic reflection survey. In the second case, active-source multistation arrays have been used at different sites along an alignment for a characterization aimed at the shallow zone. In the third example, a combination of active-source and passive tests at several sites has been used, accounting for the required depth of exploration and the availability of testing space in an highly urbanized context.

1.3.1 La Salle

The town of La Salle (Val d'Aosta) is located on a wide triangular alluvial fan lying on the left side of the Dora Baltea river. The fan is about 1.5 km long and almost 2.5 km wide at the base (Fig. 1.5). The maximum thickness of the Quaternary

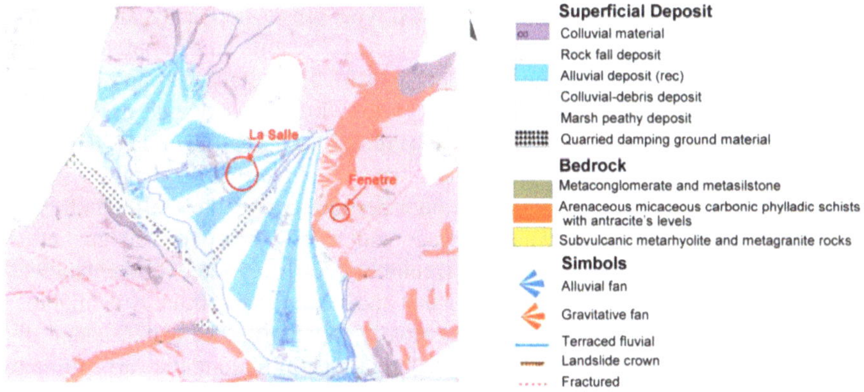

Fig. 1.5 Geological map of the La Salle area (Socco et al. 2008)

deposits is expected to be around 200 m. The fan is mainly composed of alluvial deposits (sands, gravels, pebbles), polygenic slivers, stones and blocks. The stratigraphic log to a depth of about 50 m shows the typical chaotic sequences of alpine alluvial fans, with no marked layering. This is a typical example of soils with a very wide granulometric distribution, which are very difficult to characterize from the geotechnical point of view. Geophysical seismic methods are among the few, if not the only, available tools to estimate mechanical parameters.

From the seismic hazard point of view, the region is subjected to moderate but frequent activity.

Within a project for the estimation of the seismic response of alpine valleys, a seismic characterization campaign was performed by collecting three groups of seismic datasets (Fig. 1.6):

- Two Down Hole (DH) tests up to a depth of 50 m;
- Surveys based on active and passive surface wave data at five sites;
- Two high resolution seismic reflection surveys along two crossing lines.

With respect to the latter, the ground roll (surface wave components) in seismic reflection records was subsequently processed and inverted with the laterally constrained algorithm to supply a pseudo-2D model of shear wave velocity (Socco et al. 2009). In the following we will refer mainly to this dataset, while the down-hole tests and the active-passive surface wave tests will be considered only for comparison with the final result of the laterally constrained inversion. More details on the full dataset are reported by Socco et al. (2008).

Data collected for seismic reflection surveys are often rich of surface wave components, which are considered detrimental as they tend to mask the arrivals of reflected waves. Surface wave components are indicated with the term ground roll and several techniques have been developed to filter them out (Sheriff and Geldart 1990). Some signal processing techniques for the evaluation of the experimental dispersion curve in surface wave analysis have been initially conceived for muting ground roll (e.g. McMechan and Yedlin 1981).

Since in seismic reflection surveys the acquisition is not optimized for surface-wave data analysis, it is necessary to check in first instance if the experimental data fulfill some requirements. Indeed, the source type, sensor frequency, and sampling in time and space may not be adequate. Figure 1.7 reports an example of seismic data collected at La Salle along Line 1 (Fig. 1.6). The source was an accelerated weight-drop system acting at the ground level, which produced an experimental dataset rich in surface wave components (Fig. 1.7). On the other side, data were collected using 14 Hz vertical geophones, hence the investigation depth in surface wave analysis is limited because of the poor performance of the sensors in the low frequency range. The good spatial coverage with a dense receiver array guarantees an adequate spatial resolution in the wavenumber domain. Moreover the data were collected with long time windows, so that the whole wave train is available, a prerequisite for the analysis of surface wave components. The experimental dispersion curves have been extracted for different locations along the survey line by

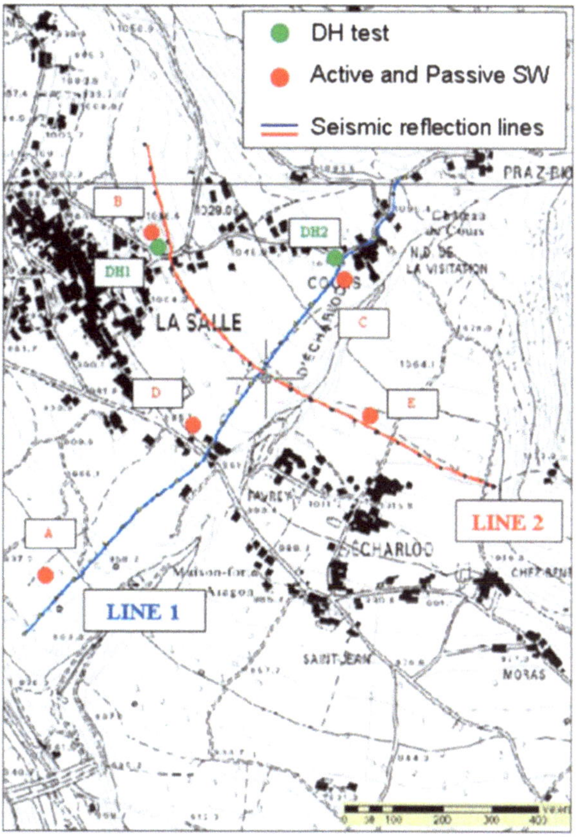

Fig. 1.6 Survey locations at La Salle: Down-Hole tests (DH) at sites DH1 and DH2; specific surface wave tests with active-source and passive data at sites A to E; seismic reflection surveys along Line 1 and Line 2 (Socco et al. 2008)

sliding a spatial window (Fig. 1.7). Details on the choice of window length and of the overlap between adjacent windows are reported in Socco et al. (2009).

Data have been analysed in the frequency-wavenumber domain with an automated procedure to get spectral maxima which are associated to the experimental dispersion curve (Fig. 1.8a). Thanks to the availability of a multifold seismic dataset in which experimental data were collected for shots at different positions, a series of experimental surface-wave dispersion curves have been extracted along the seismic line (Fig. 1.8b). Statistical analysis of the extracted experimental dispersion curves for the same spatial window (Fig. 1.8b) leads to an estimate of the uncertainties associated to the best estimate of the dispersion curve (Fig. 1.8d). The latter has been obtained from the stacked frequency-wavenumber spectrum (Fig. 1.8c) which includes information from all the shots.

The set of experimental dispersion curves obtained for the spatial windows along Line 1 is reported in Fig. 1.9. The curves are enclosed in a limited range of variation, confirming that the lateral variability at the site is limited as expected in the light of the geological setting. Indeed the alluvial fan has a chaotic structure at the scale of a representative elementary volume, but at a wider scale it can be considered a single homogeneous body.

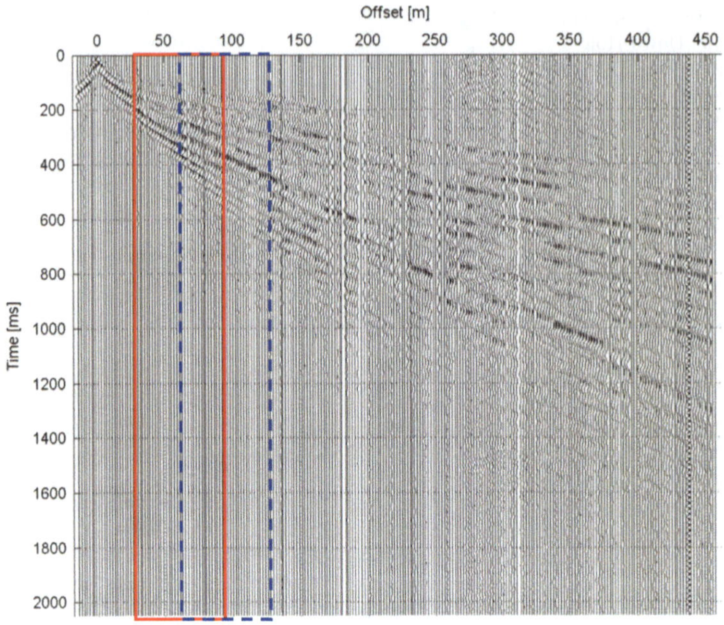

Fig. 1.7 Example of experimental data at La Salle site along Line 1 for a given shot location. The *two rectangles* represent examples of the sliding window used to extract surface wave dispersion data along the survey line

The peculiarity of the alluvial fan has another very important consequence on the solution of the inverse problem. Indeed in such a geological setting, there are no well defined boundaries between layers and the variation of shear wave velocity with depth is mainly associated to the increase of confining stress. As the solution of the forward problem of Rayleigh wave propagation assumes a layered structure with linear elastic homogeneous layers, the inversion has been performed assuming a large number of constant thickness layers and introducing a regularization criterion. With this strategy the unknowns are only the shear wave velocities. The solution for each model is linked to adjacent models by the spatial constraints which provide further regularization of the solution, which is desired considering the large number of unknowns. Layer thicknesses have been set a priori and they increase with depth to account for the variable resolution of surface wave methods. Close to the ground surface the resolution is very high, as a large amount of information is available in the high frequency range.

The solution of laterally constrained inversion for Line 1 is reported in Fig. 1.10. The solution is composed by a set of 1D shear wave velocity profiles. The parameters of each model are constrained by the experimental dispersion curve of Rayleigh waves and by the link to adjacent models (Fig. 1.4). For each profile, also the normalised residual of the misfit function (i.e. the norm of the difference between the experimental and theoretical dispersion curves) is reported. An independent inversion of each experimental dataset would have produced lower values

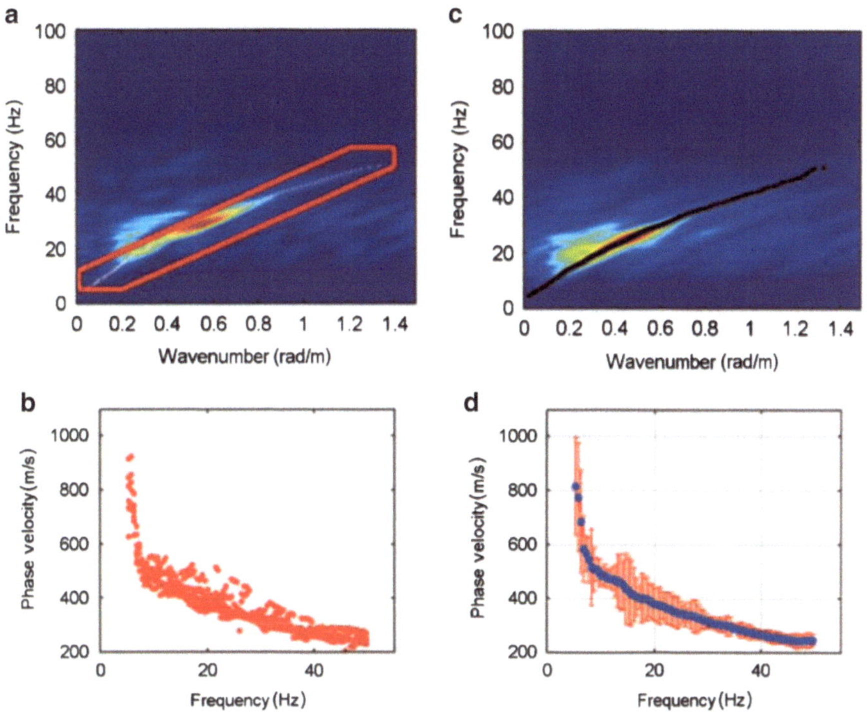

Fig. 1.8 Example processing results for a given spatial window: (**a**) f-k spectrum for a single shot. The *red line* encloses the search region for spectral maxima; the latter are indicated by the *white dots*; (**b**) Dispersion curves obtained from different shot locations; (**c**) stacked f-k spectrum obtained combining the data from different shot locations. The *black dots* represent the maxima corresponding to the best estimate of the dispersion curve; (**d**) best estimate of the experimental dispersion curve (*blue dots*) and its uncertainties (*red bars*), which are estimated as the standard deviation of the dispersion curves in panel (b) (Socco et al. 2009)

Fig. 1.9 La Salle site, experimental dispersion curves for Line 1. The curves are superimposed, but each curve corresponds to a different spatial position of the sampling window along the line (Socco et al. 2009)

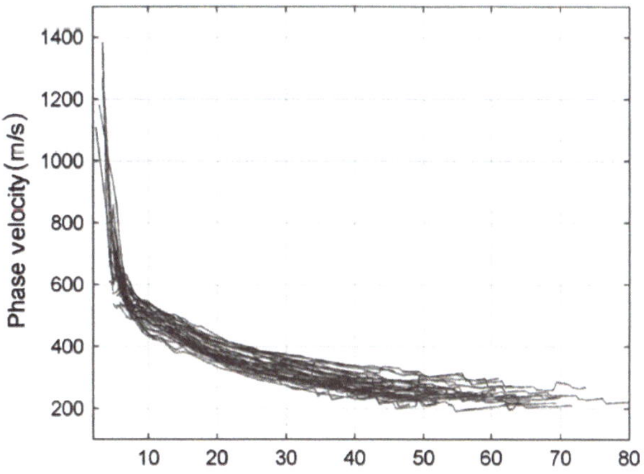

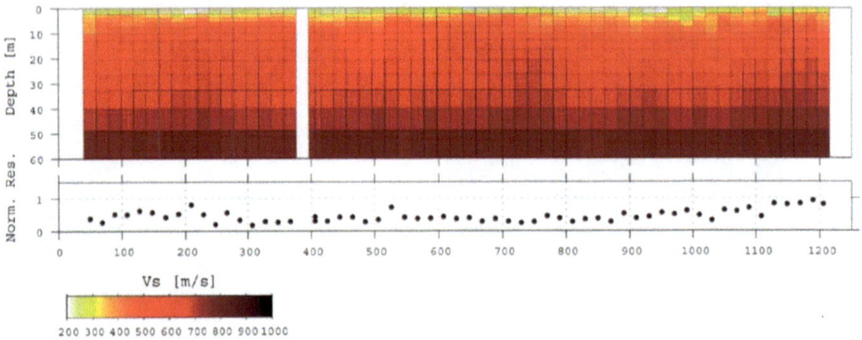

Fig. 1.10 Laterally and vertically constrained inversion results for Line 1: for each shear wave velocity profile, the normalised residuals between theoretical and experimental dispersion curves are also reported (Socco et al. 2008)

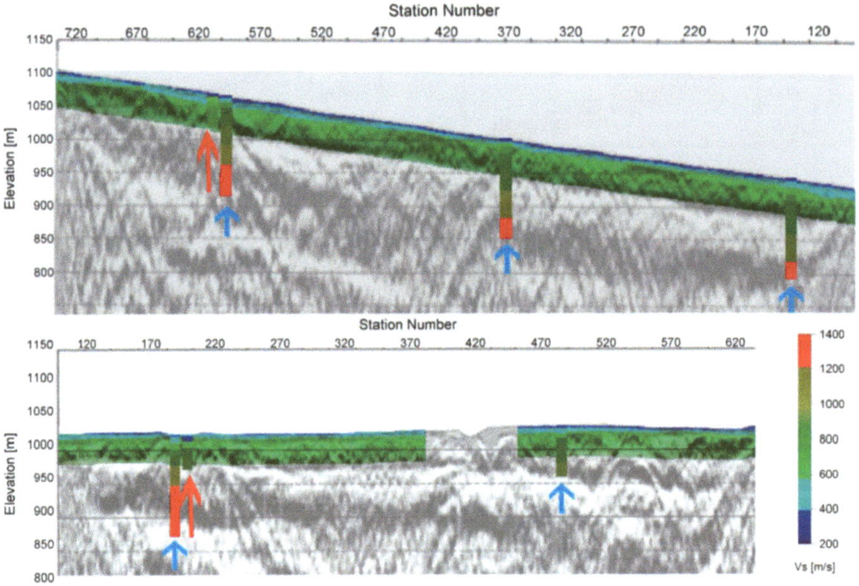

Fig. 1.11 Shear wave velocity model obtained through LCI of surface wave data extracted from seismic reflection records compared with seismic reflection results. Also 1D Vs profiles obtained with DHTs (*red arrows*) and SW active and passive tests (*cyan arrows*) are reported. *Top*: Line 1; *Bottom*: Line 2. (Socco et al. 2008)

of the normalised residuals. The lateral constrained inversion accounts for the whole ensemble of available data and produces a robust estimate of the pseudo-2D shear wave velocity model.

Finally the set of 1D models has been used to obtain a 2D map of shear wave velocity. The maps for Line 1 and Line 2 are reported in Fig. 1.11, superposed on

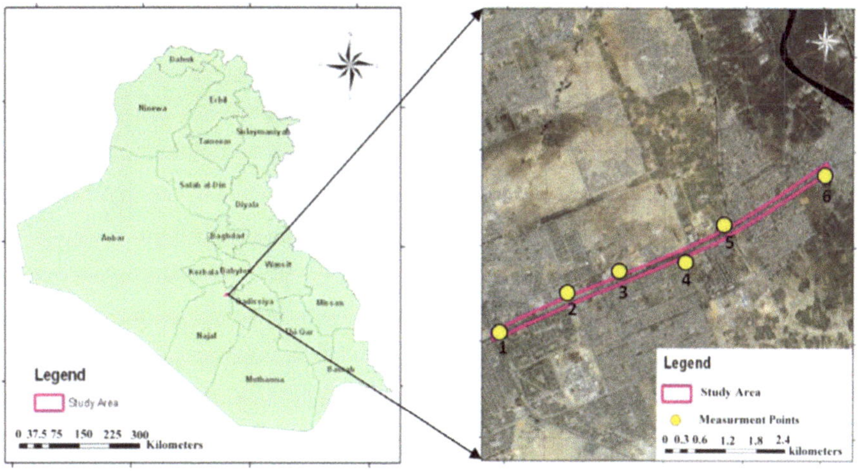

Fig. 1.12 Map of Iraq showing the study area and the position of the testing sites (Shakir et al. 2013)

the processed reflection sections. In the same graph, the 1D shear wave velocity profiles from down-hole tests and active-passive surface wave tests are reported (Socco et al. 2008). The 2D shear wave velocity model is in good agreement with punctual information obtained with other tests. The depth of investigation is limited because of the limited availability of low frequency data, caused by the use of vertical geophones with high natural frequency. The combination of active-source and passive surface wave tests allow the characterization to a larger depth. A good correspondence is found between the position of the last interface in 1D shear wave velocity profiles and the reflections from a deep interface. Combining the two pieces of information it can be postulated that the competent bedrock below the alluvial fan is at about 110 m depth.

1.3.2 Najaf

The second case history is related to site characterization along the proposed route for a subway in Najaf city (IRAQ). The location of the study area is reported in Fig. 1.12, while Fig. 1.13 reports the schematic stratigraphy as obtained from borehole logs for the sites indicated in Fig. 1.12. In this case, the geologic setting is composed by a sequence of depositional layers with different characteristics.

Surface wave data have been collected with a multichannel active-source setup at each site. Data have been processed in frequency-wavenumber domain to estimate the experimental dispersion curve (Fig. 1.14). At each site two different setups have been used: a long receiver spread with 12 geophones and inter-receiver distance 12.5 m has been used to collect low frequency data, which are necessary to

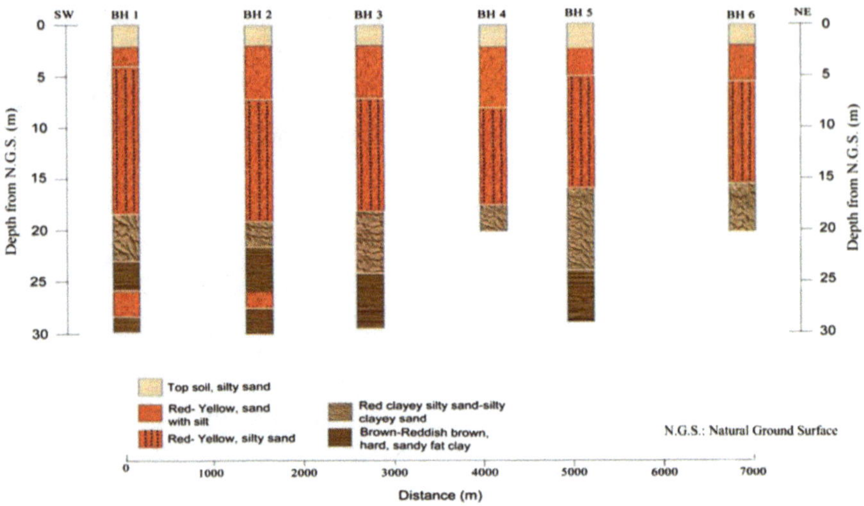

Fig. 1.13 Stratigraphic section of the study area based on both borehole logs and laboratory test results (Shakir et al. 2013)

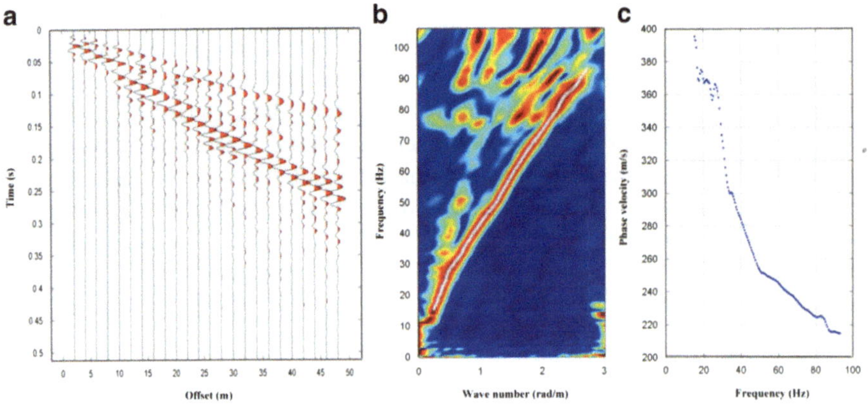

Fig. 1.14 Example of experimental data processing to estimates the dispersion curve: (**a**) Seismic raw data; (**b**) frequency-wavenumber spectrum with identification of the maxima; (**c**) experimental dispersion curve (Shakir et al. 2013)

obtain a sufficient exploration depth; a short receiver spread composed by 24 geophones with 2 m spacing has been used to extend the experimental information in the high frequency range, as required to increase the resolution close to the ground surface.

The experimental data at the six sites have been used for a laterally constrained inversion. The results are reported in Fig. 1.15. The comparison of the 1D shear wave velocity profiles (Fig. 1.15) with the stratigraphy from borehole logs (Fig. 1.13) indicates that the expected layering is overall well reconstructed,

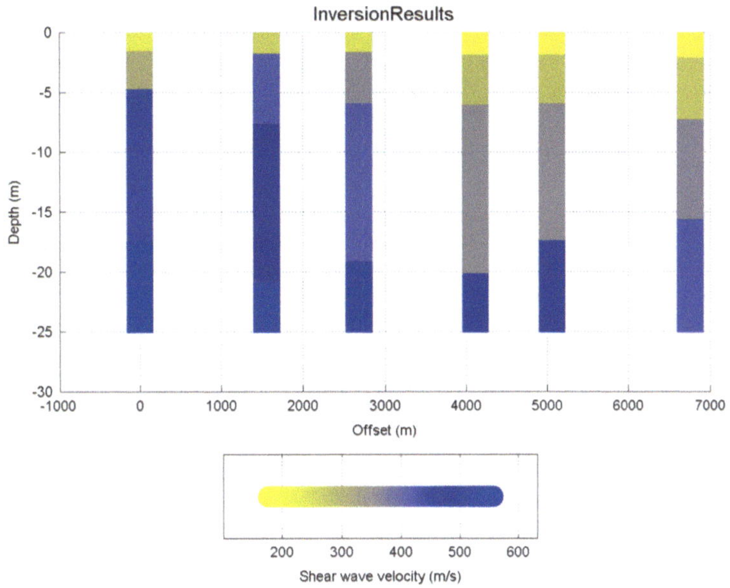

Fig. 1.15 Resulting pseudo-2D S-wave velocity model at Najaf (Shakir et al. 2013)

especially with reference to the interface with the bedrock formation which becomes shallower and shallower moving towards East. The shear wave velocity profile for site 2 is anomalous as it shows much higher velocities. Further investigation are required to assess if this is really an anomaly due to specific site conditions or the specific survey is affected by measurement errors.

1.3.3 Tarcento

The town of Tarcento is located in the North-East part of Italy (Fig. 1.16a), in the zone struck by the destructive Friuli seismic sequence in 1976. Most of the buildings are located on an alluvial basin, which for its shape and size is prone to develop significant 3D local effects. Indeed the basin is closed on three sides (North, South and East). The bedrock outcrops along the Torre creek (Fig. 1.16b). The thickness of the sediments increases from East to West up to about 100 m.

Surface wave tests have been performed at 16 sites in the town (Fig. 1.16b). Since the area is highly urbanized, the choice of the sites has been constrained by the necessity of adequate testing space. Where sufficient space was available, a combination of active-source acquisition with linear multistation array and passive measurements with circular receiver array has been implemented to increase the investigation depth.

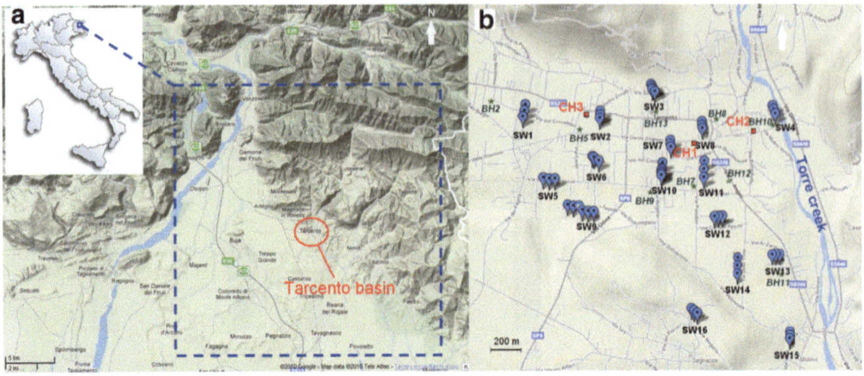

Fig. 1.16 Tarcento basin: (**a**) location; (**b**) localization of the mechanical soundings (*BH* borehole), the cross-hole tests (CH) and the 16 sites investigated with surface wave tests (SW) (Piatti et al. 2013)

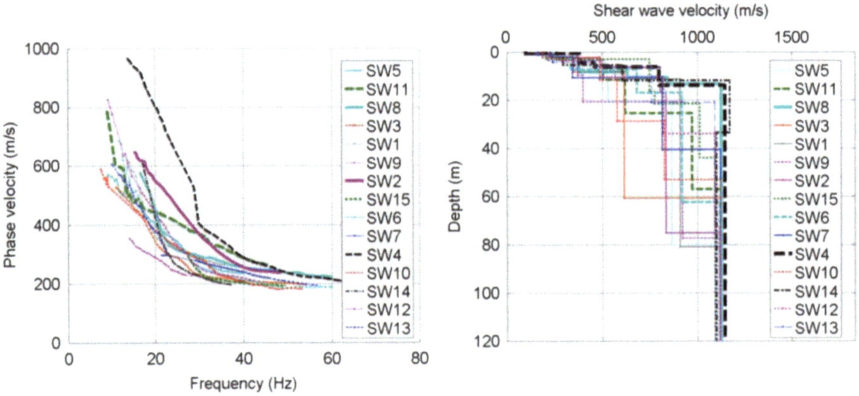

Fig. 1.17 Tarcento basin: (*left*) experimental dispersion curves for all the investigated sites within the basin; (*right*) shear wave velocity profiles obtained with LCI for each site (Piatti et al. 2013)

Results of three cross-hole tests (at the locations reported in Fig. 1.16b) were available from a previous investigation (Brambati and Faccioli 1980). These have been used to validate the results obtained with surface wave analysis. Borehole logs at several locations in the town were also available from previous studies and projects in the area. They have been used together with literature information (Brambati and Faccioli 1980) to estimate the position of the bedrock for the whole area. The bedrock position has been included in the inversion process as an a-priori information.

The experimental dispersion curves for the 15 sites within the alluvial basin are reported in Fig. 1.17a. The high variability from site to site testifies the importance of lateral variations in this geological setting, because of the marked 3D subsurface geometry. Considering this aspect, the lateral constraints in the inversion have been

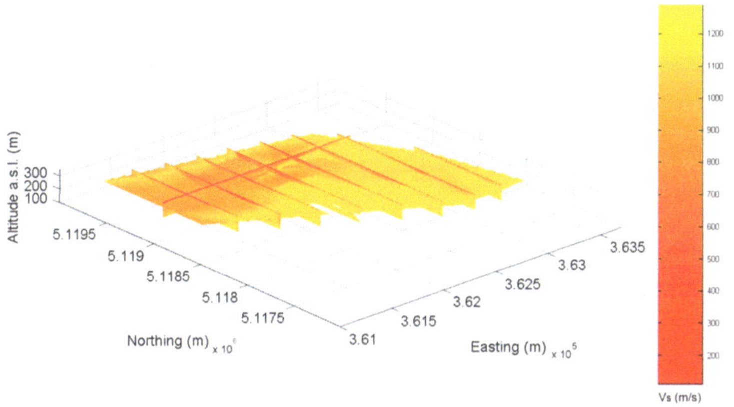

Fig. 1.18 3D shear wave velocity model of Tarcento basin (Piatti et al. 2013)

imposed only for some parameters: the velocity of layer above the bedrock and the velocity of the bedrock itself. The results of the laterally constrained inversion for the whole set of dispersion curves are reported in Fig. 1.17b, showing the variability of the shear wave velocity profiles, mainly due to the different position of the bedrock at each site. The effect of the lateral constraints is evident on the velocity of the bedrock. Indeed, a strong constraint was set to allow the information at sites with shallow bedrock to propagate to sites with bedrock too deep to be characterized with the available surface wave data.

The 1D shear wave velocity profiles have been subsequently interpolated to obtain a pseudo-3D model of shear wave velocity (Fig. 1.18). Boundary conditions were set assuming the outcrop at the edge of the basin, except along the western boundary, where the information from the closest site were used. The 3D model (Fig. 1.18) shows the presence of shallow layers that correspond to sands and gravels of the alluvial basin, with the bedrock deepening towards West.

1.4 Closing Remarks

Laterally constrained inversion provides a robust framework for the interpretation of surface wave data to get pseudo-2D/3D shear wave velocity models. Indeed the minimization of a single misfit function provide a better constrained solution than conventional inversion algorithms. Moreover it is possible to introduce a-priori information and independent data from other surveys. A crucial point is the tuning of the constraints which need to take into account the expected lateral variability. In this respect, a careful inspection of available information and some preliminary analyses to check the effects of the constraints are suggested.

The examples reported in the chapter show the capabilities of the method in different geological contexts and with different types of seismic datasets.

Acknowledgements This chapter summarizes the work carried on during the last 5 years at Politecnico di Torino with the contributions of several individuals. In particular the Authors are grateful to Roger Wisèn, Cesare Comina, Daniele Boiero, Claudio Piatti, Paolo Bergamo, Ammar M. Shakir, Flora Garofalo, Gianpiero Bianchi.

References

Aki K, Richards PG (2002) Quantitative seismology: theory and methods, 2nd edn. University Science, Sausalito

Arai H, Tokimatsu K (2005) S-wave velocity profiling by joint inversion of microtremor dispersion curve and horizontal-to-vertical (H/V) spectrum. Bull Seism Soc Am 95:1766–1778

Auken E, Christiansen AV (2004) Layered and laterally constrained 2D inversion of resistivity data. Geophysics 69:752–761

Bergamo P (2012) Surface wave analysis in laterally varying media. Ph.D. dissertation, Politecnico di Torino

Boiero D, Socco LV (2010) Retrieving lateral variations from surface wave dispersion curves analysis. Geophys Prosp 58:977–996

Brambati A, Faccioli E (1980) Studio di Microzonizzazione dell'area di Tarcento (Friuli). Università degli Studi di Trieste. Trieste (Italy). Unpublished (in Italian)

Foti S (2013) Combined use of geophysical methods in site characterization, Keynote lecture. In: Coutinho RQ, Mayne PW (eds) Geotechnical and geophysical site characterization, vol. 4. CRC Press, Leiden, pp 43–613

Foti S, Comina C, Boiero D, Socco LV (2009) Non uniqueness in surface wave inversion and consequences on seismic site response analyses. Soil Dyn Earthq Eng 29(6):982–993

Foti S, Parolai S, Albarello D, Picozzi M (2011) Application of surface wave methods for seismic site characterization. Surf Geophys 32:777–825

Garofalo F, Sauvin G, Socco LV, Lecomte I (2012) Joint inversion of surface-wave dispersion, P-wave refraction and apparent resistivity data. In: 17th EAGE Near Surface Geophysics, Expanded abstracts, Paris, A22, 5 p

Hayashi K, Matsuoka T, Hatakeyama H (2005) Joint analysis of a surface-wave method and micro-gravity survey. J Environ Eng Geophys 10:175–184

Hering A, Misiek R, Gyulai A, Ormos T, Dobroka M, Dresen L (1995) A joint inversion algorithm to process geoelectric and surface wave seismic data. Part I: Basic ideas. Geophys Prospect 43:135–156

Ivanov J, Miller RD, Lacombe P, Johnson CD, Lane JW Jr (2006) Delineating a shallow fault zone and dipping bedrock strata using multichannel analysis of surface waves with a land streamer. Geophysics 71(5):A39–A42

Jones RB (1958) In-situ measurement of the dynamic properties of soil by vibration methods. Geotechnique 8:1–21

McMechan GA, Yedlin MJ (1981) Analysis of dispersive wave by wave field transformation. Geophysics 46:869–874

Nazarian S, Stokoe II KH (1984) In situ shear wave velocities from spectral analysis of surface waves. In: Proceedings of the 8th conference on earthquake engineering, San Francisco, vol. 3. Prentice-Hall, pp 31–38

Piatti C, Boiero D, Foti S, Socco LV (2012) Constrained 1D joint inversion of seismic surface waves and P-wave refraction traveltimes. Geophys Prosp. doi:10.1111/j.1365-2478.2012.01071.x

Piatti C, Foti S, Socco LV, Boiero D (2013) Building 3D shear wave velocity models using surface wave testing: the Tarcento basin case history. Bull Seism Soc Am 103:1038–1047
Romanowicz B (2002) Inversion of surface waves: a review, International handbook of earthquake and engineering seismology. Academic Press, New York
Scherbaum F, Hinzen K-G, Ohrnberger M (2003) Determination of shallow shearwave velocity profiles in Cologne, Germany area using ambient vibrations. Geophys J Int 152:597–612
Shakir AM, Foti S, Garofalo F, Hijab BR, Laftah AA (2013) Laterally constrained inversion of surface wave data at Najaf City (IRAQ). Soil Dyn Earthq Eng 45:89–95
Sheriff RE, Geldart LP (1990) Applied geophysics. University Press, Cambridge
Socco LV, Boiero D, Comina C, Foti S, Wisén R (2008) Seismic characterisation of an alpine site. Near Surf Geophys 6:253–265
Socco LV, Boiero D, Foti S, Wisén R (2009) Laterally constrained inversion of ground roll from seismic reflection records. Geophysics 74(6):G35–G45
Socco LV, Boiero D, Foti S, Piatti C (2010a) Advances in surface and body wave integration. In: Miller RD, Bradford JD, Holliger K (eds) Advances in near-surface seismology and ground-penetrating radar, SEG monograph series. SEG, Tulsa, pp 55–73
Socco LV, Foti S, Boiero D (2010b) Surface wave analysis for building near surface velocity models: established approaches and new perspectives. Geophysics SEG, 75:A83–A102
Strobbia C, Foti S (2006) Multi-offset phase analysis of surface wave data (MOPA). J Appl Geophys 59:300–313
Tran KT, Hiltunen DR (2012) Shear wave velocity via inversion of full waveforms. In: Coutinho RQ, Mayne PW (eds) Geotechnical and geophysical site characterization 4. CRC Press, Leiden, pp 709–716
van der Veen M, Spitzer R, Green AG, Wild P (2001) Design and application of a towed landstreamer for cost-effective 2D and pseudo-3D shallow seismic data acquisition. Geophysics 66:482–500
Vangkilde-Pedersen T, Dahl JF, Ringgaard J (2006) Five years of experience with landstreamer vibroseis and comparison with conventional seismic data acquisition. In: Proceedings of the symposium on the application of geophysics to engineering and environmental problems (SAGEEP'06), Seattle, Washington, USA, 2–6 Apr 2006, pp 1086–1093
Wisén R, Christiansen AV (2005) Laterally and mutually constrained inversion of surface wave seismic data and resistivity data. J Environ Eng Geophys 10:251–262

Chapter 2
Site Classification and Spectral Amplification for Seismic Code Provisions

Anastasios Anastasiadis and Evi Riga

Abstract A global dataset of more than 3,000 ground motion records from 536 sites from Greece, Italy, Turkey, USA and Japan is used to propose elastic acceleration response spectra and soil amplification factors for a new site classification system, which uses parameters such as the thickness of soil deposits, the average shear wave velocity to the seismic bedrock and the fundamental period of the site. The dataset is also used to derive soil amplification factors for the soil classes of Eurocode 8 (EC8). Uncertainties in the estimation of soil amplification factors are captured using a logic-tree approach, which allows the use of alternative models and methods in an effective way. The results indicate that estimated soil factors for EC8 class C are significantly higher than the ones proposed in the provisions. The performance of the proposed classification system is compared to that of EC8 classification system in terms of an inter-category error, which represents the average dispersion of data within all categories of a given classification scheme and the results indicate some improvement. Error terms for the new classification system are lower than the error terms for EC8 classification system at all periods.

2.1 Introduction

Modern seismic codes, including Eurocode 8 (CEN 2004) in Europe and the International Building Code (ICC 2012) in USA have acknowledged the significant role of local site conditions and have incorporated their influence into their provisions. Both codes account for site effects through the suggestion of appropriate elastic response spectra, based on different soil categories. The main parameter used for site classification is $V_{s,30}$, i.e. the average value of shear wave velocity in the upper 30 m of the soil profile. This parameter was first proposed by Borcherdt

A. Anastasiadis (✉) • E. Riga, M.Sc.
Department of Civil Engineering, Aristotle University of Thessaloniki, Thessaloniki, Greece
e-mail: anas@civil.auth.gr

M. Maugeri and C. Soccodato (eds.), *Earthquake Geotechnical Engineering Design*, Geotechnical, Geological and Earthquake Engineering 28,
DOI 10.1007/978-3-319-03182-8_2, © Springer International Publishing Switzerland 2014

and Glassmoyer (1992) and Borcherdt (1994). The depth of 30 m was selected since it is a typical depth of geotechnical sampling borings and, thus, of site characterization. It is also believed that in most cases the main amplification is due to the trapping waves on the surface layers. However, several recent works (Steidl 2000; Wald and Mori 2000; Stewart et al. 2003; Di Giacomo et al. 2005; Mucciarelli and Gallipoli 2006 among others) have pointed the inability of $V_{s,30}$ to accurately predict amplification in several cases.

A more refined soil and site characterization was performed by the Lab. of Soil Mechanics and Foundation Engineering of the Aristotle University Thessaloniki (Pitilakis et al. 2004), based on a more comprehensive approach using a large database of high quality records in very-well known soil conditions and numerous 1D computations of ground response with different soil profiles in terms of impedance contrast, soil type and relative thickness, depth of the rigid or non-rigid bedrock etc. and of course input motion characteristics. The selected soil models and the applied numerical code were validated with real recordings at about 100 well-documented sites in Greece and worldwide. The classification of subsoil conditions includes parameters like thickness of the soil deposits, depth to bedrock, fundamental period of the site, stratigraphy, soil type, average V_s value determined in the entire thickness and strength properties (N-SPT and S_u).

The aim of the present paper is to discuss some specific aspects regarding the site classification schemes of contemporary seismic codes, together with an improved site categorization proposal and the corresponding spectral amplification according to EC8 earthquake design scenarios based on previous work (Pitilakis et al. 2004, 2006) which is the result of a long research effort in Aristotle University in Greece (Pitilakis et al. 2012a, b, 2013).

2.2 Site Effects and Seismic Codes

Site effects, in the general case, involve a combination of effects due to the characteristics and stratigraphy of subsoil materials (impendence contrast) and/or to irregular geometry of the site (topography). The main frequency dependent expressions used to characterize the effect of a site in the expected ground motion are the amplification factors representing the ratio of observed intensity measures with regard to measures of a reference site, generally for 'rock-site' conditions. The intensity measures, comprising both the effect of subsoil conditions and topography, are affected by source and path effects and it's not always valid to separate them from the observed measures. Further attention is focused on the geotechnical and geophysical definition of the surficial formations due to its effect in amplification in high frequency range and the evaluation of sediments-bedrock interface, because geometry and velocity contrast influences the amplification at low frequencies. During the last 30 years, numerous observational studies have striven to evaluate the importance of the different factors involved in site response of soils and the reader is referred to reviews of Aki (1988), Bard and Bouchon (1980a, b),

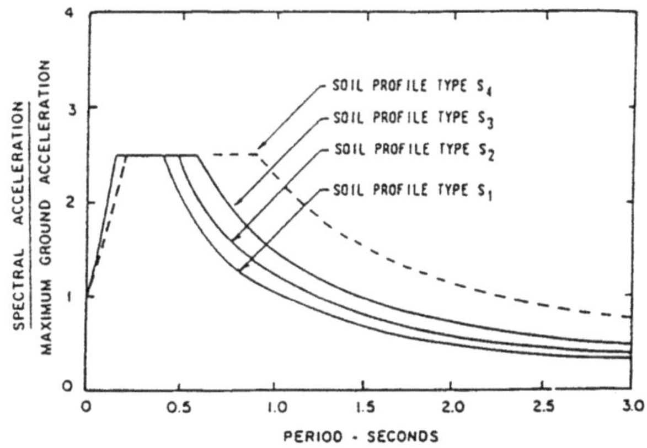

Fig. 2.1 Spectral shapes contained in seismic codes prior to NEHRP 1994 (From Martin and Dobry 1994)

Chávez-García and Bard (1994), Chávez-García and Faccioli (2000), Pitilakis (2004), Chávez-García (2007, 2011). However, as the complexity of the local geology increases together with the number of parameters affecting the response, the site effects become more complex and the possibility of extrapolating the observed results to a general application adopting into simple expressions in seismic codes decreases significantly.

Seismic code provisions incorporate the influence of site conditions on definition of seismic action via seismic intensity, different soil categories and appropriate elastic response spectra. Seed et al. (1976a, b), based on statistical study of more than 100 records from the area of California, developed average spectral shapes for various soil conditions, which still constitute the basic reference in many seismic codes worldwide. The developed spectral shapes by Seed and his co-workers (1976a, b) and similar studies by Mohraz (1976) were incorporated in the U.S. seismic codes (ATC 1978), providing three simplified spectral shapes, each one for a specific soil profile type, and associated site coefficients S which amplify only the long period range of the spectrum (Fig. 2.1 and Table 2.1). Soil profile S4 and the corresponding S factor representing deep soft clay deposits were added later in 1998 after the 1985 Mexico City earthquake (Seed et al. 1988). Each soil profile category comprises both qualitative and quantitative criteria including type, thickness and average value of shear wave velocity of soil deposits overlying bedrock (Table 2.1). The acceleration at each soil type was assumed to be equal or close to the rock acceleration based on the studies by Seed et al. (1976a) and ATC-3. Although the contribution of local geology is significant in the definition of seismic actions in seismic codes prior to NEHRP 1994, this factor had been included in many seismic codes worldwide by a generic and rather incomplete way. For example, until very recently (2011) the Greek seismic code, EAK 2000, had incorporated the soil categorization following primarily qualitative criteria, neglecting main parameters which affect site response such as the shear wave velocity and the thickness of soil deposits, as well as the intensity level of the expected ground shaking.

Table 2.1 Soil profile types and site coefficients contained in seismic codes prior to NEHRP 1994 (From Martin and Dobry 1994)

Soil profile type	Description	Site coefficient S
S_1	A soil profile with either (1) rock of any characteristic, either shale-like or crystalline in nature, that has a shear wave velocity greater than 2,500 ft per second or (2) stiff soil conditions where the soil depth is less than 200 ft and the soil types overlying the rock are stable deposits of sands, gravels, or still clays	1.0
S_2	A soil profile with deep cohesionless or stiff clay conditions where the soil depth exceeds 200 ft and the soil types overlying rock are stable deposits of sands, gravels, or stiff clays	1.2
S_3	A soil profile containing 20–40 ft in thickness of soft-to-medium-stiff clays with or without intervening layers of cohesionless soils	1.5
S_4	A soil profile characterized by a shear wave velocity of less than 500 ft per second containing more than 40 ft of soft clays or silts	2.0

Observations and recordings after the 1985 Mexico City, 1989 Loma Prieta, 1994 Northridge and 1995 Kobe earthquakes, in addition to numerous studies regarding the effect of nonlinear soil response, showed the effect of the level of shaking amplification at soft sites and especially at the short period range (Idriss 1990, 1991). Contemporary seismic codes (IBC2000 and EC8) based on these results of extended research studies and actual recordings, have largely accepted the significant role of site effects and attempt to incorporate their influence by an amplification function dependent on soil class, frequency range and lately upon the shaking intensity of bedrock. These provisions, sharing in general the same principles and design procedures, lead to a more rationalistic soil categorization by quantitative criteria, recognizing that spectra amplification is a nonlinear function of the intensity of the expected motion and of soil conditions. The first aim of the improved provisions achieved incorporation of a part of 1D soil dynamic characteristics such as the variation of shear wave velocity of the upper 30 m or strength parameters such as the values from standard penetration test and characteristics of the stratigraphy. The second aim in IBC2000 refers to the incorporation of non-linear behaviour achieved in the 1997 NEHRP provisions via parameters F_a and F_v, for short and high period ranges respectively, dependent upon the soil category and the intensity of the expected motion, incorporating the pioneering work of Borcherdt (1994), Crouse and McGuire (1996), Dobry et al. (1994), Martin and Dobry (1994), Seed et al. (1994a, b).

In EC8 the elastic response spectra are proposed for two different levels of seismic action, Type 1 and Type 2. Type 1 spectra have more energy in long-period motions and are proposed for use in regions having high seismic activity and stronger earthquakes; it is estimated that Type 1 spectra should be used when the earthquakes that contribute to the seismic hazard have a surface wave magnitude M_s greater than 5.5. Type 2 spectra are recommended for $M_s \leq 5.5$, having larger normalized spectral amplitudes at short periods. Ground motion amplification to account for local soil effects is established through a frequency-independent

Table 2.2 Ground types defined in EC8

Ground type	Description of stratigraphic profile	Parameters: Vs,30 (m/s)	N-SPT	Su (kPa)
A	Rock or other rock-like geological formation, including at most 5 m of weaker material at the surface	>800	–	–
B	Deposits of very dense sand, gravel, or very stiff clay, at least several tens of meters in thickness, characterized by a gradual increase of mechanical properties with depth	360–800	>50	>250
C	Deep deposits of dense or medium-dense sand, gravel or stiff clay with thickness from several tens to many hundreds of meters	180–360	15–50	70–250
D	Deposits of loose-to-medium cohesionless soil (with or without some soft cohesive layers), or of predominantly soft-to-firm cohesive soil	<180	<15	<70
E	A soil profile consisting of a surface alluvium layer with Vs values of type C or D and thickness varying between about 5 and 20 m, underlain by stiffer material with Vs > 800 m/s			
S1	Deposits consisting, or containing a layer at least 10 m thick, of soft clays/silts with a high plasticity index (PI > 40) and high water content	<100 (indicative)	–	10–20
S2	Deposits of liquefiable soils, of sensitive clays, or any other soil profile not included in types A – E or S1			

amplification factor S, called 'soil factor', which increases uniformly the normalized elastic response spectra of each type of seismic action in all periods, incorporating the pioneering work of Faccioli (1992) and Rey et al. (2002).

Both in IBC2000 and EC8 the $V_{s,30}$ parameter is used along with N-SPT values, plasticity index PI and undrained shear strength S_u to define different soil profile types, while extra special ground types are also proposed for special soils (Tables 2.2 and 2.3). The above criteria of site classification have been incorporated in many seismic codes worldwide, while in many countries (Japan, Turkey, New Zealand, etc.) soil classification criteria stemming from microzonation studies and the local geology have been included in their codes. Reader is referred to a World list of regulations for seismic design of the International Association for Earthquake engineering www.iaee.or.jp/worldlist.html.

It is important to note that the correlation of the elastic with the design spectra is inevitably related to the level of energy absorption which is anticipated to take place through cycles of inelastic behaviour and damage of the structural members. The latter is usually described as the behaviour factor q (in Eurocodes, the Greek Seismic Code and a number of other codes) and the response modification factor R in the U.S. codes. It is therefore necessary, before proceeding to the critical and comparable evaluation of the code-defined spectra, to account for the different assumptions on the reduction of forces prescribed in different codes and proceed to

Table 2.3 Site class definitions in IBC2000

Site class	Description of stratigraphic profile	Parameters		
		Vs,30 (m/s)	N-SPT	Su (kPa)
A	Hard rock	>1500	–	–
B	Rock	760–1500	–	–
C	Very dense soil and soft rock	360–760	>50	Su≥100
D	Stiff soil profile	180–360	15≤N≤50	50≤Su≤100
E	Soft soil profile	<180	<15	Su≤50
E	Any profile with more than 10 ft (3 m) of soil having the following characteristics: 1. Plasticity index PI > 20, 2. Moisture content w ≥ 40 %, and 3. Undrained shear strength Su < 500psf (25 kPa)			
F	Any profile containing soils having one or more of the following characteristics: 1. Soils vulnerable to potential failure or collapse under seismic loading such as liquefiable soils, quick and highly sensitive clays, collapsible weakly cemented soils 2. Peats and/or organic clays [H > 10 ft (3 m) of peat and/or highly organic clay where H = thickness of soils] 3. Very high plasticity clays [H > 25 ft (8 m) with PI > 75] 4. Very thick soft/medium stiff clays [(H > 120 ft (36 m)] with Su < 1000 psf (50 kPa)			

the appropriate calibrating calculations (Pitilakis 2004). This is an ample field and it's beyond the aim of the present paper.

2.2.1 Spectral Amplification of EC8

The elastic response spectra proposed by EC8 for the horizontal components of seismic action are a function of site category according to soil and site characterization type (Fig. 2.2) and are defined from the following equations and four fixed period values given in Table 2.4.

$$0 \le T \le T_B \quad S_e(T) = a_g \cdot S\left[1 + \frac{T}{T_B} \cdot (n \cdot 2,5 - 1)\right] \tag{2.1}$$

$$T_B \le T \le T_C \quad S_e(T) = a_g \cdot S \cdot n \cdot 2,5 \tag{2.2}$$

$$T_C \le T \le T_D \quad S_e(T) = a_g \cdot S \cdot n \cdot 2,5 \cdot \left[\frac{T_C}{T}\right] \tag{2.3}$$

$$T_D \le T \le 4.0 \quad S_e(T) = a_g \cdot S \cdot n \cdot 2,5 \cdot \left[\frac{T_C \cdot T_D}{T^2}\right] \tag{2.4}$$

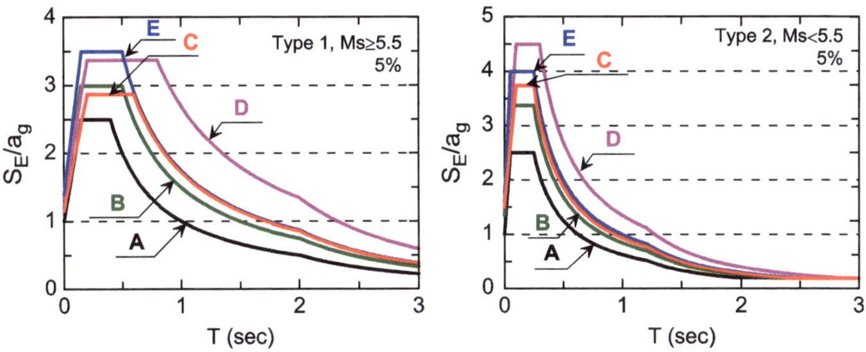

Fig. 2.2 Recommended Type 1 (*left*) and Type 2 (*right*) elastic response spectrum for soils A to E according to Eurocode 8 (EN1998)

Table 2.4 Soil factors and periods proposed by EC8 (EN1998)

	Type 1 spectrum – $M_S \geq 5.5$				Type 2 spectrum – $M_S < 5.5$			
Ground type	S	T_B(s)	T_C(s)	T_D(s)	S	T_B(s)	T_C(s)	T_D(s)
A	1.0	0.15	0.4	2.0	1.0	0.05	0.25	1.2
B	1.2	0.15	0.5	2.0	1.35	0.05	0.25	1.2
C	1.15	0.20	0.6	2.0	1.5	0.1	0.25	1.2
D	1.35	0.20	0.8	2.0	1.8	0.1	0.30	1.2
E	1.4	0.15	0.5	2.0	1.6	0.05	0.25	1.2

where:

$S_e(T)$: the elastic spectra
S: is the soil factor
a_g: the design peak ground acceleration value defined as $a_{g,Rock} \cdot \gamma_I$
n: is the damping factor with n = 1 for 5 % viscous damping

The shape of EC8 spectrum follows the Newmark and Hall (1973) criteria, comprising three branches controlled by acceleration at short period range (T_B-T_C), velocity at intermediate periods (T_C-T_D) and displacement at long period range (T_D-4s). The horizontal components of seismic action for each site category depends on PGA value, where the four controlling period values are fixed for each type of earthquake design scenario (Type I and Type II) devolved to the National Annexes.

In addition, in EC8 (CEN 2004) the elastic response spectra for vertical seismic motion, as well as the displacement spectra are defined. The vertical spectrum depends on horizontal peak acceleration value and the effect of source to site distance is not included. This paper aims at presenting an updated classification scheme and only the horizontal elastic spectra shapes and for this reason the reader is referred to reviews of Bommer et al. (2000, 2010), Bommer and Pinho (2006), Tolis and Faccioli (1999), Pitilakis et al. (2007), Faccioli and Villani (2009), Bommer (2011).

2.2.2 *Site Classification in Codes Based on the $V_{s,30}$*

The site classification according to the contemporary seismic provisions IBC2000 and EC8 is now based mainly on the $V_{s,30}$, the effective average shear wave velocity of the upper 30 m. $V_{s,30}$ is not the arithmetic average of the shear wave velocity to a depth of 30 m but is the velocity that produces the same travel time through the upper 30 m of the subsoil (Boore 2004). The restriction to the depth of 30 m makes it in many cases feasible in practice in order to obtain the required measurements, since geotechnical inspection frequently extends in depths of 25–40 m. Definitely, this parameter has provided researchers with an unambiguous and measurable quantity for empirical studies of future earthquake data aiming at verifying, refining or modifying parameters as more data become available (Dobry et al. 2000). For example, at the Euroseistest valley (http://euroseis.civil.auth.gr/; EUROSEISTEST, EV5V-CT.93-0281 1993–1995) a physical laboratory (Test Site), located at a distance of 30 km from Thessaloniki in northern Greece has been running continuously from 1981. The structure of the valley in 1D, 2D (see Fig. 2.3) and 3D is very well known (EUROSEISMOD, ENV4-CT.96-0255 1996–1999) through a combined geological, geotechnical and geophysical surveys, as well as, empirical and theoretical studies of site effects based on high quality data of many weak and moderate earthquakes, (Pitilakis et al. 1999; Raptakis et al. 1998, 2000, 2005; Chávez-García et al. 2000; Makra et al. 2005; Makra and Raptakis 2008; Manakou et al. 2010) (Fig. 2.4).

The variation of V_s of the upper 30 m obtained from crosshole and downhole (CH, DH), wave inversion (SWI), P and SH refraction (REF) and analyses of weak to moderate recordings from the DH array at TST site at the center of the valley with regard to the mean V_s of the reference model of the valley, is depicted in Fig. 2.5. The intra-method variability is of the order of 3–10 %, with the lower values obtained from the CH and SWI and the higher from DH and the analysis of DH seismic array records, and consistent with findings of Moss (2008) and Asten and Boore (2005). On the other hand, the mean N-SPT value obtained at closely-spaced boreholes varies from 9 to 19, while this variation is reduced when the corrected values for 60 % of the maximum free-fall energy are used (Fig. 2.5). Thus, the N-SPT values proposed in EC8 should be corrected to a hammer energy ratio of 60 %, N(60), due to the fact that in many countries the "rope and pulley" technique is still applied and in many cases available data have to be corrected before V_s estimation. In addition, S_u values proposed in EC8 as an alternative parameter for site classification for site classes B, C and D are different compared to those of IBC2000 even for subsoil characterized by the same value of stiffness (e.g. in site class B, $S_{u,\,EC8} > 250$ kPa instead of $S_{u,\,IBC2000} > 100$ kPa). Hence, the mean N-SPT values of the upper 30 m, as an alternative parameter for site classification, in some cases, may not be used in a direct way but through appropriate empirical correlations of the shear wave velocity.

The main downside of $V_{s,30}$ is that it assumes that the properties of the subsoil below 30 m are the same everywhere and the underlying bedrock has the same

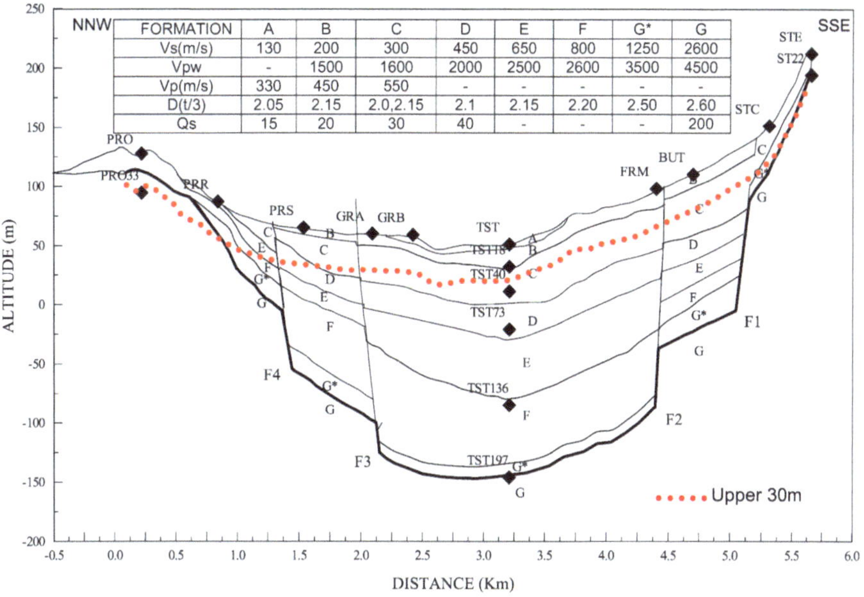

Fig. 2.3 Profitis – Stivos 2D cross section (Raptakis et al. 2000). A to G represents different soil and rock categories, V_s S-wave velocity, V_p and V_{pw} the P-wave velocity of unsaturated and saturated soils, Q_s quality factor and D density, where *dotted line* depicts the upper 30 m

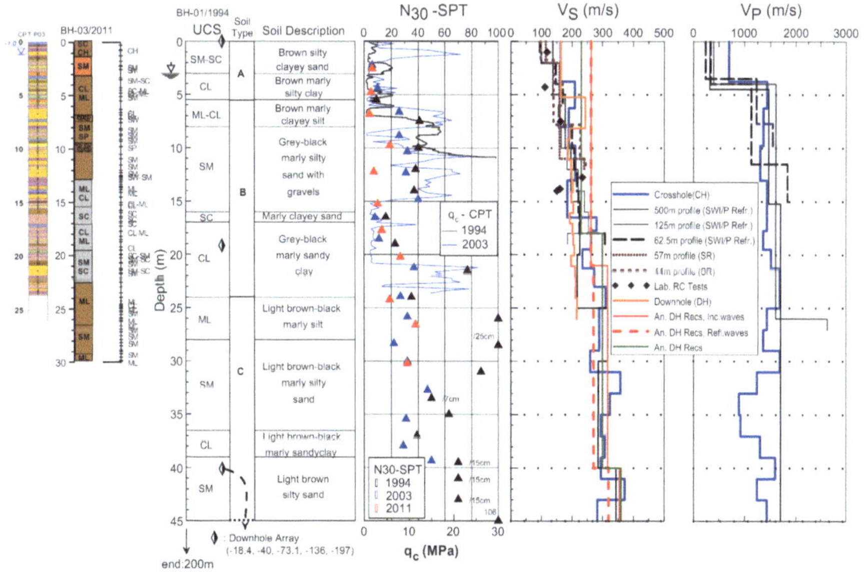

Fig. 2.4 Borehole and CPT logs, SPT variation with depth and Vs, Vp resulted from numerous methods at the upper 40 m at the Euroseistest DH array (TST) (Pitilakis et al. 1999; Raptakis et al. 2000)

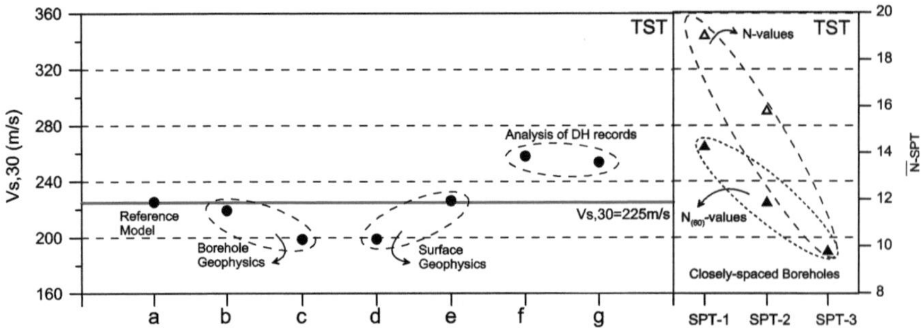

Fig. 2.5 (*Left*): $V_{s,30}$ values obtained from various methods and analyses: (**a**) reference model of the valley cross-section (Raptakis et al. 2000), (**b**) Borehole geophysics comprised CH (Raptakis et al. 2000), (**c**) DH tests at closely spaced boreholes, (**d**) surface wave inversion of moderate seismic profiles (L ~ 125 m), (Raptakis et al. 2000), (**e**) surface wave inversion of seismic profiles L ~ 500 m, source: explosion, (Raptakis et al. 2000), (**f**) analysis of DH records (Makra and Raptakis 2008), (**g**) analysis of DH records at TST station (EuroseisRisk report 2005), (*right*): Mean N and N(60) SPT values of the upper 30 m obtained from closely-spaced boreholes at TST site

velocity values. This is consistent with results of Joyner et al. (1981), Aki (1988) and Dobry et al. (1994) showing that the amplification is a function of the wave velocity of the shallow subsoil and is less independent of the soil deposits thickness.

Many references are available that discuss in detail variant aspects of this issue. Boore et al. (1997) proposed the use of average V_s at depth of one-quarter of the wavelength associated with the period of interest, which can be greater than 30 m and is consistent with the theory.

The amplification is caused by impedance contrast determined by the properties of soil deposits and the underlying bedrock, and taking into account only the properties of the soil, the actual amplification will be very different from the predicted one. The amplification at ground is correlated in many cases with stiffness, as well as impedance contrasts at depths greater than 30 m; sometimes the overall amplification is a combination of both shallow and deep effects (Dobry and Iai 2000). If we take into account only the stiffness of the top 30 m, the actual response and the amplification will be significantly different from that predicted. Moreover, in both classification schemes the qualitative geotechnical description is restricted only to stiffness characterization (e.g. in IBC2000 soft, stiff) and in EC8 includes in the same class deposits of loose-to-medium cohesionless soil and predominantly soft-to-firm cohesive soil.

For example, examining the results in terms of response spectra ratios stemming from Euroseistest downhole array at TST station from two local weak to moderate events (Fig. 2.6c, d), a clear amplification in high frequencies range ($T < 0.5$ s) followed by large amplification at higher period ranges (T: 1.2–2 s) is observed between surface and 136 m depth. Comparing the spectra ratios between surface and 73 m depth, as well as the rest spectra ratios at 40 and 18 m, no amplification at long period range is observed.

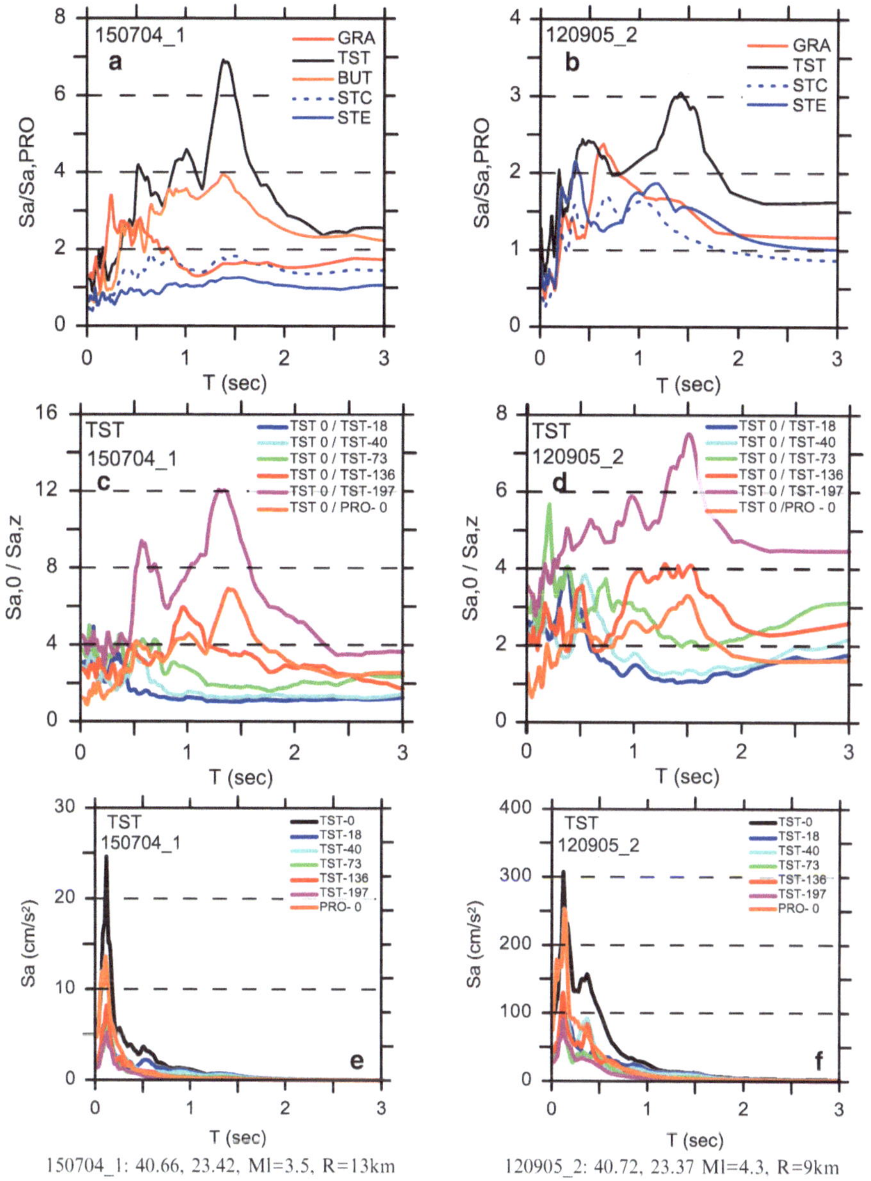

Fig. 2.6 Mean response spectra ratios at surface stations (**a**, **b**) with regard to a nearby site at outcropping rock conditions (PRO), and response spectra ratios (**c**, **d**) and response spectra (**e**, **f**) at the Euroseistest TST vertical DH array from two local weak events

The 1D resonance effect at the fundamental frequency of the site ($f_0 = 0.7$ s) and the effect of long period waves, mainly surface waves, generated at the lateral discontinuities, disappear when only the uppermost layers are taken into account together with a 1D SH wave pattern (the case of Euroseistest valley), (Pitilakis 2004).

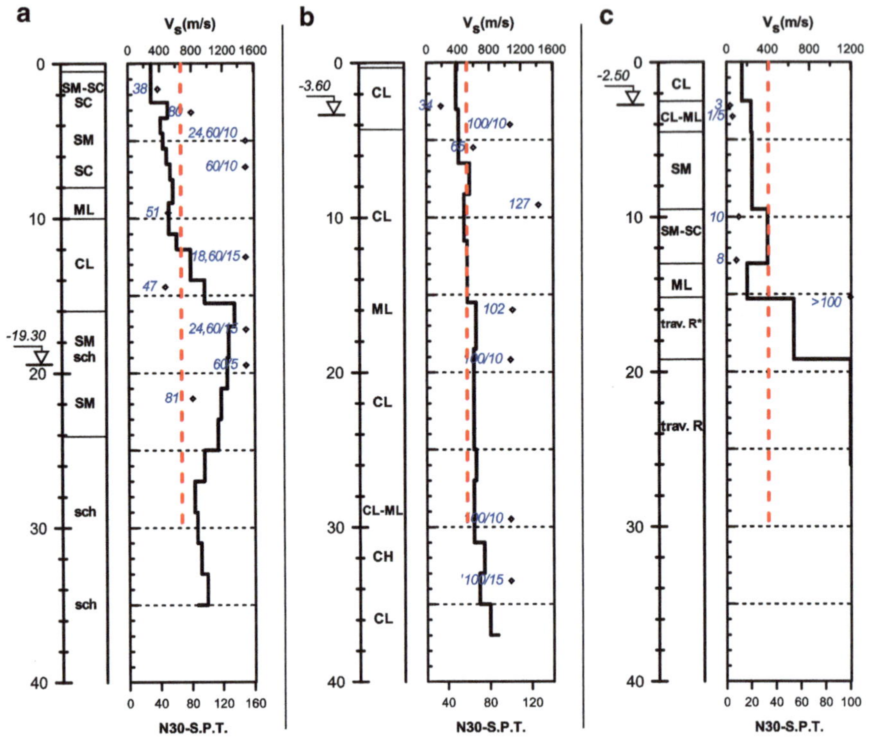

Fig. 2.7 Representative soil profiles of category B-EC8 (from left $V_{s,30}$: B1: 670 m/s, B2: 519 m/s, B3: 400 m/s)

Furthermore, significant differences in spectra amplification at surface stations with regard to values at nearby site at outcropping rock conditions (PRO, site category A according to EC8) are observed. Sites TST, BUT, GRA are classified as site category C according to EC8 (with $V_{s,30}$: 226, 252, 270 m/s respectively) but their response is controlled mainly by 1D resonance, as well as, the effect of the valley structure.

The use of $V_{s,30}$ for site classification in practice resulted in the execution and improvement of numerous available in-situ tests and methods and the engineering community became acquainted with the application of shear wave velocity. In engineering practice the use of one basic parameter ($V_{s,30}$), as well as the limited number of codes site classes disregarding the soil thickness as a basis for site classification is misleading in many cases. For example, EC8 site category B contains a huge variety of soil deposits from very dense sands-gravels to very stiff – hard clays having thickness of several tens of meters and $V_{s,30}$ from 360 to 800 m/s.

Figures 2.7, 2.8 and 2.9 present representative soil profiles from strong motion network sites in Greece (Figs. 2.7 and 2.8) and microzonation studies (Fig. 2.9)

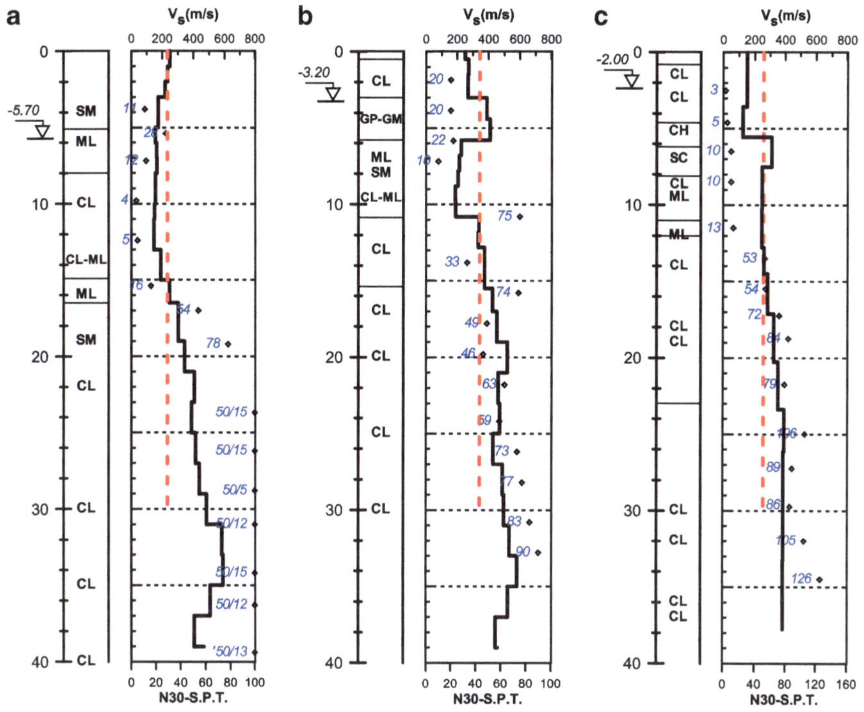

Fig. 2.8 Representative soil profiles of category C-EC8 (from left $V_{s,30}$: 235, 346, 258 m/s)

where a detailed geotechnical and geophysical survey was conducted comprising classical geotechnical tests, CH and DH tests, as well as resonant column tests. It is evident that the response of soil profiles of Fig. 2.7 can differ vastly. On the other hand, in many cases, the use of one main parameter for site classification led to overlooking the required steps for classifying a site (e.g. according to FEMA450-1/2003), especially in sites consisting of soft clayey layers or highly organic clays, where a site specific study is required. The soil profiles of Fig. 2.8 are classified as C site-category according to EC8, but the existence of soft clayey-sandy layers having thickness more than 5 m, requires site specific evaluation according to UBC criteria. Finally, in many European coastal areas where the local geology consists of elevated marine terraces, plio-pleistocene calcarenites, cemented gravelly soils, sandstones, etc. (Fig. 2.9a, b) or weathered igneous, metamorphic formations, the surficial layers are characterized by a gradual decrease of mechanical properties and Vs values with depth at depths up to 30 m. In conclusion it is clear that the $V_{s,30}$ hypothesis for site characterization as the basic criterion, disregarding geotechnical criteria and the thickness of soil deposits, may lead to erroneous estimates of ground motion in many cases.

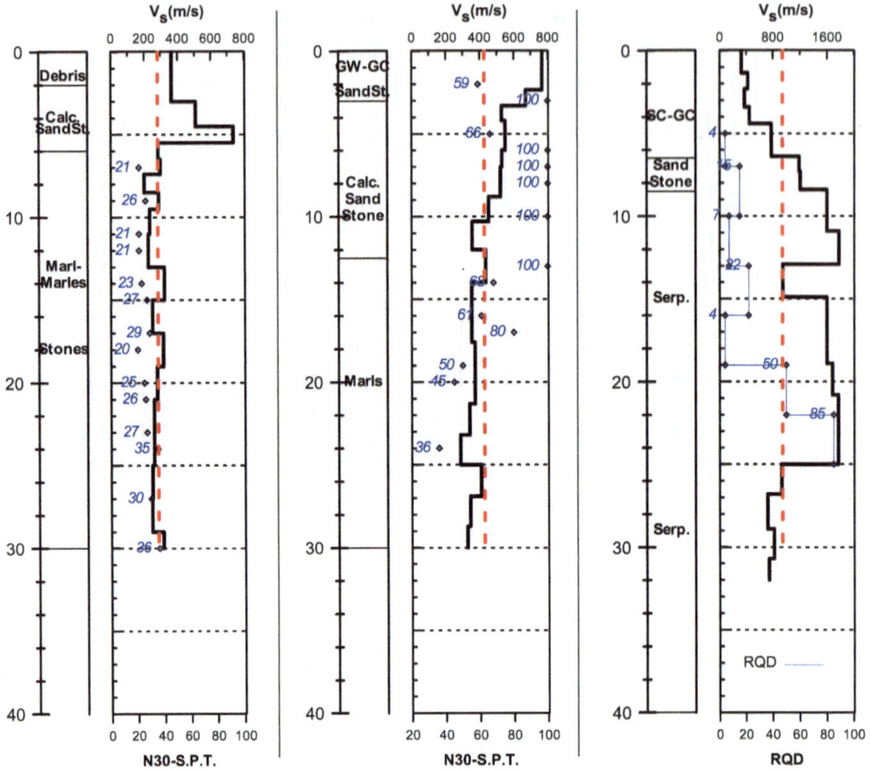

Fig. 2.9 Representative soil profiles at special geological site conditions (from left $V_{s,30}$: 283, 427, 916 m/s)

2.2.3 *Alternative Site Classification Methods and Amplification*

Several researchers used experimental techniques in order to study site effects and to examine the impact to the site classification schemes. The Japanese Road Association (1980–1990) used both the predominant soil period (T_0) and the $V_{s,30}$ in their classification. Steidl (2000) pointed out a poor correlation between site class and site amplification and proposed that a depth to basement parameter might be more useful in order to predict the ground motion amplification. Rodriguez-Marek et al. (2001) proposed a classification scheme based on the analysis of recordings stemming from the 1989 Loma Prieta and 1994 Northridge earthquakes and associated period-dependent spectral amplification factors for different site categories. The proposed classification based on site period and average shear wave velocity resulted in reduction of standard error compared to simple rock/soil categorization. Stewart et al. (2003) studied the amplification factors for spectral acceleration in active tectonically regions from California, Turkey and Japan, concluding that detailed surface geology and shear wave velocity provide an effective means of soil-site categorization at small periods, but at longer periods none of the schemes is optimized relative to both the dispersion and distinction criteria. Park and Hashash (2004) studied the response in the deep basin of the Mississippi embayment and pointed out that NEHRP provisions may not be appropriate for thick sediments because the short period coefficients are highly overestimated, whereas the long period coefficients are underestimated and new depth dependent site coefficients were developed following the NEHRP format. Choi and Stewart (2005), based on the PEER strong motion database, studied the existence of nonlinearity in site amplification, pointing out that amplification increases with decreasing $V_{s,30}$ at low and mid-period range while $V_{s,30}$ as site metric is more effective than NEHRP or geotechnical classification scheme proposed by Rodriguez-Marek et al. (2001) at most periods, and roughly equally effective as surface geology characterization. Lang and Schwartz (2006) proposed an implementation of NEHRP site classification scheme based on HVSR measurements comprising the thickness of soil deposits and shear wave velocities. Zhao et al. (2006b) proposed an empirical classification scheme based on results of HVSR technique on strong motion stations in Japan, pointing out the advantages of using HVSR versus shear wave velocity profiles. Fukushima et al. (2007) based on H/V spectral ratios classified stations and studied the effect on empirical ground motion models. Cadet et al. (2008), based on the analysis of 500 sites from the Kik-Net network, proposed an alternative site classification and associated spectral shapes based on a two parameter scheme, comprising the average shear wave velocity $V_{s,z}$ over z meters (5–30 m) and the site fundamental frequency f_0, concluding that the largest variance reduction is obtained using both the Vs,z and f_0 parameters, while a simple classification based only on f_0 leads to satisfactory results. Kokusho and Sato (2008) studied the spectral amplification on recorded data from Kik-Net downhole arrays, as well as, by theoretical analyses, pointing out that good and unique correlation was found between the peak amplifications and S-wave velocity ratios defined by S-wave velocity in base layer divided by average S-wave velocity V_s, for different sites and different earthquakes and the $V_{s,30}$ did not correlate well with the obtained amplifications. Assimaki et al. (2008) studied site amplification and attenuation based on weak-motion recordings at 38 stations of KiK-Net network, pointing out that the currently employed site classification criteria need to be re-evaluated accordingly, to reflect more realistically the anticipated average response of near-surface formations. Gallipoli and Mucciarelli (2009) studied 45 downhole profiles of 30 m at sites where earthquake recordings were available and pointed out that $V_{s,10}$ could predict site classification with the same performance of $V_{s,30}$. Lee and Trifunac (2010) also pointed out that the effect of deeper than 30 m layers is very important and the $V_{s,30}$ parameter is not adequate in order to scale the strong motion amplitudes. Cadet et al. (2012), based on empirically derived amplification factors from a large subset of Kik-Net recordings, corrected for the varying depths and impedance, examined the correlation of several site parameters such as $V_{s,z}$ (z: 5, 10, 20, 30 m) and the fundamental frequency f_0, and pointed out that the best results are obtained with f_0 and $V_{s,30}$ while the best single parameter was the fundamental frequency. Luzi et al. (2011), based on results at 126 recording sites of the Italian accelerometric network, proposed two soil classifications schemes in which the fundamental frequency becomes an alternative or a complement to $V_{s,30}$ parameter, concluding that a classification based on fundamental frequency leads to satisfactory results compared with classification based on $V_{s,30}$. Di Alessandro et al. (2012), extending the classification introduced by Zhao et al. (2006b) based on H/V spectral ratios, studied its impact on empirical ground motion equations comparing with that of conventional classification (rock/soil). Although the approach resulted in a small reduction of the overall deviation, the use of H/V ratios in site classification does capture the signature of sites with flat frequency-response, as well as deep and shallow-soil profiles, characterized by long- and short-period of resonance, respectively.

2.2.4 Estimation of f_0 Based on HVSR Technique

During the last decades various methods have been introduced for site effects evaluation and used for site classification. Among them the application of instrumental techniques (SSR, GIS, HVSR, etc.) is straightforward correlated to the improvement of high quality, digital recordings and the drastic reduction of the requisite equipment. SSR and HVSR are the most commonly used experimental techniques for the estimation of site amplification due to local soil conditions (Field et al. 1992; Steidl 1993; Lermo and Chávez-García 1993; Field and Jacob 1995; Raptakis et al. 1998, 2000, 2004a, b). Among the experimental-empirical techniques that do not require a reference site, the horizontal to vertical spectral ratio (HVSR) technique of ambient vibration record presented by Nakamura (1989) was applied in many cases and was the subject of heated debates. The method is simple, very cheap and fast and can produce reliable results regarding the fundamental period of the site (Chávez-García 2011; Pitilakis 2004). On the other hand, in many cases numerous results of HVSR ratios at urban areas specifically in the framework of microzonation and risk studies, not correlated with available geotechnical information and not validated with simple 1D models in a straightforward way, led to disregarding the important effect of well known parameters in site response observations and classification: e.g. the thickness of soil deposits, the distinction between loose sandy and soft clayey soils, the effect of impedance contrast in intermediate layers and all the issues regarding the non-linear behaviour of soil deposits. In addition, the exact application of the H/V spectral ratio of ambient noise requires the examination of special requirements regarding the site and the parameters that could affect ambient noise measurements such as local geology, environmental conditions, data acquisition system, processing techniques, etc. (Bard 1999; Atakan et al. 2004; guidelines in http://sesame-fp5.obs.ujf-grenoble.fr/index.htm). In conclusion, the HVSR technique is reliable in site classification in order to estimate the fundamental frequency of the soil profile and should be applied with caution regarding the aforementioned remarks.

2.3 Design Spectra and Amplification Factors

It is widely accepted that there are several more factors that may affect the seismic ground motion and site response, such as the impedance contrast between surface and underlying stiff deposits or rock basement, the soil type and stratigraphy, the material non linearity and damping and its potential variation with the intensity of the ground motion; all these should ideally be taken into account in an improved site classification system. Based on these fundamental observations, a new improved soil and site classification system is presented, suitable for seismic code provisions, based on a new worldwide database of strong ground motion records. The primary source of records was the database compiled in the framework of European Seventh Framework Programme (FP7) project 'Seismic Hazard Harmonization in Europe' (SHARE). The present work constitutes a further step to previous work (Pitilakis et al. 2004, 2006), in which a soil classification system that includes soil type, stratigraphy, depth to bedrock and soil stiffness as key parameters was presented, and amplification factors were assessed, based exclusively on 1D theoretical analyses of various representative models of realistic site conditions. At this stage, the strong motion dataset is used to redifine the limits of the parameters used to define the different soil classes and to propose new elastic response spectra. Normalized response spectra are proposed based on the conceptual assumption that they should fit as close as possible to the 84th percentile of the normalized spectra of the strong-motion records, while amplification factors are estimated following a logic tree approach, which allows the use of alternative models, capturing in this way effectively the uncertainties of each model. Although the proposed classification system should ideally be used to replace the existing classification scheme in EC8, soil amplification factors are also proposed for the existing soil classes of EC8 using again the logic tree approach.

2.3.1 Description of Databases

In the framework of SHARE, an extended strong motion database was compiled, by unifying existing databases, such as the European strong motion database, the Turkish national strong-motion database, the Next Generation Attenuation database, the Kik-Net database, the global worldwide database compiled by Cauzzi and Faccioli (2008), the Internet Site for European Strong-motion Data and the Italian Accelerometric Archive database (Yenier et al. 2010). This database, hereinafter called SHARE database, covers earthquakes back to 1930s and contains a total of 14193 corrected records from 2448 events. The version of the database that was used is v3.1 (March 2010) and contains 13,500 records. For each record, information is provided on the earthquake (magnitude, location, focal mechanism) and on the accelerograph station. The soil and site documentation of the stations of the SHARE database includes the $V_{s,30}$ values and site classification according to EC8.

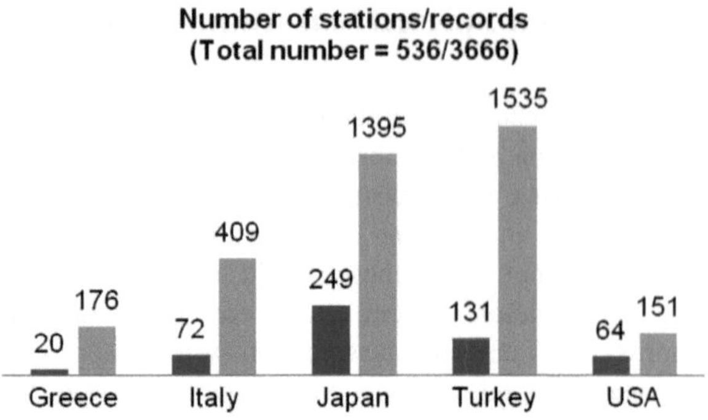

Fig. 2.10 Geographic distribution of selected stations (in *black*) and records (in *grey*)

As a result, strong motion records of SHARE database were used to estimate amplification factors for the soil classification system of Eurocode 8, which is based mainly on $V_{s,30}$.

However, the soil and site documentation of the stations included in SHARE database was rather insufficient for the proposal of the improved soil classification system and corresponding elastic response spectra. To overcome this obstacle, a complementary database was compiled within SHARE (hereinafter SHARE-AUTH database), which contains 3,666 records from 536 recording stations from Greece, Italy, Japan, Turkey and USA. All sites in SHARE-AUTH database dispose accurately measured shear wave velocity profiles, which were carefully collected from a number of sources, such as the Italian Accelerometric Archive (Working Group ITACA 2010, http://itaca.mi.ingv.it), Kik-Net (http://www.kik.bosai.go.jp/), K-Net (http://www.k-net.bosai.go.jp/), the Turkish national strong-motion database (Akkar et al. 2010), ROSRINE program (http://gees.usc.edu/ROSRINE/), Dr. David Boore's personal webpage (http://www.daveboore.com/) and USGS open-file reports. The V_s profiles of the stations of SHARE-AUTH database have been measured with a variety of surveying methods; for most of the sites (70 % of the total sample) V_s profiles have been obtained from borehole measurements (e.g. cross-hole, down-hole), while for the rest of the sites, for example the Turkish stations (24 % of the total sample), the V_s profiles have been evaluated using surface geophysical surveys. As a result, SHARE-AUTH database constitutes a very reliable set of empirical measurements for estimation of influence of local site conditions. The geographic distribution of the stations and records of SHARE-AUTH database is presented in Fig. 2.10. Magnitude M and peak ground acceleration PGA distributions with respect to the epicentral distance R, for the records in SHARE-AUTH database are shown in Fig. 2.11a, b respectively. It is observed that there are significantly fewer records with PGA values exceeding 200 cm/s^2 and many weak motion records with records with peak values less than 20 cm/s^2.

For the sites of SHARE-AUTH database $V_{s,30}$ values were calculated using Eq. (2.5):

$$V_{s,30} = 30 / \sum_{i=1,N} h_i / V_i \tag{2.5}$$

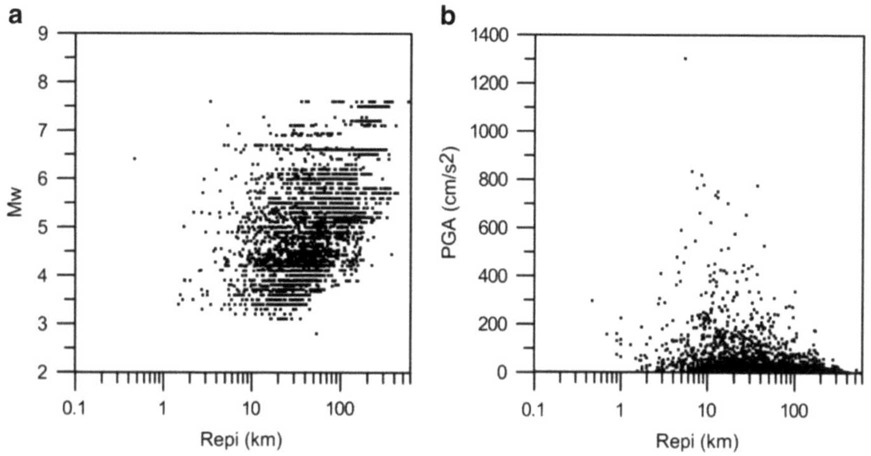

Fig. 2.11 Data coverage of SHARE-AUTH database records in terms of (**a**) moment magnitude – epicentral distance R and (**b**) *PGA* peak ground acceleration – epicentral distance R

where h_i and V_i denote respectively the thickness and shear-wave velocity of the i-th layer, in a total of N, existing in the top 30 m. However, for some sites (mostly K-net sites) the available shear-wave velocity models did not extend down to 30 m.

In that case, $V_{s,30}$ was estimated using Eq. (2.6), which assumes a constant shear wave velocity from the bottom of the available velocity model to the depth of 30 m:

$$V_{s,30} = 30/[tt(d) + (30 - d)/V_b] \tag{2.6}$$

where d is the depth to the bottom of the velocity model, tt(d) is the travel time to depth d and V_b is the velocity at the bottom of the model (Boore 2004).

For all 536 stations of the SHARE-AUTH database, the thickness of the soil deposits H (i.e. depth to "seismic" bedrock – $V_s > 800$ m/s), the average shear wave velocity $V_{s,av}$ of the entire soil deposit (Eq. 2.7) and the fundamental period T_0 of the soil deposit (Eq. 2.8) were estimated in addition to $V_{s,30}$:

$$V_{s,av} = H/\sum_{i=1,N_h} h_i/V_i \tag{2.7}$$

$$T_0 = 4H/V_{s,av} \tag{2.8}$$

where h_i and V_i denote respectively the thickness and shear-wave velocity of the i-th layer, in a total of N_b layers from top until "seismic" bedrock. For those sites, the shear wave velocity profiles of which did not extend to the depth of "seismic" bedrock, depth of bedrock and Vs profile until bedrock were estimated using three different methods. For the Turkish sites, general linear fit curves were developed for EC8 soil classes B and C. For the sites which could be characterized as rather uniform based on their borehole log (41 sites), the slopes of the general fit lines were used to extend the shear wave velocity at the bottom of each model until it reached the value of 800 m/s. For the sites characterized as non-uniform or for which no borehole log was available (45 sites), site-specific linear fit curves were developed for each site and were used to estimate the V_s profiles until the depth which corresponds to a shear wave velocity equal to 800 m/s. Site-specific linear fits were also developed for nine USA sites. For the Japanese K-net sites, the horizontal-to-vertical Fourier spectrum ratio (HVSR) technique was used. This technique was introduced by Nakamura (1989) to analyze ambient seismic noise records, and was later applied by Lermo and Chávez-García (1993) to earthquake ground motion. HVSR was applied only to stations with at least 3 ground motion records (81 stations) to estimate the resonant frequency of each station. The depth of "seismic" bedrock was then estimated, assuming a constant shear wave velocity from bottom of V_s model until the "seismic" bedrock. HVSR technique was also applied to 20 Italian sites. An extensive list of the stations of SHARE-AUTH database is provided in Pitilakis et al. (2012a).

The following criteria were applied to both SHARE and SHARE-AUTH databases for the selection of the ground motion records used in this work:

- Surface wave magnitude $M_s \geq 4$. For the records for which M_s was not available, it was estimated from M_w using the empirical relation proposed by Scordilis (2006).
- Available spectral values at least up to 2.5 s.
- Peak Ground Acceleration PGA $\geq$ 20 cm/s^2, considering the fact that design spectra should be derived basically from records from strong earthquakes.

The application of the aforementioned criteria resulted to a dataset extracted from SHARE database, consisting of 3500 3-component accelerograms and to a dataset extracted from SHARE-AUTH database, consisting of 715 3-component accelerograms. The former dataset was used for the validation of Eurocode 8 amplification factors and design spectra, while the latter was used for the proposal of the new soil classification system and corresponding normalized spectra and soil amplification factors. It should be stressed, however, that EC8 spectra were validated against two more datasets extracted from SHARE database, with different ranges of PGA, one including all records regardless of PGA and a second one containing strong motion records with PGA $\geq$ 150 cm/s^2, as representative for high seismicity. The results of these two datasets are presented in Pitilakis et al. (2012b).

2.3.2 Proposed Soil Classification Scheme

A new, more detailed, soil classification system that includes soil type, stratigraphy, depth to bedrock, soil stiffness, fundamental period of soil deposit and average shear wave velocity of the entire soil deposit as key parameters was presented by

Pitilakis et al. (2004, 2006), based exclusively on theoretical 1D ground response analyses of various representative models of realistic site conditions and was later modified by Pitilakis et al. (2012a, 2013) by adjusting the limits of the parameters describing each soil class.

According to the proposed classification system, which is described in detail in Table 2.5, the sites are classified into six basic categories (A, B, C, D, E and X) based on their qualitative description and stiffness characteristics. This form is generally compatible with the categories proposed in EC8, introducing at the same time some extra subclasses, which allow to take into consideration the influence of the depth of bedrock. It should be stressed however, that there is no complete correspondence between the proposed soil classes and the main EC8 soil classes. For instance, a site which is classified as C1, with the new classification, may be a B site according to EC8. The new improved soil classification system is more convenient and practical from geotechnical point of view. At the same time it introduces as main classification parameter the predominant period of the site (To), which is a fundamental factor for site amplification.

The classification of the 536 sites of the SHARE-AUTH database according to the new classification scheme is given in Fig. 2.12. Due to the insufficient number of data for the classification of D sites to subclasses D1, D2 and D3, the three subclasses were unified to one single class (D).

2.3.2.1 Shape of Horizontal Elastic Response Spectra

Regarding the equations describing the proposed response spectra for the soil classes of the new classification system, the general form of the equations proposed by EC8 was adopted, allowing however for a differentiation of spectral amplification parameter, which in EC8 is constant and equal to 2.5. The equations describing the elastic acceleration spectra for 5 % damping are the following:

$$0 \leq T \leq T_B : \frac{S_a(T)}{PGA_{rock}} = S \cdot \left[1 + \frac{T}{T_B} \cdot (\beta - 1)\right] \tag{2.9}$$

$$T_B \leq T \leq T_C : \frac{S_a(T)}{PGA_{rock}} = S \cdot \beta \tag{2.10}$$

$$T_C \leq T \leq T_D : \frac{S_a(T)}{PGA_{rock}} = S \cdot \beta \cdot \frac{T_C}{T} \tag{2.11}$$

$$T_D \leq T : \frac{S_a(T)}{PGA_{rock}} = S \cdot \beta \cdot T_C \cdot \left(\frac{T_D}{T^2}\right) \tag{2.12}$$

where PGA_{rock} is the design ground acceleration at rock-site conditions, S is the soil amplification factor, T_B and T_C are the limits of the constant spectral acceleration branch, T_D is the value defining the beginning of the constant spectral displacement

Table 2.5 Proposed soil and site characterization

Soil class	Description	T_O (s)	Remarks
A1	Rock formations		$V_s \geq 1500$ m/s
A2	Slightly weathered/segmented rock formations (thickness of weathered layer <5.0 m)	≤0.2	Surface weathered layer: $V_s \geq 200$ m/s Rock Formations: $V_s \geq 800$ m/s
	Geologic formations resembling rock formations in their mechanical properties and their composition (e.g. conglomerates)		$V_s \geq 800$ m/s
B1	Highly weathered rock formations whose weathered layer has a considerable thickness (5.0–30.0 m)	≤0.5	Weathered layer: $V_s \geq 300$ m/s
	Soft rock formations of great thickness or formations which resemble these in their mechanical properties (e.g. stiff marls)		V_s: 400–800 m/s *N*-SPT > 50, *Su* > 200 kPa
	Soil formations of very dense sand – sand gravel and/or very stiff/to hard clay, of homogenous nature and small thickness (up to 30.0 m)		V_s: 400–800 m/s *N*-SPT > 50, $S_u > 200$ kPa
B2	Soil formations of very dense sand – sand gravel and/or very stiff/to hard clay, of homogenous nature and medium thickness (30.0–60.0 m), whose mechanical properties increase with depth	≤0.8	V_s: 400–800 m/s *N*-SPT > 50, $S_u > 200$ kPa
C1	Soil formations of dense to very dense sand – sand gravel and/or stiff to very stiff clay, of great thickness (>60.0 m), whose mechanical properties and strength are constant and/or increase with depth	≤1.5	V_s: 400–800 m/s *N*-SPT > 50, $S_u > 200$ kPa
C2	Soil formations of medium dense sand – sand gravel and/or medium stiffness clay (*PI* > 15, fines percentage > 30 %) of medium thickness (20.0–60.0 m)	≤1.5	V_s: 200–450 m/s *N*-SPT > 20, $S_u > 70$ kPa
C3	Category C2 soil formations of great thickness (>60.0 m), homogenous or stratified that are not interrupted by any other soil formation with a thickness of more than 5.0 m and of lower strength and V_s velocity	≤1.8	V_s:200–450 m/s *N*-SPT > 20, $S_u > 70$ kPa
D1	Recent soil deposits of substantial thickness (up to 60.0 m), with the prevailing formations being soft clays of high plasticity index (*PI* > 40), high water content and low values of strength parameters	≤2.0	$V_s \leq 300$ m/s *N*-SPT < 25, $S_u < 70$ KPa
D2	Recent soil deposits of substantial thickness (up to 60.0 m), with prevailing fairly loose sandy to sandy-silty formations with a substantial fines percentage (not to be considered susceptible to liquefaction)	≤2.0	$V_s \leq 300$ m/s *N*-SPT < 25
D3	Soil formations of great overall thickness (>60.0 m), interrupted by layers of category D1 or D2 soils of a small thickness (5.0–15.0 m), up to the depth of ~40 m, within soils (sandy and/or clayey, category C) of evidently greater strength, with $V_s \geq 300$ m/s	≤3.0	V_s: 150–600 m/s
E	Surface soil formations of small thickness (5.0–20.0 m), small strength and stiffness, likely to be classified as category C and D according to its geotechnical properties, which overlie category A formations ($V_s \geq 800$ m/s)	≤0.7	Surface soil layers: $V_s \leq 400$ m/s

(continued)

Table 2.5 (continued)

Soil class	Description	T_O (s)	Remarks
X	Loose fine sandy-silty soils beneath the water table, susceptible to liquefaction (unless a special study proves no such danger, or if the soil's mechanical properties are improved), Soils near obvious tectonic faults, Steep slopes covered with loose lateral deposits, Loose granular or soft silty-clayey soils, provided they have been proven to be hazardous in terms of dynamic compaction or loss of strength. Recent loose landfills, Soils with a very high percentage in organic material, Soils requiring site-specific evaluations		

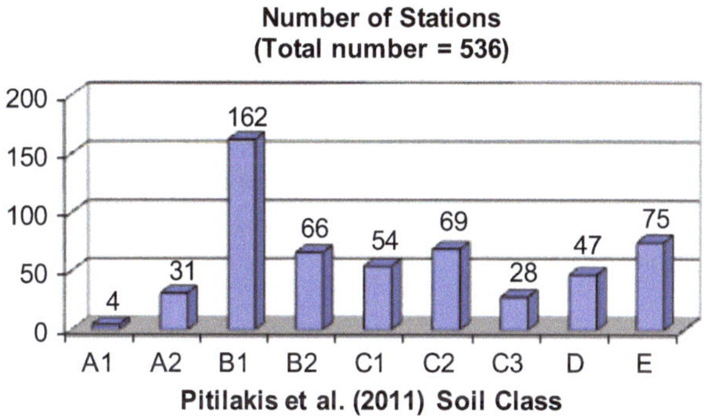

Fig. 2.12 Classification of the sites of SHARE-AUTH database according to the new soil classification scheme

range of the spectrum and β is the spectral amplification parameter. Parameters S, T_B, T_C, T_D and β depend on soil class and level of seismicity. For reasons of homogeneity, the two levels of seismicity proposed by EC8 were adopted (Type 2 spectrum if seismic hazard has been assessed mostly from earthquakes with surface-wave magnitude $M_s \leq 5.5$, otherwise Type 1 spectrum). Parameters T_B, T_C, T_D and β, which define the shape of the response spectra, were determined by fitting the general spectral equations (Eqs. (2.9), (2.10), (2.11), and (2.12) divided by soil amplification factor S) to the empirical data. Five percent damped normalized acceleration response spectra resulting from the selected strong-motion records, i.e. the acceleration spectra divided by PGA, were plotted for each soil class and for the two levels of seismicity. For each soil class and level of seismicity, the median normalized acceleration spectra, as well as the 16th and 84th percentiles were calculated. The specific percentiles were selected since, in the case of normal distribution of data, they represent the values of average minus one standard deviation and average plus one standard deviation respectively. The proposed normalized acceleration response spectra resulted by trying to fit the general spectral functions as close as possible to the 84th percentile, in order to increase

Table 2.6 Parameters of proposed normalized acceleration response spectra

	Type 2 (Ms ≤ 5.5)				Type 1 (Ms > 5.5)			
Soil class	T_B (s)	T_C (s)	T_D (s)	β	T_B (s)	T_C (s)	T_D (s)	β
ς	0.05	0.3	1.2	2.5	0.1	0.4	2	2.5
B1	0.05	0.25	1.2	2.75	0.1	0.4	2	2.75
B2	0.05	0.3	1.2	2.5	0.1	0.5	2	2.5
C1	0.1	0.25	1.2	2.5	0.1	0.6	2	2.5
C2	0.1	0.4	1.2	2.5	0.1	0.6	2	2.5
C3	0.1	0.5	1.2	2.5	0.1	0.9	2	2.5
D	0.1	0.7	1.2	2.5	0.1	0.7	2	2.5
E	0.05	0.2	1.2	2.75	0.1	0.35	2	2.75

the confidence limits, compared to fitting the proposed spectra to the median curve, which is not conservative enough for all soil classes. Moreover, it was considered important that normalized acceleration response spectra should be derived based on a common rationale for all soil classes, which is not the case for the normalized response spectra of Eurocode 8. Parameters T_B, T_C, T_D and β for the soil classes of the proposed classification system and for both seismicity types are given in Table 2.6. Figures 2.13, 2.14, 2.15, 2.16, 2.17, 2.18, 2.19, and 2.20 illustrate for the different soil classes the range of the computed normalized acceleration spectra between 16 % and 84 % percentiles and the proposed design normalized acceleration spectra.

2.3.2.2 Soil Amplification Factors

A logic tree approach was used for the estimation of soil amplification factors to the reference rock basement motion. The logic tree approach (Scherbaum et al. 2005; Bommer and Scherbaum 2008) allows the use of alternative models, each of which is assigned a weighting factor that is interpreted as the relative likelihood of that model being correct. In this way, the epistemic uncertainties associated with the different models can be captured in an efficient way.

The logic tree that was implemented, which is shown in Fig. 2.21, combines two state-of-the-art methods with equal weights. The first method (Approach 1) calculates period-dependent amplification factors using Ground Motion Prediction Equations (GMPEs) for the estimation of reference acceleration spectral values, while the second method (Approach 2) calculates a constant period-independent amplification factor with respect to the rock sites of the database.

2.3.2.3 Approach 1

In this approach, amplification is evaluated by normalizing the spectra of recorded motions by a reference (rock) spectrum obtained from GMPEs. Amplification

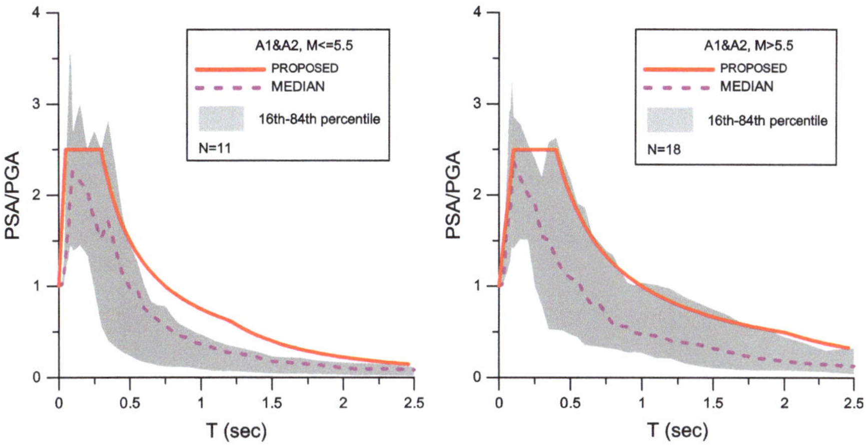

Fig. 2.13 Normalized elastic acceleration response spectra for soil class A (A1 + A2) of the proposed classification system for Type 2 seismicity (*left*) and Type 1 seismicity (*right*)

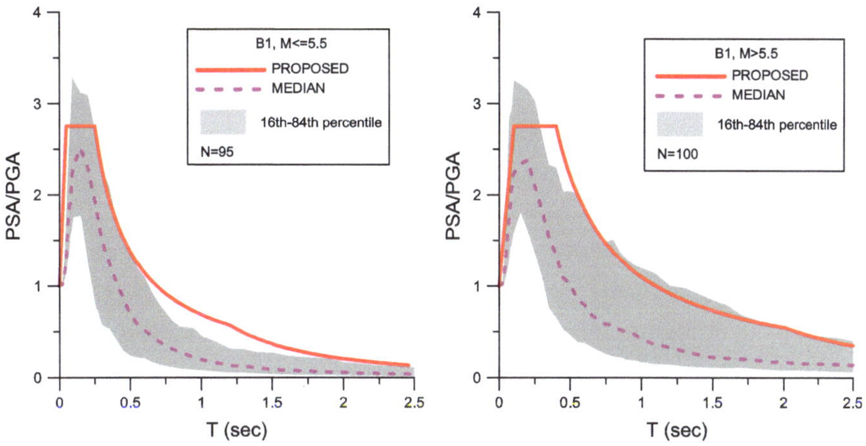

Fig. 2.14 Normalized elastic acceleration response spectra for soil class B1 of the proposed classification system for Type 2 seismicity (*left*) and Type 1 seismicity (*right*)

factor for ground motion j within site class i, S_{ij}, at spectral period T, was evaluated from the geometric mean of 5 % damped acceleration response spectra for the two horizontal components of shaking, GM_{ij}, and the reference ground motion for the site, $(GM_r)_{ij}$, using Eq. (2.13) (Choi and Stewart 2005):

$$S_{ij}(T) = GM_{ij}/(GM_r)_{ij} \tag{2.13}$$

GM_{ij} and $(GM_r)_{ij}$ were computed at the same spectral period, which was varied from 0 to 2.5 s.

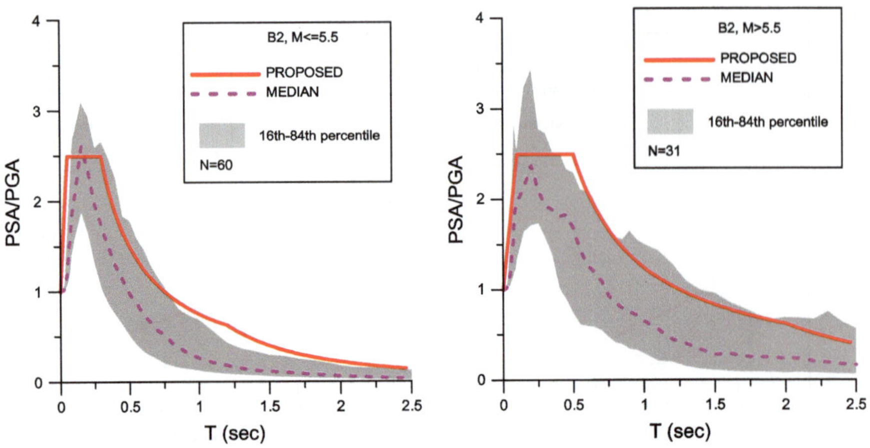

Fig. 2.15 Normalized elastic acceleration response spectra for soil class B2 of the proposed classification system for Type 2 seismicity (*left*) and Type 1 seismicity (*right*)

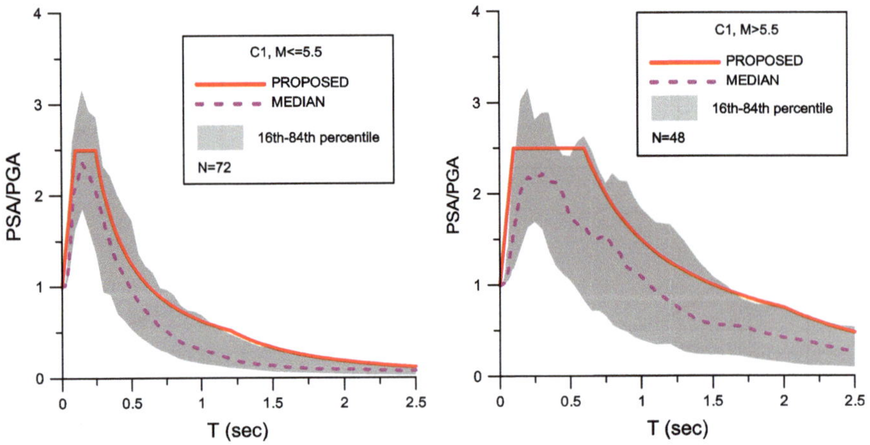

Fig. 2.16 Normalized elastic acceleration response spectra for soil class C1 of the proposed classification system for Type 2 seismicity (*left*) and Type 1 seismicity (*right*)

For each dataset the reference motion parameter $(GM_r)_{ij}$ was estimated as the weighted average of the rock predictions of the four GMPEs proposed in SHARE for active shallow crustal regions (Delavaud et al. 2012), since the majority of the stations are in active regions. The proposed models and corresponding weights are the following:

- Akkar and Bommer (2010), hereinafter referred to as A&B, with a weighting factor equal to 0.35
- Cauzzi and Faccioli (2008), hereinafter referred to as C&F, with a weighting factor equal to 0.35

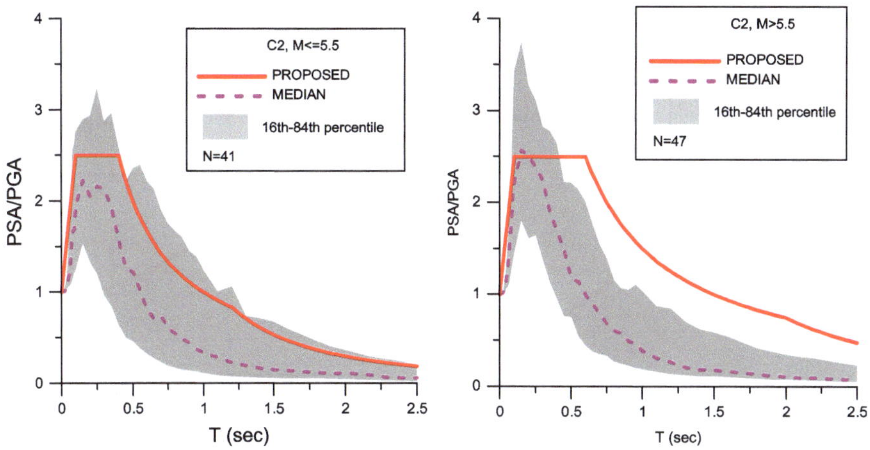

Fig. 2.17 Normalized elastic acceleration response spectra for soil class C2 of the proposed classification system for Type 2 seismicity (*left*) and Type 1 seismicity (*right*)

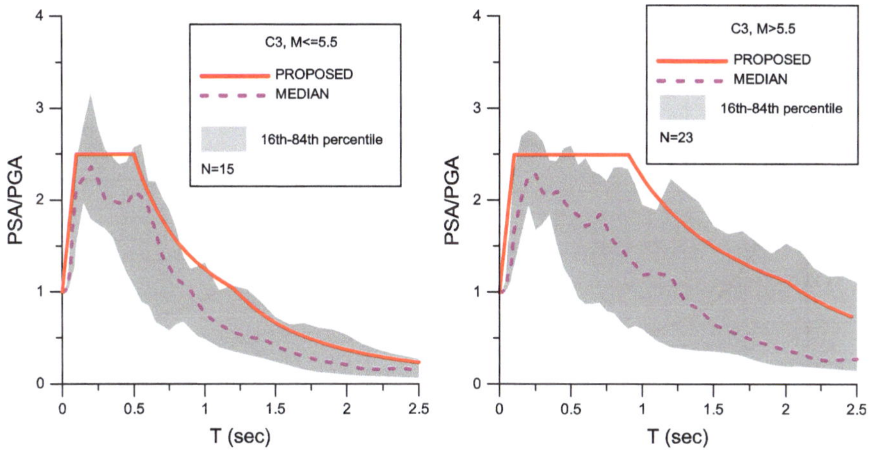

Fig. 2.18 Normalized elastic acceleration response spectra for soil class C3 of the proposed classification system for Type 2 seismicity (*left*) and Type 1 seismicity (*right*)

- Zhao et al. (2006a), hereinafter referred to as Zh, with a weighting factor equal to 0.10
- Chiou and Youngs (2008), hereinafter referred to as C&Y, with a weighting factor equal to 0.20

The four GMPEs require the knowledge of different parameters for the prediction of ground motion. For example, regarding the distance measures, A&B use Joyner-Boore distance R_{jb}, C&F the hypocentral distance R_{hyp} and Zh and C&Y the shortest distance to the rupture plane R_{rup}. These distance measures were not

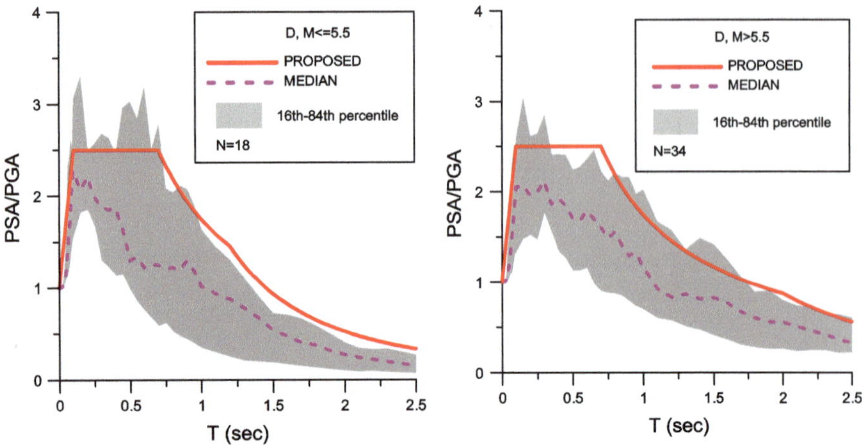

Fig. 2.19 Normalized elastic acceleration response spectra for soil class D (D1, D2 and D3) of the proposed classification system for Type 2 seismicity (*left*) and Type 1 seismicity (*right*)

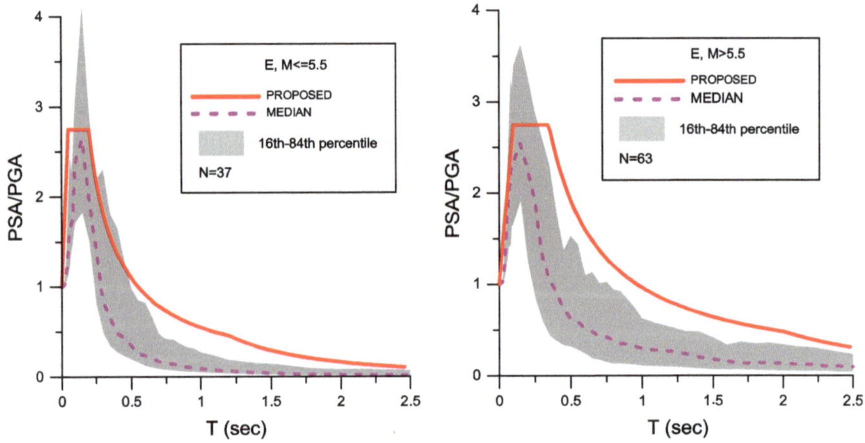

Fig. 2.20 Normalized elastic acceleration response spectra for soil class E of the proposed classification system for Type 2 seismicity (*left*) and Type 1 seismicity (*right*)

available for all records in the database. As a result, each GMPE could be implemented for a variable number of records, and in any case, only for those records for which all necessary data were either available or could be estimated in a reliable manner. Taking into account all four GMPEs and using the aforementioned weights, the reference ground motion $(GM_r)_{ij}$ at a certain period was calculated for each ground motion j within site class i with Eq. (2.14):

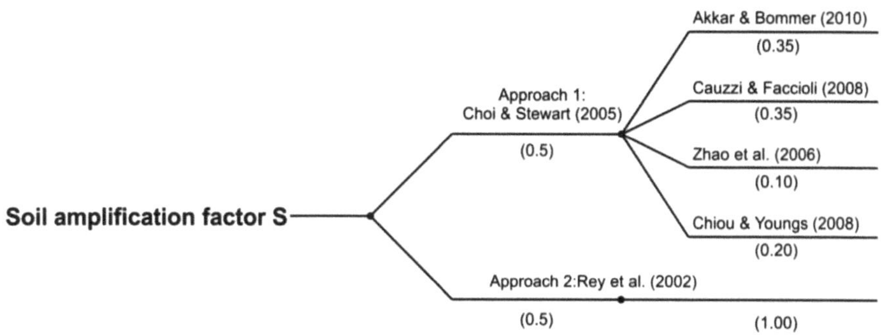

Fig. 2.21 Logic tree for estimation of soil amplification factors

$$\begin{aligned}(\mathrm{GM_r})_{\mathrm{ij}}(\mathrm{T}) &= 0.35\cdot(\mathrm{GM_r})_{\mathrm{ij,AB}} + 0.35\cdot(\mathrm{GM_r})_{\mathrm{ij,CF}} \\ &\quad + 0.10\cdot(\mathrm{GM_r})_{\mathrm{ij,Zh}} + 0.20\cdot(\mathrm{GM_r})_{\mathrm{ij,CY}}\end{aligned} \tag{2.14}$$

where $(\mathrm{GM_r})_{\mathrm{ij},\mathrm{AB}}$, $(\mathrm{GM_r})_{\mathrm{ij},\mathrm{CF}}$, $(\mathrm{GM_r})_{\mathrm{ij},\mathrm{Zh}}$, $(\mathrm{GM_r})_{\mathrm{ij},\mathrm{CY}}$, are the reference spectral accelerations at period T, calculated using the A&B, C&F, Zh and C&Y GMPEs respectively.

The computation of reference spectral acceleration with Eq. (2.14) requires that reference spectral acceleration at period T can be estimated with all four GMPEs. However, each GMPE could be implemented for a different subset of data, since each GMPE requires the knowledge of different parameters. As a result, the computation of reference spectral acceleration with all four GMPEs, and thus the estimation of amplification factors with Approach 1, was not feasible for all 715 selected records of SHARE-AUTH database, but only for a subset, hereinafter referred to as subset of common records (Table 2.7). A representative plot showing the median amplification factors for soil class B2-Type 2, estimated from the subset of common records using Eq. (2.14), is illustrated in Fig. 2.22a.

It is obvious that the restriction of using only those strong motion records, for which all GMPEs can be applied, limits the dataset significantly. As a result, for some soil classes (e.g. B1) there are adequate data, while for other soil classes (e.g. soil classes D and E) there are very few or even no available data. In order to overcome this obstacle, we decided, in addition to estimating amplification factors using only the subset of common records, to calculate the median amplification factors by implementing each one of the A&B, C&F, Zh and C&Y GMPEs separately to as many records as possible and then to apply the weights to these median values. In this way, each GMPE is applied to a different subset and amplification factors are given by the following equation:

$$S(T) = \frac{1}{\dfrac{0.35}{S_{AB}} + \dfrac{0.35}{S_{CF}} + \dfrac{0.10}{S_{Zh}} + \dfrac{0.20}{S_{CY}}} \tag{2.15}$$

Table 2.7 Number of strong motion records from SHARE-AUTH database for which reference spectral acceleration could be estimated with all GMPEs (subset of common records)

Soil class	Type 2 ($4 \leq M_s \leq 5.5$)	Type 1 ($M_s > 5.5$)
B1	33	41
B2	29	9
C1	5	18
C2	15	9
C3	5	10
D	1	7
E	–	4

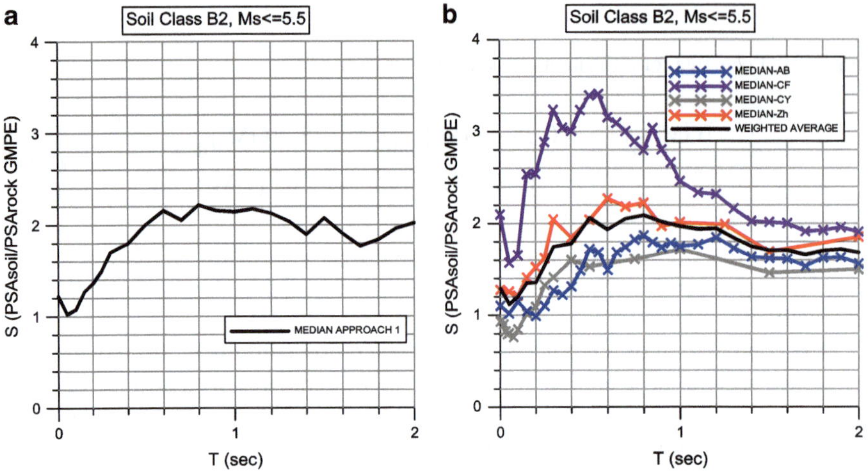

Fig. 2.22 Amplification factors estimated with Approach 1 for soil class B2-Type 2 seismicity, using (**a**) the subset of common records and (**b**) the different subsets

where $\overline{S_{AB}}$, $\overline{S_{CF}}$, $\overline{S_{Zh}}$ and $\overline{S_{CY}}$ are the median amplification factors, calculated using solely each one of the A&B, C&F, Zh and C&Y GMPEs respectively. Table 2.8 presents the number of strong motion records from SHARE-AUTH database for which each one of the four GMPEs could be implemented separately. A sample plot for soil class B2-Type 2 showing the median amplification factors calculated using each one of the four GMPEs separately (different subsets), as well as the weighted average as derived from Eq. (2.15), is illustrated in Fig. 2.22b.

Finally, in order to estimate a single period-independent amplification factor for each soil class and each seismicity level, appropriate for incorporation into the elastic response spectra equations, the amplification factors were averaged over a range of periods from T = 0 to T = 2.0s, since Zh and C&Y GMPEs provide no intermediate values between 2.0 and 2.5s. It should be stressed, however, that such an averaging eliminates the period-dependent character of the amplification. The resulting amplification factors do not represent only the amplification related to the increase of ordinates of soil spectra with respect to rock spectra, but also the amplification due to the change in shape of PGA-normalized response spectra, since average spectra of softer soils tend to have a larger and shifted towards longer periods plateau compared to rock spectra (Rey et al. 2002). In order to assure that

Table 2.8 Number of strong motion records from SHARE-AUTH database for which each GMPE could be implemented (different datasets)

Soil class	A&B		C&F		Zh		C&Y	
	$M_s \leq 5.5$	$M_s > 5.5$	$M_s \leq 5.5$	$M_s > 5.5$	$M_s \leq 5.5$	$M_s > 5.5$	$M_s \leq 5.5$	$M_s > 5.5$
B1	52	49	76	96	76	96	41	52
B2	41	14	41	25	41	25	31	14
C1	11	20	21	44	21	44	5	27
C2	17	12	26	38	26	38	17	11
C3	11	12	14	20	14	20	15	11
D	7	10	17	34	17	34	1	8
E	8	4	25	46	25	46	–	13

Table 2.9 Spectral shape ratios SR for the new classification scheme

Soil class	$M_s \leq 5.5$	$M_s > 5.5$
B1	0.97	1.10
B2	1.00	1.16
C1	0.87	1.29
C2	1.20	1.29
C3	1.38	1.62
D	1.67	1.41
E	0.82	1.00

the proposed soil factors represent only the amplification related to the increase of ordinates of soil spectra with respect to rock spectra, the period-averaged amplification factors were divided by the spectral shape ratio values (SR), which represent the amplification due to the change in shape of PGA-normalized response spectra, since average spectra of softer soils tend to have a larger and shifted towards longer periods plateau compared to rock spectra. The spectral shape ratios SR for the soil classes of the proposed soil classification scheme are given in Table 2.9.

Table 2.10 summarizes the soil factors obtained for the new soil classes with Approach 1, using on the one hand the subset of common records and on the other hand the different subsets. Soil factor values for classes with sufficient strong motion data in the common dataset (e.g. B1 and B2) obtained from the different subsets are very close to the ones obtained from the subset of common records. This justifies the decision to use the different datasets for the cases where there were only few or even no available strong motion data in the common dataset.

2.3.2.4 Approach 2

This approach, proposed by Rey et al. (2002), is based on a systematic analysis of response spectra ordinates as a function of magnitude and site conditions. Period-independent soil amplification factors are calculated for each soil class and for

Table 2.10 Soil factors for the new soil classes with Approach 1 obtained from the subset of common records, compared to those obtained from the different datasets

	$M_s \leq 5.5$		$M_s > 5.5$	
Soil class	Approach 1 (common dataset)	Approach 1 (different datasets)	Approach 1 (common dataset)	Approach 1 (different datasets)
B1	1.28	1.25	1.03	1.05
B2	1.89	1.77	1.36	1.33
C1	2.36	2.02	2.19	1.95
C2	2.08	1.86	1.35	1.26
C3	2.29	2.59	1.57	1.56
D	1.98	2.19	1.69	2.03
E	–	1.54	0.93	1.10

different magnitude intervals (M.I.) of $M_s = 0.5$, with respect to the rock sites of the database using Eq. (2.16):

$$S = \frac{I_{soil}}{I_{rock}} \cdot \frac{1}{SR} \tag{2.16}$$

SR is the spectral shape ratio, I_{soil} and I_{rock} are the spectrum intensities for soil and rock respectively, originally defined by Housner (1952) for spectral velocities and here adapted for spectral accelerations (Eq. 2.17):

$$I = \int_{0.05}^{2.5} \overline{R \cdot S_a(T)}\, dt \tag{2.17}$$

where $\overline{R \cdot S_a(T)}$ denotes the log-average of distance-normalized 5 % spectral ordinates RSa(T) for each soil class and magnitude interval. I_{soil}/I_{rock} ratios provide a scaling factor for site effect that represents an average amplification globally affecting the whole spectrum (Rey et al. 2002).

Figure 2.23 illustrates the log average of distance-normalized response spectra for soil class C1, with respect to soil class A, for six magnitude intervals, ranging from $M_s = 4.0$ to $M_s = 7.0$. It is observed that the log average curves for soil type C1 have a general trend to lie above the corresponding curves for soil A, so the estimation of an average amplification through a single period-independent factor is justified (Rey et al. 2002). Table 2.11 presents the calculated I_{C1}/I_A ratios for all M.I. The table also contains the number of available strong motion records (third and seventh columns) and the range of PGA (fourth and eighth columns, with the number in the parenthesis representing the median PGA for soil class C1 records) for each M.I. It is obvious that for some magnitude intervals there were not adequate data, so the corresponding I_{C1}/I_A ratios could not be considered reliable. For example, for the M.I. ranging from 5.5 to 6.0, the log-average curve for soil class A, which has been estimated from only two strong-motion records, lies above the corresponding curve for soil class C1, which has been estimated from five

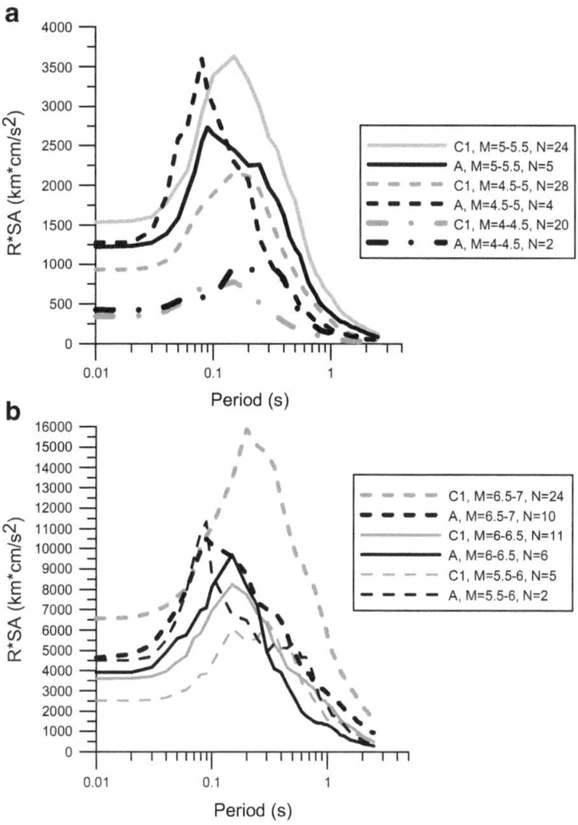

Fig. 2.23 Log-average, distance-normalized acceleration response spectra for sites of soil class C1 (*grey lines*) and rock sites (*black lines*) for (**a**) Type 2 and (**b**) Type 1 magnitude intervals

Table 2.11 I_{soil}/I_A ratio for soil class C1 and all magnitude intervals

$M_s \leq 5.5$				$M_s > 5.5$			
M_s	I_{C1}/I_A	n(C1)/n(A)	PGA (cm/s^2)	M_s	I_{C1}/I_A	n(C1)/n(A)	PGA (cm/s^2)
4.0–4.5	0.53	20/2	<69 (29)	5.5–6.0	0.86	5/2	≤251 (84)
4.5–5.0	1.16	28/4	≤215 (55)	**6.0–6.5**	1.36	11/6	≤445 (114)
5.0–5.5	1.40	24/5	≤332 (80)	**6.5–7.0**	1.84	24/10	≤1302 (377)

strong motion records (Fig. 2.23). The magnitude intervals with a satisfactory number of available strong motions for both soil C1 and rock are depicted in bold. Only the ratios corresponding to these intervals were used to estimate a mean I_{C1}/I_A ratio for each spectrum type, which was then inserted in Eq. (2.16) to derive the soil amplification factor S.

Table 2.12 gives the I_{soil}/I_A ratios for the soil classes of the new classification scheme and for the two seismicity contexts. I_{soil}/I_A coefficients of Table 2.12 were calculated as the mean values of the coefficients from the magnitude intervals considered as more reliable.

Table 2.12 I_{soil}/I_A ratios for the new soil classes and both seismicity contexts

	$M_s \leq 5.5$		$M_s > 5.5$	
Soil class	Selected M.I.	I_{soil}/I_A	Selected M.I.	I_{soil}/I_A
B1	4.5–5.0, 5.0–5.5	0.99	6.0–6.5, 6.5–7.0	1.12
B2	4.5–5.0, 5.0–5.5	1.17	6.0–6.5, 6.5–7.0	1.70
C1	4.5–5.0, 5.0–5.5	1.28	6.0–6.5, 6.5–7.0	1.60
C2	4.5–5.0, 5.0–5.5	1.66	6.0–6.5, 6.5–7.0	1.55
C3	4.5–5.0, 5.0–5.5	2.22	6.0–6.5, 6.5–7.0	1.91
D	4.5–5.0, 5.0–5.5	3.78	6.0–6.5, 6.5–7.0	2.58
E	4.5–5.0, 5.0–5.5	1.18	6.0–6.5, 6.5–7.0	0.96

The values of the soil factors for the new classification system obtained with the two different approaches are summarized in Table 2.13. As far as Approach 1 is concerned, for the cases where there were only few available strong motion data in the subset of common records, the results obtained from the different subsets are presented. Since both approaches were assigned a weighting factor equal to 0.5, the weighted average values are the mean of the values obtained with the two approaches considered.

The finally proposed S factors for the soil classes of the new classification scheme were determined by approximating and rounding (usually to slightly higher values) the weighted average of the values obtained from the two approaches. In case of excessively or "unrealistically" high values (i.e. for soil class D, Type 2 earthquakes) the proposed factor is lowered to more realistic values. For soil class E the proposed amplification factors were calculated from Kik-Net surface and bedrock strong-motion records. The final proposed S factors for the new soil classification system are included in Table 2.13.

The elastic acceleration response spectra for Type 2 and Type 1 seismicity, normalized by design ground acceleration at rock-site conditions a_g, are illustrated in Fig. 2.24a, b respectively.

2.3.3 Validation of Eurocode 8

Keeping the EC8 site classification criteria unchanged, amplification factors were estimated for the existing soil classes of EC8 using the selected strong motion records from SHARE database and the logic tree presented in Fig. 2.21, in order to compare the empirical amplification factors to those specified in the EC8 provisions. The same database was used for the validation of EC8 normalized acceleration response spectra (Pitilakis et al. 2012a, b). Median normalized response spectra and 16th and 84th percentiles were estimated for all EC8 soil classes. Two representative plots for EC8 soil class C and both seismicity types are given in Fig. 2.25. In general, EC8 spectra match rather well the empirical, since they are in most cases within the 16 and 84 % percentiles. However, they do not seem to have been derived based on a common rationale for all soil classes. For example, in some

Table 2.13 Soil factors for the soil classes of the proposed classification system obtained with Approaches 1 and 2, weighted average and proposed values

	$M_s \leq 5.5$				$M_s > 5.5$			
Soil class	Approach 1	Approach 2	Weighted average	Proposed	Approach 1	Approach 2	Weighted average	Proposed
B1	1.28	1.02	1.15	1.2	1.03	1.02	1.02	1.1
B2	1.89	1.17	1.53	1.5	1.36	1.47	1.41	1.4
C1	2.02	1.47	1.75	1.8	2.19	1.24	1.72	1.7
C2	2.08	1.38	1.73	1.7	1.35	1.20	1.28	1.3
C3	2.59	1.61	2.10	2.1	1.57	1.18	1.38	1.4
D	2.19	2.26	2.23	2.0	2.03	1.83	1.93	1.8
E	1.54	1.44	1.49	1.8[a]	1.10	0.96	1.03	1.4[a]

[a]Estimated from Kik-Net surface and bedrock strong-motion records

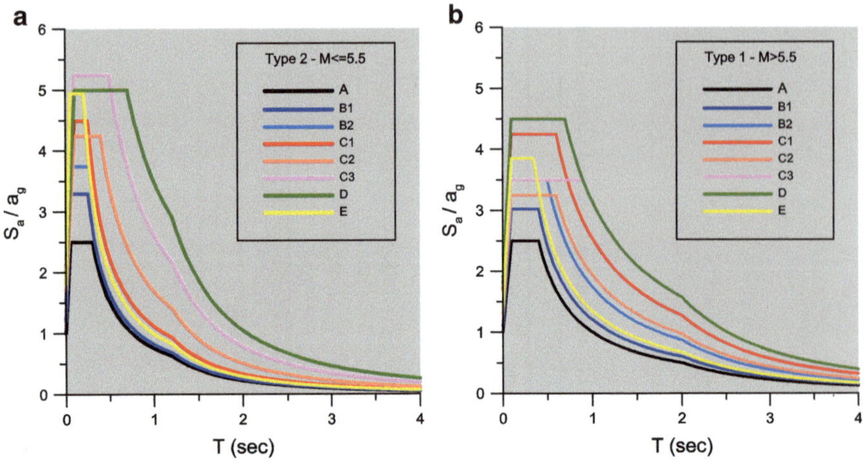

Fig. 2.24 (**a**) Type 2 and (**b**) Type 1 elastic acceleration response spectra for the proposed classification system

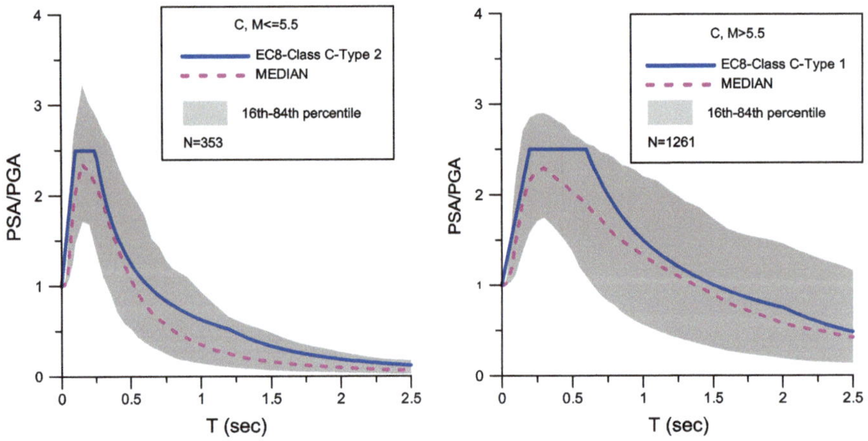

Fig. 2.25 Normalized elastic acceleration response spectra for EC8 soil class C for Type 2 seismicity (*left*) and Type 1 seismicity (*right*). *Solid lines* correspond to EC8 proposed spectra, *shaded areas* correspond to the region between the 16th and 84th percentile and *dashed lines* correspond to median empirical spectra

cases (e.g. soil classes A, B and C) EC8 spectra lie close to the median normalized curve, while in other cases (e.g. soil class E) EC8 spectra lie closer to or even above the 84th percentile of the empirical normalized spectra. EC8 spectra also seem to be better representing short periods than long periods. For example for soil class C-Type 1, the proposed spectrum is far below the 84th percentile for spectral periods greater than 0.7s, while it is much closer to the 84th percentile for shorter periods.

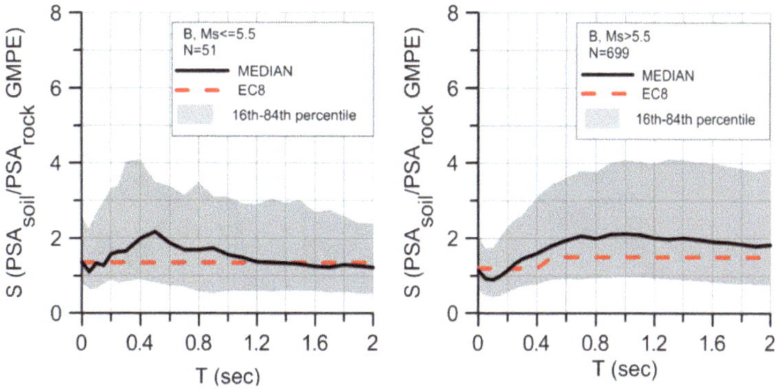

Fig. 2.26 Amplification factors estimated with Approach 1 for EC8 soil class B, for Type 2 (*left*) and Type 1 seismicity (*right*), compared to the corresponding EC8 spectra

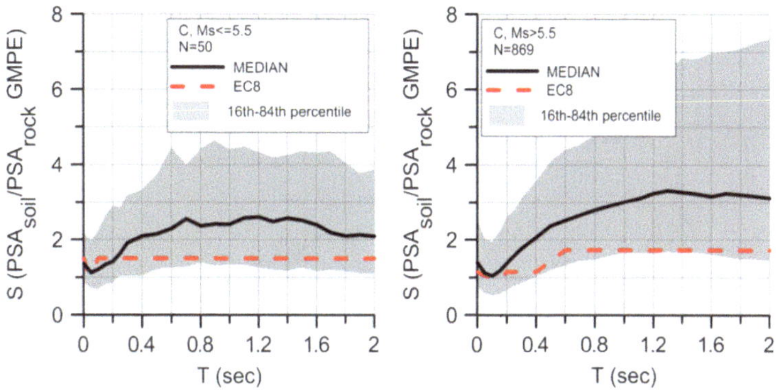

Fig. 2.27 Amplification factors estimated with Approach 1 for EC8 soil class C, for Type 2 (*left*) and Type 1 seismicity (*right*), compared to the corresponding EC8 spectra

Period-dependent amplification factors obtained with Approach 1 are presented in Figs. 2.26, 2.27, 2.28, and 2.29. For soil classes B and C the median amplification factors derived from the subset of common records are shown, along with the 16th and 84th percentiles (Figs. 2.26 and 2.27). The range of the amplification between these two percentiles, which is equivalent to the ± one standard deviation, is remarkably wide. For soil classes D and E the results from the different subsets are presented, illustrating the medians of the amplification factors derived by using each one of the four GMPEs separately for the estimation of reference ground motion, as well as the weighted average as derived with Eq. (2.15) (Figs. 2.28 and 2.29). In all cases the corresponding EC8 acceleration response spectra divided by the spectral values for soil class A are also provided for direct comparison.

The median amplification factors for soil class B and both types of seismicity do not exhibit a strong period-dependency, ranging between one and two for a broad period range between 0 and 2 s. Moreover they are in good agreement with the EC8

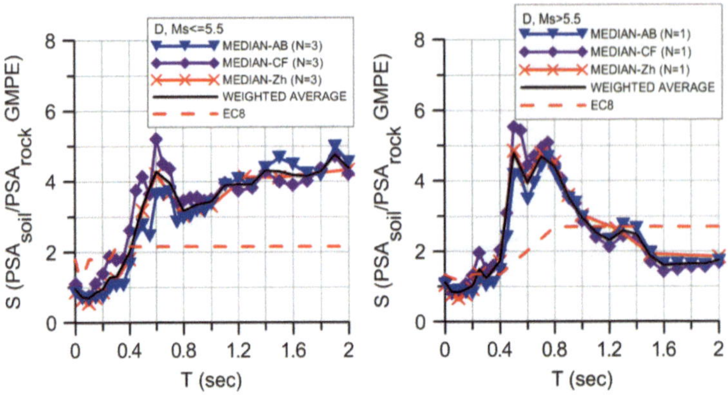

Fig. 2.28 Median amplification factors estimated with Approach 1 for soil class D and PSA_{rock} estimated with all GMPEs, for Type 2 (*left*) and Type 1 seismicity (*right*)

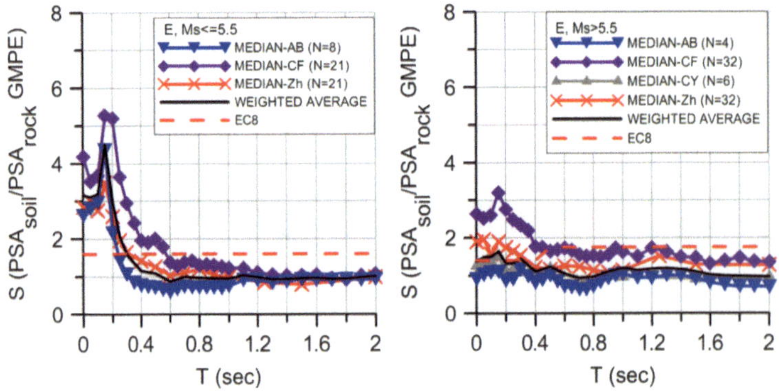

Fig. 2.29 Median amplification factors estimated with Approach 1 for soil class E and PSA_{rock} estimated with all GMPEs, for Type 2 (*left*) and Type 1 seismicity (*right*)

proposed spectra, with the latter lying very close to the median for Type 2 and slightly below the median for Type 1 seismicity.

For soil class C on the other hand, amplification factors are influenced by the spectral period, especially for Type 1 seismicity. In particular, for Type 2 seismicity the peak amplification is observed for spectral periods ranging from 0.7 to 1.4 s, while for Type 1 amplification is constantly increasing for all periods greater than 0.1 s. EC8 spectra seem unable to represent accurately the amplification potential of this soil class, since they are very close to the 16th percentile for almost the whole period range. A much stronger tendency for increasing amplification with period is observed for soil class D-Type 2, with peak amplification exceeding the value of 4.5, while for Type 1 the only available record presents a plateau between 0.5 and 0.9 s with amplification levels around 4.5. For soil class E-Type 2, amplification peaks at very low periods (around 0.15 s) reaching a value of 4.5 for the weighted

Table 2.14 I_{soil}/I_A ratios for EC8 soil classes and both seismicity contexts

	$M_s \leq 5.5$		$M_s > 5.5$	
Soil class	Selected M.I.	I_{soil}/I_A	Selected M.I.	I_{soil}/I_A
B	4.0–4.5, 4.5–5.0, 5.0–5.5	1.37	5.5–6.0, 6.0–6.5, 6.5–7.0	1.25
C	4.0–4.5, 4.5–5.0, 5.0–5.5	2.10	5.5–6.0, 6.0–6.5, 6.5–7.0	1.88
D	5.0–5.5	2.26	5.5–6.0	1.41
E	4.0–4.5, 4.5–5.0	1.96	5.5–6.0, 6.0–6.5, 6.5–7.0	0.96

Table 2.15 EC8 soil factors and soil factors obtained with Approaches 1 and 2 and their weighted average

	$M_s \leq 5.5$				$M_s > 5.5$			
Soil class	Approach 1	Approach 2	Weighted average	EC8	Approach 1	Approach 2	Weighted average	EC8
B	1.51	1.37	1.44	1.35	1.53	1.08	1.31	1.20
C	2.19	2.12	2.16	1.50	2.06	1.46	1.76	1.15
D	2.92	2.00	2.46	1.80	1.56	0.92	1.24	1.35
E	1.30	1.96	1.63	1.60	0.97	0.83	0.90	1.40

average, and then decreases sharply, remaining around 1.0 for periods greater than 0.4 s. For soil class E-Type 1, amplification ranges between 1 and 1.5 regardless of period value.

Table 2.14 gives the I_{soil}/I_A ratios calculated with Approach 2 for soil classes B, C, D and E and the two seismicity contexts of EC8. I_{soil}/I_A coefficients of Table 2.14 were calculated as the mean values of the coefficients from those magnitude intervals considered as more reliable.

The values of the soil factors for EC8 soil classes obtained with the two approaches are summarized in Table 2.15. The S factors of the current version of Eurocode 8 are also included.

Comparing the estimated S factors with those proposed in EC8, we observe that the S factors estimated for soil class B and both seismicity types are in general close to the EC8 soil factors. A slight increase of soil factor values for both seismicity types could be justified. For soil class C, however, the values adopted in EC8 are clearly smaller and need to be increased. For soil class D-Type 2, the estimated soil factor is much higher than the one proposed in EC8, while for Type 1 it is close to the present EC8 S factor. However, soil factors for soil class D have been derived from a limited number of records and should be interpreted with caution. For soil class E-Type 2, the obtained empirical S factors are close to the soil factors proposed in EC8, while for Type 1, the obtained S factors are much lower, and in fact close to unity. This is due on one hand to the poor seismological documentation of soil class E stations of the database, which led to a scarce number of records that could be utilized for the estimation of amplification on soil class E stations, and on the other hand on the averaging process that both methods use in order to compute a period-independent S factor, which can be applied to the whole spectrum. It is expected that in the near future the available records for soil classes D and E will be increased to allow for a better estimation of the amplification factors.

2.3.4 Comparison of the Performance of the New Classification System and EC8 Classification System

The performance of the proposed soil classification system was examined and compared to the performance of EC8 soil classification system in terms of an inter-category error term (Stewart et al. 2003). This error term represents the average dispersion of data within all categories of a given classification scheme. In this way, the ability of each classification scheme to capture site-to-site variations of spectral acceleration can be quantified. The inter-category error term is calculated with the following equation:

$$\sigma_R = \sqrt{\frac{\sum_{i=1}^{M_C}\sum_{i=1}^{N_i}\left(\varepsilon_{ij}-\varepsilon_i\right)^2}{\left(\sum_{i=1}^{M_C} N_i\right) - df}} \tag{2.18}$$

where M_c is the number of categories in the scheme and df is the total number of degrees-of-freedom.

Residuals ε_{ij}, which have a mean value ε_i, are calculated between the amplification prediction, which is derived from least-square regression analyses, and the actual amplification, as was calculated from Approach 1. It is reminded that in Approach 1, period-dependent amplification factors for ground motion j within site class i, S_{ij}, were evaluated for each strong-motion record, by dividing the geometric mean of 5 % damped acceleration response spectra for the two horizontal components of shaking with the reference ground motion, which was estimated using the weighted average of the four GMPEs.

Soil amplification factors calculated for the subset of common records of SHARE-AUTH dataset were sorted into the site categories defined by both the EC8 and the proposed classification scheme. A total amount of 191 strong motion records was used. For each scheme, regression analyses were performed to relate amplification factors S_{ij} at a certain period with a parameter G_{ij} which represents the amplitude of reference ground motion as follows:

$$\ln\left(S_{ij}\right) = a_i + b_i \ln\left(G_{ij}\right) + \varepsilon_{ij} \tag{2.19}$$

where a_i and bi are the regression coefficients. Peak reference ground acceleration (PGA_r) was selected as G_{ij}, as in Stewart et al. (2003).

The residuals ε_{ij} were calculated with the following equation:

$$\varepsilon_{ij} = \ln\left(S_{ij}\right)_{data}\text{-}\ln\left(S_{ij}\right)_{model} \tag{2.20}$$

Example results for EC8 classification scheme are given in Fig. 2.30, which illustrates the amplification factors at spectral period T = 0.3 s for soil classes B and C. Results of regression analyses performed according to Eq. (2.19) (solid lines), as well as median regression ± standard error (dashed lines) are also plotted.

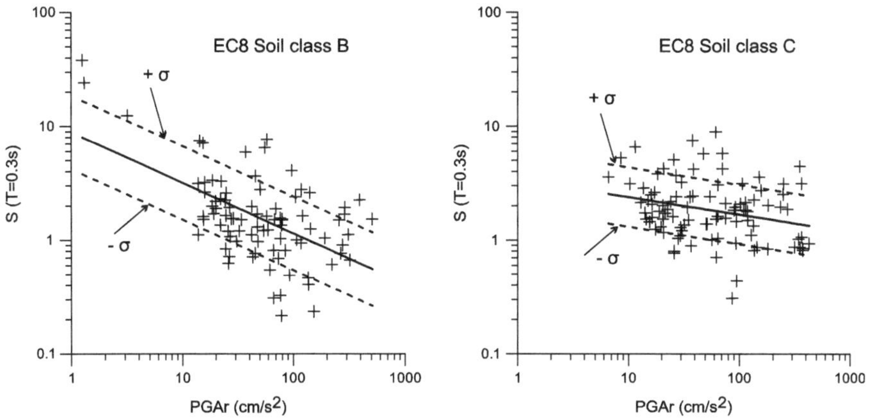

Fig. 2.30 Regression results for EC8 soil classes B and C at spectral period T = 0.3 s

Table 2.16 Regression coefficients for EC8 soil classes B and C

Soil class	Period	a	b	σ
B	PGA	2.090	−0.468	0.660
	0.3 s	2.171	−0.443	0.744
	1.0 s	1.606	−0.326	0.830
C	PGA	0.907	−0.138	0.534
	0.3 s	1.218	−0.152	0.599
	1.0 s	1.294	−0.090	0.754

The regression coefficients and standard error terms for three different spectral periods (T = 0, 0.3, 1.0s) are listed in Table 2.16.

Example results for the proposed classification scheme are given in Fig. 2.31, which illustrates the amplification factors at spectral period T = 0.3 s for soil classes B1, B2, C1 and C2. Results of regression analyses performed according to Eq. (2.19) (solid lines), as well as median regression ± standard error (dashed lines) are also plotted. The regression coefficients and standard error terms for three different spectral periods (T = 0, 0.3, 1.0s) are listed in Table 2.17.

Inter-category error terms σ_R for the two classification systems were calculated with Eq. (2.18) and are plotted as a function of period in Fig. 2.22. We observe that for the new classification system, σ_R error terms at all periods are lower than the error terms for EC8 classification system. The differences are amplified for longer periods (T > 0.4 s). Although the performance of the proposed classification system is better than that of the EC8 classification system, the level of improvement does not seem very impressive at first glance. For this reason we decided to additionally quantify the improvement caused when classifying the records of the same dataset to the soil classes of EC8, compared to classifying them based on a simplified criterion (two soil classes, one with $V_{s,30}$ < 400 m/s and a second one with $V_{s,30}$ > 400 m/s). The inter-category error for the simplified classification in two soil classes is also given in Fig. 2.32. We observe that the improvement caused

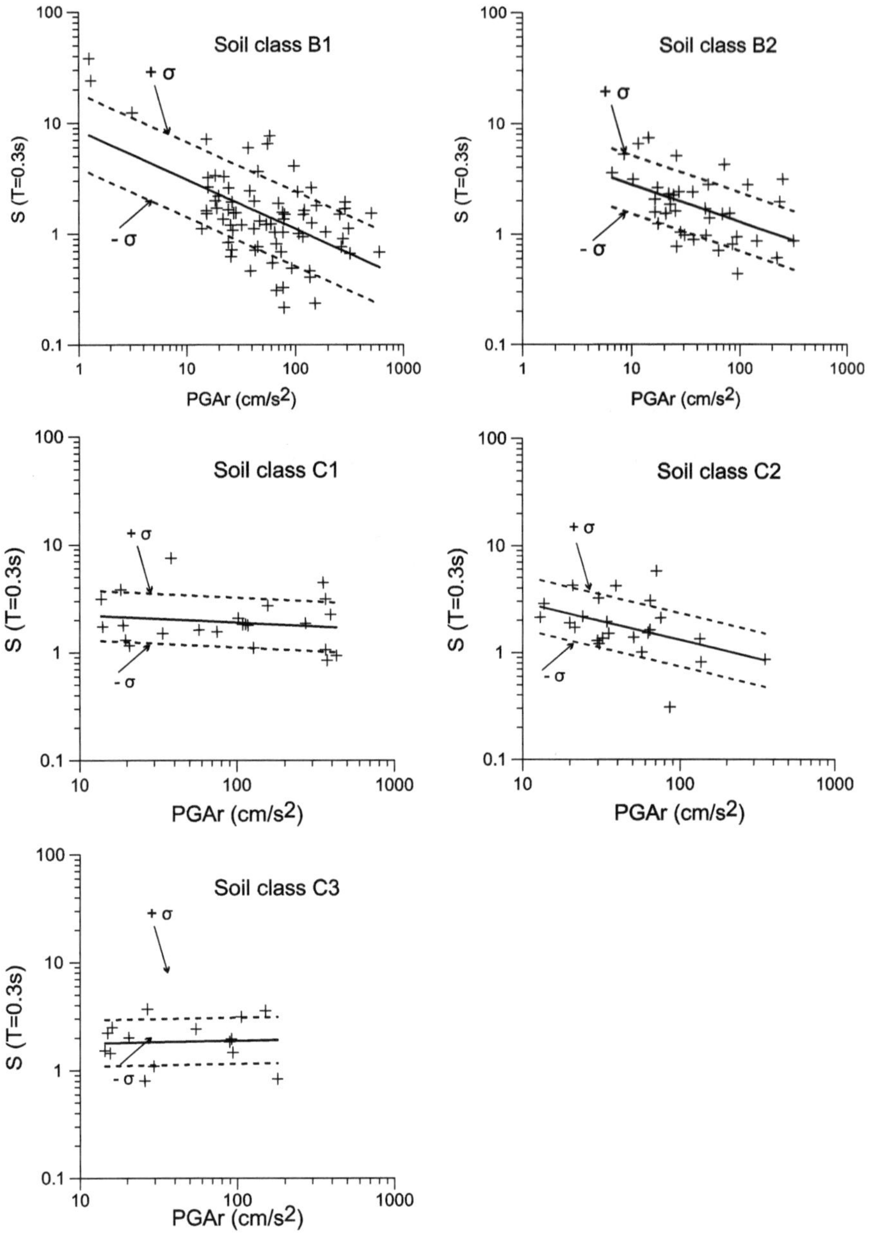

Fig. 2.31 Regression results for new soil classes at spectral period T = 0.3 s

when using the EC8 classification system instead of the simplified two-class system is almost negligible and, in any case, less significant than the one caused when using the proposed classification system instead of that of EC8. This shows that the nature of the problem allows for no major improvements. As a result, the improvement

Table 2.17 Regression coefficients for new soil classes B1, B2, C1, C2 and C3

Soil class	Period	a	b	σ
B1	PGA	2.272	−0.508	0.674
	0.3 s	2.143	−0.444	0.776
	1.0 s	1.566	−0.353	0.830
B2	PGA	1.280	−0.296	0.515
	0.3 s	1.801	−0.337	0.607
	1.0 s	1.110	−0.113	0.678
C1	PGA	0.991	−0.123	0.422
	0.3 s	0.966	−0.071	0.530
	1.0 s	1.540	−0.094	0.413
C2	PGA	1.022	−0.199	0.532
	0.3 s	1.871	−0.347	0.573
	1.0 s	1.944	−0.248	0.830
C3	PGA	0.471	−0.013	0.507
	0.3 s	0.518	0.025	0.492
	1.0 s	1.515	−0.103	0.588

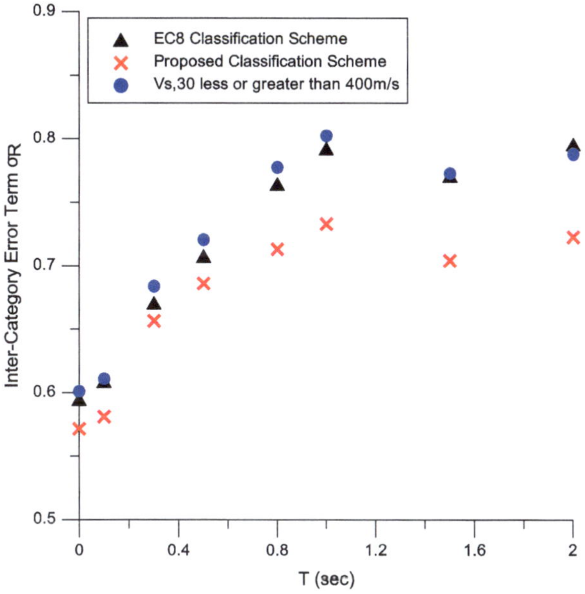

Fig. 2.32 Comparison of inter-category error term (σ_R) for the EC8, the proposed and the simplified classification systems as a function of period

brought by the new classification system could be considered as satisfactory as one could expect.

The dataset of the 191 well-documented strong motion records, which was used to estimate the inter-category error terms, was further used to compare the obtained soil amplification factors to site factors from recent studies of empirical ground motion data performed by a number of investigators for the NEHRP site classes.

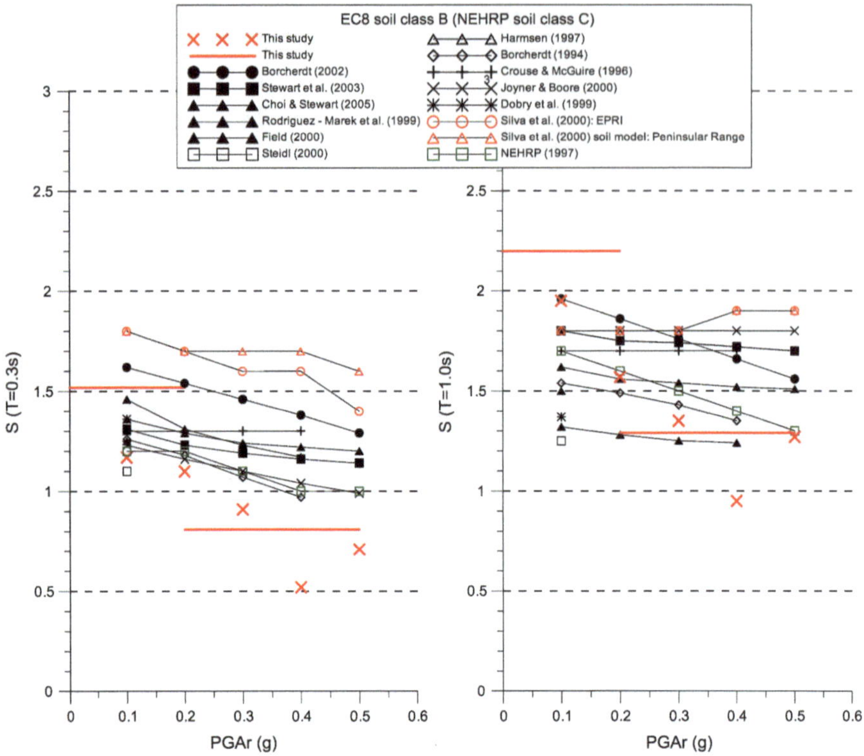

Fig. 2.33 Soil amplification factors for EC8 soil class B (NEHRP soil class C) obtained from recent studies of empirical ground motion data

Site amplification factors are plotted versus peak rock acceleration PGA_r for two periods of vibration, 0.3 and 1.0 s, in Figs. 2.33 and 2.34 respectively. Red x symbols from this study refer to the mean amplification factors corresponding to the specific $PGA_r \pm 0.05$ g (e.g. amplification for PGA_r equal to 0.3 g has been estimated from records with PGAr ranging from 0.25 to 0.35 g). Red continuous lines from this study refer to the mean soil amplification factors derived from all records with $PGA_r < 0.2$ g or >0.2 g.

The results of this study are generally in good agreement with the results from previous studies, particularly for EC8 soil class C. The obtained soil amplification factors for soil class B and period T = 0.3 s are quite low for large PGA_r values (>0.4 g), a result which is attributed to the very few available data for this PGA_r range.

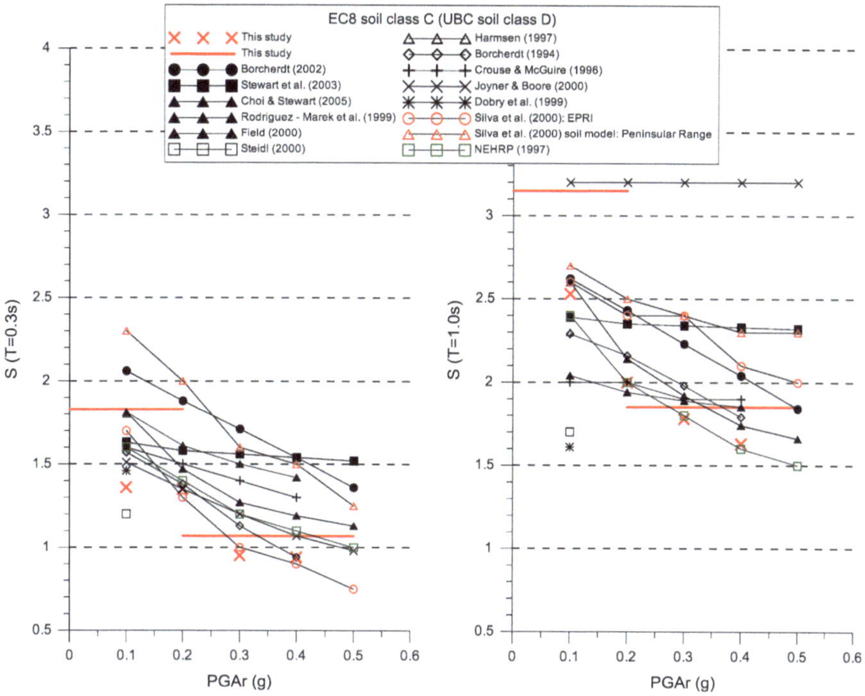

Fig. 2.34 Soil amplification factors for EC8 soil class C (NEHRP soil class D) obtained from recent studies of empirical ground motion data

2.4 Conclusions

In the frame of improvement of EC8, a new soil and site classification system for EC8-Part 5 was presented, aiming at eliminating the drawbacks of using $V_{s,30}$ as a key parameter for the definition of different soil classes and description of amplification potential for different soil conditions. The new classification system, which uses parameters such as the thickness of soil deposits, the average shear wave velocity to the seismic bedrock and the fundamental period of the site, exhibits an improved performance in terms of inter-category error compared to the classification system of Eurocode 8. Site amplification factors, as well as normalized acceleration response spectra were proposed for each soil class of the new classification system and the two seismicity contexts used in EC8 provisions, using a new worldwide strong motion database, which includes records from sites with accurately measured shear wave velocity profiles. New soil amplification factors were also estimated for the existing EC8 site classes, using a larger dataset. The comparison between the estimated values and the ones proposed in EC8 shows that for soil class C, the proposed values are much lower than the ones estimated in the present work for both spectrum types and need to be increased. For soil class B the estimated values are close to the EC8 soil factors, while for soil classes D and E a need for more data is identified.

Acknowledgements The work presented herein was performed in the framework of the EU collaborative project SHARE (FP7, Reference code 226967, http://www.share-eu.org/).

References

Aki K (1988) Local site effects on ground motion. In: Von Thun JL (ed) Earthquake engineering and soil dynamics II: recent advances in ground motion evaluation. Geotechnical special publication no 20, American Society of Civil Engineering, pp 103–155

Akkar S, Bommer JJ (2010) Empirical equations for the prediction of PGA, PGV and spectral accelerations in Europe, the Mediterranean region and the Middle East. Seismol Res Lett 81 (2):195–206

Akkar S, Çağnan Z, Yenier E, Erdogan E, Sandıkkaya MA, Gülkan P (2010) The recently compiled Turkish strong-motion database: preliminary investigation for seismological parameters. J Seismol 14(3):457–479

Applied Technology Council (ATC) (1978) Tentative provisions for the development of seismic regulations for buildings, ATC3-06 report, Redwood City, CA

Assimaki D, Li W, Steidl JH, Tsuda K (2008) Site amplification and attenuation via downhole array seismogram inversion: a comparative study of the 2003 Miyagi-Oki aftershock sequence. Bull Seismol Soc Am 98(1):301–330

Asten MW, Boore DM (2005) Comparison of shear-velocity profiles of unconsolidated sediments near the Coyote borehole (CCOC) measured with fourteen invasive and non-invasive methods, U.S. geological survey open-file report 2005-1169

Atakan K, Duval A-M, Theodulidis N, Bard P-Y, SESAME Team (2004) The H/V spectral ratio technique: experimental conditions, data processing and empirical reliability assessment. In: Proceedings of the 13th world conference on earthquake engineering, Paper no 2268, Vancouver, Canada, 1–6 Aug 2004

Bard PY (1999) Microtremor measurements: a tool for site effect estimation? In: Irikura K, Kudo K, Okada H, Sasatani T (eds) The effects of surface geology on seismic motion. Balkema, Rotterdam, pp 1251–1279

Bard PY, Bouchon M (1980a) The seismic response of sediment-filled valleys. Part 1. The case of incident SH waves. Bull Seismol Soc Am 70(4):1263–1286

Bard PY, Bouchon M (1980b) The seismic response of sediment-filled valleys. Part 2. The case of incident P and SV waves. Bull Seismol Soc Am 70(4):1921–1941

Bommer JJ (2011) Ground-motion models for defining seismic actions in Eurocode 8. In Akkar S, Gülkan P, van Eck T (eds) Earthquake data in engineering seismology – predictive models, data management and networks. Geotechnical, geological and earthquake engineering, vol 14, Springer

Bommer JJ, Pinho R (2006) Adapting earthquake actions in Eurocode 8 for performance-based seismic design. Earthq Eng Struct Dyn 35(1):39–55

Bommer JJ, Scherbaum F (2008) The use and misuse of logic trees in probabilistic seismic hazard analysis. Earthquake Spectra 24(4):997–1009

Bommer JJ, Elnashai AS, Weir AG (2000) Compatible acceleration and displacement spectra for seismic design codes. In: Proceedings of the 12th world conference on earthquake engineering, Auckland, New Zealand

Bommer JJ, Stafford PJ, Akkar S (2010) Current empirical ground-motion prediction equations for Europe and their application to Eurocode 8. Bull Earthq Eng 8(1):5–26

Boore DM (2004) Estimating $\overline{V}_{s,30}$ (or NEHRP site classes) from shallow velocity models (depths < 30 m). Bull Seismol Soc Am 94(2):591–597

Boore DM, Joyner WB, Fumal TE (1997) Equations for estimating horizontal response spectra and peak acceleration from western North American earthquakes: a summary of recent work. Seismol Res Lett 68(1):128–153

Borcherdt RD (1994) Estimates of site-dependent response spectra for design (methodology and justification). Earthquake Spectra 10(4):617–653

Borcherdt RD, Glassmoyer G (1992) On the characteristics of local geology and their influence on ground motions generated by the Loma Prieta earthquake in the San Francisco Bay region, California. Bull Seismol Soc Am 82(2):603–641

Cadet H, Bard PY, Duval AM (2008) A new proposal for site classification based on ambient vibration measurements and the Kiknet strong motion data set. In: Proceedings of the 14th world conference on earthquake engineering, Beijing, 12–17 Oct 2008

Cadet H, Bard PY, Rodriguez-Marek A (2012) Site effect assessment using KiK-net data: part 1. A simple correction procedure for surface/downhole spectral ratios. Bull Earthq Eng 10 (2):421–448

Cauzzi C, Faccioli E (2008) Broadband (0.05 to 20 s) prediction of displacement response spectra based on worldwide digital records. J Seismol 12(4):453–475

CEN (European Committee for Standardisation) (2004) Eurocode 8: design of structures for earthquake resistance, part 1: general rules, seismic actions and rules for buildings. EN 1998-1:2004, Brussels, Belgium

Chávez-García FJ (2007) Site effects: from observation and modelling to accounting for them in building codes. In Pitilakis KD (ed) Earthquake geotechnical engineering. In: Proceedings of the 4th international conference of earthquake geotechnical engineering-invited lectures, Thessaloniki, Greece, pp 53–72

Chávez-García FJ (2011) Site effects due to topography and to soft soil layers: progress made and pending issues: a personal perspective. In Vertugo R (ed) Earthquake geotechnical engineering. In: Proceedings of the 5th international conference of earthquake geotechnology engineering-invited lectures, Santiago, Chile, 10–13 Jan 2011

Chávez-García FJ, Bard P-Y (1994) Site effects in Mexico City eight years after the September 1985 Michoacan earthquakes. Soil Dyn Earthq Eng 13(4):229–247

Chávez-García FJ, Faccioli E (2000) Complex site effects and building codes: making the leap. J Seismol 4(1):23–40

Chávez-García FJ, Raptakis D, Makra K, Pitilakis K (2000) Site effects at Euroseistest – II. Results from 2D numerical modeling and comparison with observations. Soil Dyn Earthq Eng 19 (1):23–39

Chiou BS-J, Youngs RR (2008) An NGA model for the average horizontal component of peak ground motion and response spectra. Earthquake Spectra 24(1):173–215

Choi Y, Stewart JP (2005) Nonlinear site amplification as function of 30 m shear wave velocity. Earthquake Spectra 21(1):1–30

Crouse CB, McGuire JW (1996) Site response studies for purpose of revising NEHRP seismic provisions. Earthquake Spectra 12(3):407–439

Delavaud E, Cotton F, Akkar S, Scherbaum F, Danciu L, Beauval C, Drouet S, Douglas J, Basili R, Sandikkaya MA, Segou M, Faccioli E, Theodoulidis N (2012) Toward a ground-motion logic tree for probabilistic seismic hazard assessment in Europe. J Seismol. doi:10.1007/s10950-012-9281-z

Di Alessandro C, Bonilla LF, Boore DM, Rovelli A, Scotti O (2012) Predominant-period site classification for response spectra prediction equations in Italy. Bull Seismol Soc Am 102 (2):680–695

Di Giacomo D, Gallipoli MR, Mucciarelli M, Parolai S, Richwalski SM (2005) Analysis and modeling of HVSR in the presence of a velocity inversion: the case of Venosa, Italy. Bull Seismol Soc Am 95(6):2364–2372

Dobry R, Iai S (2000) Recent developments in understanding of earthquake site response and associated seismic code implementation. GeoENG 2000, vol 1. Melbourne, pp 186–219, 19–24 Nov 2000

Dobry R, Martin GM, Parra E, Bhattacharya A (1994) Development of site dependent ratios of elastic response spectra (RRS) and site categories for building seismic codes. In: Proceedings NCEER, SEAOC, BSSC workshop on site response during earthquakes and seismic code provisions, University of Southern California

Dobry R, Borcherdt RD, Crouse CB, Idriss IM, Joyner WB, Martin GR, Power MS, Rinne EE, Seed RB (2000) New site coefficients and site classification system used in recent building seismic code provisions. Earthquake Spectra 16(1):41–67
EUROSEISMOD (ENV4-CT.96-0255) (1996–1999) Development and experimental validation of advanced modeling techniques in engineering seismology and earthquake engineering. European Commission, Directorate General XII for Science, Research and Development, Environment and Climate Programme (Climate and natural hazards), Program Coordinator: K. Pitilakis
EUROSEISRISK (EVG1-CT-2001-00040) (2002–2005) Seismic hazard assessment, site effects and soil structure interaction in an instrumented basin. Directorate General XII for Science, Research and Development (Environment Programme: Global change and natural disasters) of European Union, Program Coordinator: K. Pitilakis
EUROSEISTEST (EV5V-CT.93-0281) (1993–1995) Volvi – Thessaloniki: a European test-site for engineering seismology, earthquake engineering and seismology. Funded by the European Commission, Directorate General XII for Science, Research and Development, Environment and Climate Programme (Climate and natural hazards unit), Program Coordinator: K. Pitilakis
Faccioli E (1992) Selected aspects of the characterisation of seismic site effects, including some recent European contributions. Invited lecture. In: Proceedings of the international symposium on the effects of surface geology on seismic motion (ESG1992), vol 1. Odawara (Japan), 25–27 Mar 1992, pp 65–96
Faccioli E, Villani M (2009) Seismic hazard mapping for Italy in terms of broadband displacement response spectra. Earthquake Spectra 25(3):515–539
Field EH, Jacob K (1995) A comparison and test of various site response estimation techniques, including three that are non reference-site dependent. Bull Seismol Soc Am 85(4):1127–1143
Field EH, Jacob KH, Hough SE (1992) Earthquake weak motion estimation: a weak motion case study. Bull Seismol Soc Am 82(6):2283–2307
Fukushima Y, Bonilla LB, Scotti O, Douglas J (2007) Site classification using horizontal-to-vertical response spectral ratios and its impact when deriving empirical ground-motion prediction equations. J Earthq Eng 11(5):712–724
Gallipoli MR, Mucciarelli M (2009) Comparison of site classification from VS30, VS10, and HVSR in Italy. Bull Seismol Soc Am 99:340–351
Housner GW (1952) Spectrum intensities of strong-motion earthquakes. In: Proceedings of the symposium on earthquakes and blast effects on structures, Earthquake Engineering Research Institute
ICC (International Code Council) (2012) International building code. International Code Council, Inc. www.iccsafe.org/Pages/default.aspx
Idriss IM (1990) Response of soft soil sites during earthquakes. Symposium to Honor Professor H.B. Seed, Berkeley, pp 273–289
Idriss IM (1991) Earthquake ground motions at soft soil sites. In: Proceedings of the 2nd international conference on recent advances in geotechnical earthquake engineering and soil dynamics, vol III, St. Luis, pp 2265–2273
Joyner WB, Warrick R, Fumal T (1981) The effect of Quaternary alluvium on strong ground motion in the Coyote Lake, California earthquake of 1979. Bull Seismol Soc Am 71 (4):1333–1349
Kokusho T, Sato K (2008) Surface-to-base amplification evaluated from KiK-net vertical array strong motion records. Soil Dyn Earthq Eng 28(9):707–716
Lang DH, Schwarz J (2006) Instrumental subsoil classification of Californian strong motion sites based on single-station measurements. In: Proceedings of the 100th anniversary earthquake conference, 8NCEE, San Francisco, CA, 18–22 Apr 2006
Lee VW, Trifunac MD (2010) Should average shear-wave velocity in the top 30 m of soil be used to describe seismic amplification? Soil Dyn Earthq Eng 30(11):1540–1549
Lermo J, Chávez-García FJ (1993) Site effect evaluation using spectral ratios with only one station. Bull Seismol Soc Am 83(5):1574–1594

Luzi L, Puglia R, Pacor F, Gallipoli MR, Bindi D, Mucciarelli M (2011) Proposal for a soil classification based on parameters alternative or complementary to Vs,30. Bull Earthq Eng 9 (6):1877–1898

Makra K, Raptakis D (2008) Analysis of strong ground motion recordings at a downhole array and correlation with simulation results. In: Proceedings of 3rd Greek conference on earthquake engineering and engineering seismology, 5–7 Nov 2008, Athens, paper n. 1965, 19p

Makra K, Chávez-García FJ, Raptakis D, Pitilakis K (2005) Parametric analysis of the seismic response of a 2D sedimentary valley: implications for code implementations of complex site effects. Soil Dyn Earthq Eng 25(4):303–315

Manakou M, Raptakis D, Chavez-Garcia FJ, Apostolidis PI, Pitilakis KD (2010) 3D soil structure of the Mygdonian basin for site response analysis. Soil Dyn Earthq Eng 30(11):1198–1211

Martin GR, Dobry R (1994) Earthquake site response and seismic code provisions. NCEER Bull 8 (4):1–6

Mohraz B (1976) Earthquake response spectra for different geologic conditions. Bull Seismol Soc Am 66(3):915–935

Moss RE (2008) Quantifying measurement uncertainty of thirty-meter shear-wave velocity. Bull Seismol Soc Am 98(3):1399–1411

Mucciarelli M, Gallipoli MR (2006) Comparison between Vs_{30} and other estimates of site amplification in Italy. In: Proceedings of the first European conference on earthquake engineering and seismology, Geneva, Switzerland, Paper no 274

Nakamura Y (1989) A method for dynamic characteristics estimation of subsurface using microtremors on the ground surface. Railway Technical Research Institute, Q Rep 30(1):25–33

Newmark NM, Hall WJ (1973) Procedures and criteria for earthquake resistant design. Building practices for disaster mitigation, Building science series, no 46, National Bureau of Standards, Washington, DC, pp 209–236

Park D, Hashash YMA (2004) Probabilistic seismic hazard analysis with non linear site effects in the Mississippi embayment. In: Proceedings of the 13th world conference on earthquake engineering, Vancouver, Paper no 1549 (on CD-Rom)

Pitilakis K (2004) Site effects. In: Ansal A (ed) Recent advances in earthquake geotechnical engineering and microzonation. Kluwer Academic, Dordrecht/Boston, pp 139–197

Pitilakis K, Raptakis D, Lontzetidis K, Tika-Vassilikou T, Jongmans D (1999) Geotechnical and geophysical description of EURO-SEISTEST using field, laboratory tests and moderate strong motion recordings. J Earthq Eng 3(3):381–409

Pitilakis K, Gazepis C, Anastasiadis A (2004) Design response spectra and soil classification for seismic code provisions. In: Proceedings of the 13th world conference on earthquake engineering, Vancouver, BC, Canada. Paper no 2904

Pitilakis K, Gazepis C, Anastasiadis A (2006) Design response spectra and soil classification for seismic code provisions. In: Proceedings of geotechnical evaluation and application of the seismic Eurocode EC8 2003–2006, ETC 12 workshop, NTUA Athens, pp 37–52

Pitilakis K, Manou D, Anastasiadis A (2007) Displacement response spectra for different soil conditions. In: Proceedings of the special section on geotechnical aspects of Eurocode 8, organized by ERTC-12 at the XIV European conference on soil mechanics and geotechnical engineering, Madrid, Spain, 24–27 Sept 2007

Pitilakis K, Anastasiadis A, Riga E (2012a) New site classification scheme and associated site amplification factors. SHARE Deliverable, Thessaloniki, Greece, 4.3

Pitilakis K, Riga E, Anastasiadis A (2012b) Design spectra and amplification factors for Eurocode 8. Bull Earthq Eng 10:1377–1400. doi:10.1007/s10518-012-9367-6

Pitilakis K, Riga E, Anastasiadis A (2013) New code site classification, amplification factors and normalized response spectra based on a worldwide ground-motion database. Bull Earthq Eng 11:925–966. doi:10.1007/s10518-013-9429-4

Raptakis D, Theodulidis N, Pitilakis K (1998) Data analysis of the Euroseistest strong motion array in Volvi (Greece): standard and horizontal to vertical ratio techniques. Earthquake Spectra 14(1):203–224

Raptakis D, Chávez-García FJ, Makra K, Pitilakis K (2000) Site effects at EUROSEISTEST – I. Determination of the valley structure and confrontation of observations with 1D analysis. Soil Dyn Earthq Eng 19(1):1–22

Raptakis D, Makra K, Anastasiadis A, Pitilakis K (2004a) Complex site effects in Thessaloniki (Greece): I. Soil structure and comparison on observations with 1D analysis. Bull Earthq Eng 2 (3):271–300

Raptakis D, Makra K, Anastasiadis A, Pitilakis K (2004b) Complex site effects in Thessaloniki (Greece): II. 2D, SH modeling and engineering insights. Bull Earthq Eng 2(3):301–327

Raptakis D, Manakou M, Chávez-García FJ, Makra K, Pitilakis K (2005) 3D configuration of Mygdonian basin and preliminary estimate of its site response. Soil Dyn Earthq Eng 25 (11):871–887

Rey J, Faccioli E, Bommer JJ (2002) Derivation of design soil coefficients (S) and response spectral shapes for Eurocode 8 using the European strong-motion database. J Seismol 6 (4):547–555

Rodriguez-Marek A, Bray JD, Abrahamson NA (2001) An empirical geotechnical seismic site response procedure. Earthquake Spectra 17(1):65–87

Scherbaum F, Bommer J, Bungum H, Cotton F, Abrahamson NA (2005) Composite ground-motion models and logic trees: methodology, sensitivities, and uncertainties. Bull Seismol Soc Am 95(5):1575–1593

Scordilis EM (2006) Empirical global relations converting M_S and m_b to moment magnitude. J Seismol 10(2):225–236

Seed HB, Murarka R, Lysmer J, Idriss IM (1976a) Relationships between maximum acceleration, maximum velocity, distance from source and local site conditions for moderately strong earthquakes. Bull Seismol Soc Am 66(4):1323–1342

Seed HB, Ugas C, Lysmer J (1976b) Site dependent spectra for earthquake-resistant design. Bull Seismol Soc Am 66(1):221–244

Seed HB, Romo MP, Sun I, Jaime A, Lysmer J (1988) The Mexico earthquake of September 19, 1985 – relationships between soil conditions and earthquake ground motions. Earthquake Spectra 4(4):687–729

Seed RB, Dickenson E, Rau A, White K, Mok C (1994a) Observations regarding seismic response analyses for soft and deep clay sites. In: Martin G (ed) 1992 NCEER/SEAOC/BSSC workshop on site response during earthquake and seismic code provisions. Special publication. NCEER-94SP01, New York

Seed RB, Dickenson SE, Mok CM (1994b) Site effects on strong shaking and seismic risk; recent developments for seismic design codes and practice. American Society of Civil Engineering Structures Congress, vol 12, pp 573–578

Steidl JH (1993) Variation of site response estimates at the UCSB dense array of portable accelerometers. Earthquake Spectra 9(2):289–302

Steidl JH (2000) Site response in southern California for probabilistic seismic hazard analysis. Bull Seismol Soc Am 90(6B):S149–S169

Stewart JP, Liu AH, Choi Y (2003) Amplification factors for spectral acceleration in tectonically active regions. Bull Seismol Soc Am 93(1):332–352

Tolis SV, Faccioli E (1999) Displacement design spectra. J Earthq Eng 3(1):107–125

Wald LA, Mori J (2000) Evaluation of methods for estimating linear site-response amplifications in the Los Angeles Region. Bull Seismol Soc Am 90(6B):S32–S42

Yenier E, Sandikkaya MA, Akkar S (2010) Report on the fundamental features of the extended strong motion databank prepared for the SHARE project (v1.0)

Zhao JX, Zhang J, Asano A, Ohno Y, Oouchi T, Takahashi T, Ogawa H, Irikura K, Thio HK, Somerville PG, Fukushima Y, Fukushima Y (2006a) Attenuation relations of strong ground motion in Japan using site classification based on predominant period. Bull Seismol Soc Am 96(3):898–913

Zhao JX, Irikura K, Zhang J, Fukushima Y, Somerville PG, Asano A, Ohno Y, Oouchi T, Takahashi T, Ogawa H (2006b) An empirical site classification method for strong motion station in Japan using H/V response spectral ratio. Bull Seismol Soc Am 96(3):914–925

Part II
Liquefaction

Chapter 3
Sand Liquefaction Observed During Recent Earthquake and Basic Laboratory Studies on Aging Effect

Takaji Kokusho, Yohta Nagao, Fumiki Ito, and Takashi Fukuyama

Abstract During the 2011 Tohoku Pacific Ocean earthquake (M9.0), liquefaction occurred extensively in reclaimed land in Kanto area more than 200 km far from the earthquake fault. The liquefied sand generally contained a lot of non-plastic fines with fines content more than 50 % in some places. Almost all sand deposits along the Tokyo bay area reclaimed in 1960s or later liquefied, while in a good contrast, those older than that did not.

In order to study the aging effect on liquefaction strength of sands containing fines, a series of basic laboratory tests combining innovative miniature cone penetration and subsequent cyclic undrained loading were carried out in a modified triaxial apparatus on sand specimens containing fines. A unique curve relating cone resistance q_t and liquefaction strength R_L was obtained for reconstituted specimens, despite the differences in relative density D_r and fines content F_c, quite contradictory to the current liquefaction potential evaluation practice. Then a small amount of cement was added to fines in the sand specimens to simulate a geological aging effect in a short time. It was found that the liquefaction strength R_L increases with increasing F_c more than the penetration resistance q_t, resulting in higher liquefaction strength under the same cone resistance. Thus it has been clarified that not the fines content itself but the aging effect by cementation, which becomes more pronounced in sands with higher F_c, can facilitate reasonable basis why liquefaction strength is modified with increasing fines content in the evaluation practice.

In addition to the accelerated tests on the reconstituted specimens with cement, intact samples recovered from in situ Pleistocene and Holocene deposits with known ages have been tested and confirmed the similar trend to the above-mentioned results.

T. Kokusho (✉) • Y. Nagao • F. Ito • T. Fukuyama
Department of Civil and Environmental Engineering, Chuo University, Tokyo, Japan
e-mail: kokusho@civil.chuo-u.ac.jp

M. Maugeri and C. Soccodato (eds.), *Earthquake Geotechnical Engineering Design*, Geotechnical, Geological and Earthquake Engineering 28,
DOI 10.1007/978-3-319-03182-8_3, © Springer International Publishing Switzerland 2014

3.1 Liquefaction in Tokyo Bay Area During M9.0 Earthquake

During the recent Tohoku Pacific Ocean earthquake (M9.0), liquefaction extensively took place along Tokyo bay area, more than 200 km far from the earthquake fault. Almost all sand deposits along the bay area liquefied this time were reclaimed in 1960s or later. In a good contrast, those older than that did not liquefy. The sand generally contained a lot of non-plastic fines, with fines content F_c more than 40 % in many places.

In Urayasu City in particular, extensive liquefaction took place in newly developed land as shown in Fig. 3.1. Almost all the area, developed after 1968 at the mouth of Edo-River by filling sea-bed soils by hydraulic methods, liquefied extensively. In a good contrast, the area of the same city existed before 1948 did not liquefy. The thick dashed line in Fig. 3.1 separating liquefied and non-liquefied area can be drawn very clearly based on the damage of sewage systems and road. There were a plenty of bore-holes and SPT loggings in this area both in the old and new areas developed by Chiba Prefectural Government (wwwp.pref.chiba.lg.jp/pbgeogis/servlet/infobank.index) as illustrated in Fig. 3.2. The N-values at four locations A, B, C and D in the map are shown, among which A, B were in the non-liquefied area and locations C, D were heavily liquefied. The water table was from GL −1.0 to −2.0 m at almost all locations. They share the common soil profiles consisting of surface sand layer (10–20 m thick) and underlying Holocene soft clay. The clay layer is 30–40 m thick underlain by Pleistocene dense sand of $N > 50$. The upper portion of the surface sand layer is a fill in the reclaimed area with N-values lower than 10, whereas the lower portion is Holocene sands with N-values around 10–20. It is clearly seen from Fig. 3.2 that the depth of sand layer with $N > 10$ gets deeper as the location moves from A to D, from inland to the shore front.

Figure 3.3 indicates the changes of SPT N-values converted from Swedish penetration tests (Inada 1960) conducted before and after the earthquake. It is

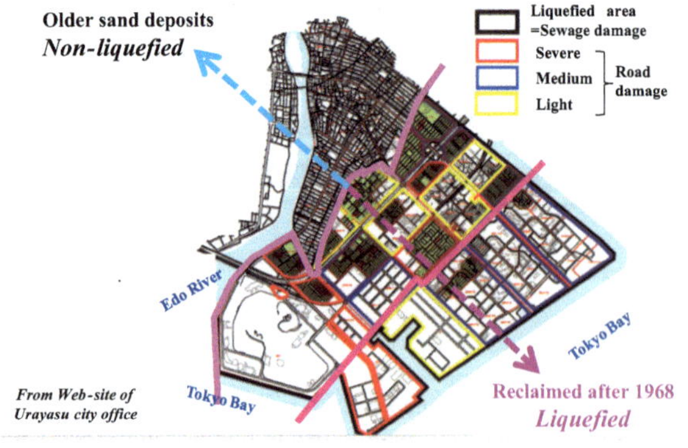

Fig. 3.1 Liquefied and non-liquefied areas in Urayasu city

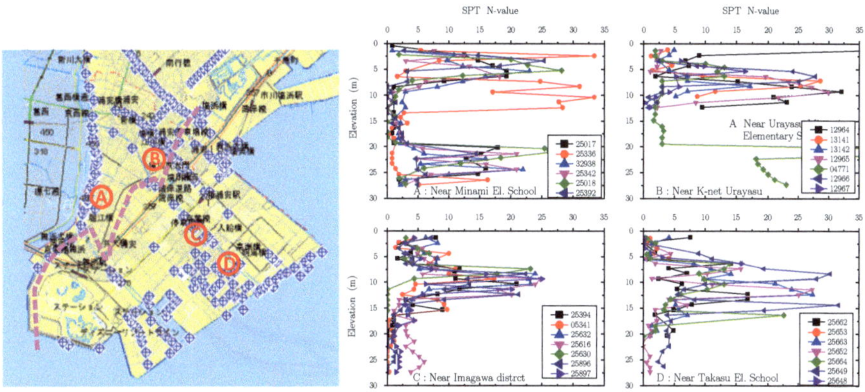

Fig. 3.2 Liquefied and non-liquefied areas in Urayasu city

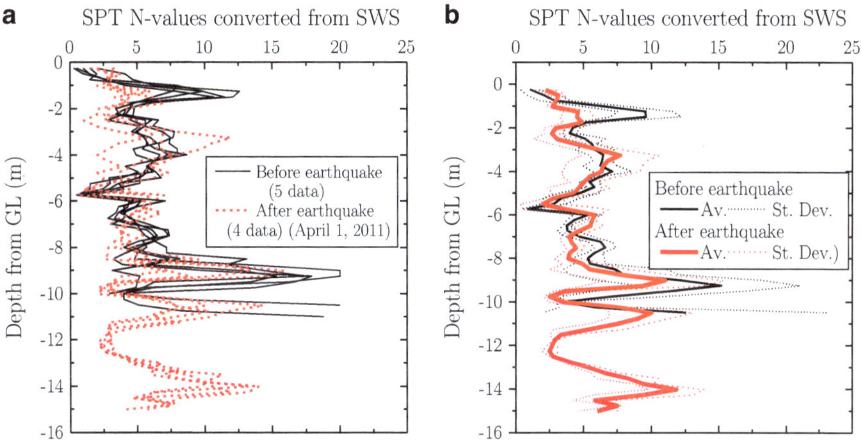

Fig. 3.3 Changes in SPT N-values converted from SWS conducted before and after EQ at location C (**a**) and their averages ∓ standard deviations (**b**)

clearly seen from the individual data (a) and also from the averages ± standard deviations (b) that the penetration resistance decreased at GL. −1 to −3 m and around GL. −7 to −8 m, which seems to correspond to fill sands by reclamation. The sea-bed sand with $N > 10$ also decreased the penetration resistance, which may or may not reflect liquefaction. Thus, loosening seems more dominant than densifying at all the depth. Though it seems highly probable that liquefaction occurred exclusively in reclaimed sand deposits shallower than −8 m, further investigations based on more penetration test data are needed. Figure 3.4 shows grain-size curves of erupted sands in the liquefied areas in Urayasu with thick lines. The erupted sands in Urayasu were extraordinarily rich of fines with fines content, $F_c = 30–70$ %, in comparison with liquefied sands of other places and the fines were all non-plastic.

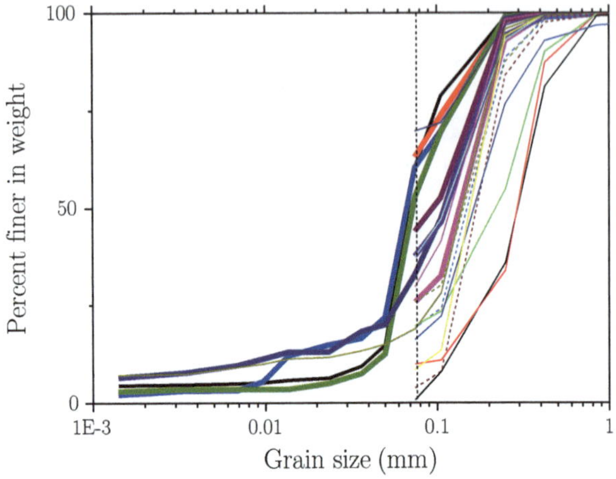

Fig. 3.4 Grain-size curves of erupted sands from liquefied sites in Kanto area (Urayasu sands by thicker lines compared with those from other places)

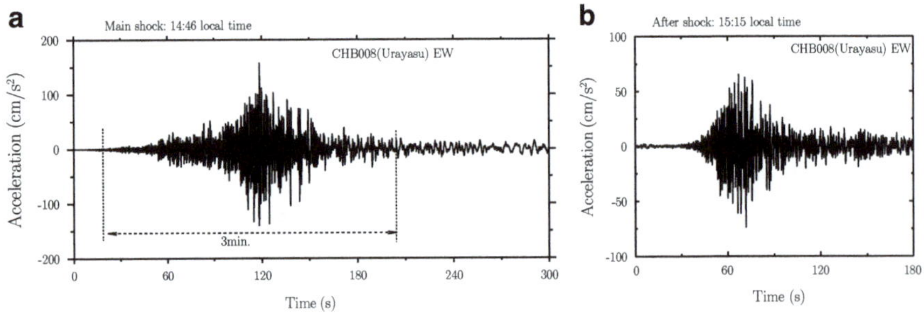

Fig. 3.5 Acceleration records at K-net Urayasu during main shock (**a**) and big aftershock (**b**)

The ground motions during the main shock of Tohoku Pacific Ocean earthquake (M9.0, 14:46 local time) were recorded at the KiK-net Urayasu station (CHB008 NIED). The duration was longer than 3 min, with the maximum acceleration 168 gal and the maximum velocity 29 kine as depicted in Fig. 3.5a. The strong aftershock (M7.4, 15:15 local time) recorded at the same station also shown in Fig. 3.5b seems to have strong effect on post-liquefaction behavior of liquefying deposits, giving large flow potential of boiled sand.

Thus, a fundamental question posed by this liquefaction case is whether or not the clear distinction between liquefaction and non-liquefaction in old and newly developed areas in Urayasu can be explained by;

(a) the difference of soil density,
(b) the difference in ground motion,
or
(c) the aging effect of fines-containing sand.

Although, further studies are needed using comprehensive datasets of geotechnical conditions and ground motions to answer this question clearly, recent case studies including the present case indicate that aging effect is likely to play an important role in occurrence of liquefaction in sand deposits. Liquefaction occurred almost exclusively in backfills and reclaimed ground of young ages during recent earthquakes worldwide. Also pointed out is that cases are increasing in which liquefied sands contained considerable amount of low-plasticity fines as in Adapazari during 1999 Kocaeli earthquake, in Christchurch during 2010 Darfield earthquake in New Zealand and in Tokyo Bay area during the recent M9.0 earthquake.

3.2 Aging Effect on Sands Containing Fines

In engineering practice, liquefaction potential is evaluated using penetration resistance in standard penetration tests (SPT) or cone penetration tests (CPT). If sands contain a measurable amount of fines, liquefaction strength is normally raised in accordance to fines content F_c in most liquefaction evaluation methods. This F_c-dependent modification of liquefaction strength may be originated from liquefaction case studies (Tokimatsu and Yoshimi 1983; Seed and De Alba 1984), in which empirical boundary curves, developed in situ, separating liquefaction/non-liquefaction to SPT N_1-values were found being strongly dependent on fines content. More directly, Suzuki et al. (1995) carried out CPT tests and soil sampling by in situ freezing technique in the same sand deposits. The study showed that the higher the fines content the greater the liquefaction strength for the same q_t-value. In contrast to their finding, however, quite a few laboratory tests using reconstituted specimens, having the same relative density D_r (e. g. Kokusho 2007) or the same void ratio (Papadopoulou and Tika 2008), have shown that R_L clearly decreases with increasing fines content of low plasticity fines from $F_c = 0$–$30\ \%$. Thus, a lack of understanding seems to remain in the current practice for liquefaction potential evaluation in the field in relation with laboratory test results for sands containing fines.

In this paper, a systematic experimental study is addressed, in which miniature cone penetrations and subsequent cyclic loading tests were carried out on the same triaxial test specimens (Kokusho et al. 2005). An innovative simple mechanism introduced in a normal cyclic triaxial apparatus by Kokusho et al. (2003) was used enabling a miniature cone to penetrate into the sand specimen at a constant speed. The results from the two sequential tests on the same specimen were compared to develop direct $q_t - R_L$ correlations for sands containing various amounts of fines. In the first part of the present paper, the aging effect by cementation is investigated in a series of laboratory tests using reconstituted specimens containing fines. A small quantity of cement was added to reconstituted specimens to simulate cementation or chemical bonding for a geological time span in a short time. Penetration resistance q_t and liquefaction strength R_L were measured in the same specimens to investigate the aging effect on the $q_t - R_L$ correlations considering fines content as a key parameter. In the latter part of the paper, intact specimens sampled from Pleistocene and Holocene deposits were tested in the same test apparatus to know the direct $q_t - R_L$ relationship of natural soils. Then, the same soils, disturbed and then reconstituted to the original density, were tested again to compare the results with the intact soils.

3.3 Test Apparatus, Soil Material and Test Procedures

In the triaxial apparatus used in this research, the specimen size was 100 mm in diameter and 200 mm in height. In the liquefaction tests, the soil specimen was loaded cyclically by a pneumatic actuator as a stress-controlled test. In order to carry out a cone penetration test in the same specimen prior to the liquefaction tests, a pedestal below the soil specimen was modified as shown in Fig. 3.6, so that a miniature cone can penetrate into the specimen from the bottom. For that goal, the pedestal consists of two parts, a circular base to which the cone rod is fixed and a movable metal cap. Through the center of the cap, the cone rod penetrates up into the overlying specimen. The annulus between the two parts is sealed by O-rings, enabling the cap to slide up and down (Kokusho et al. 2003). Before the test, the pedestal cap is set up at the highest position by filling water in the reservoir, and the specimen is constructed on it.

By opening a valve at the start of cone penetration, the water in the reservoir is squeezed by the cell pressure, resulting in settlement of the total specimen at the top of the pedestal. Due to the settlement, the cone penetrates into the specimen by 25 mm (from the initial to final height, 45–70 mm).

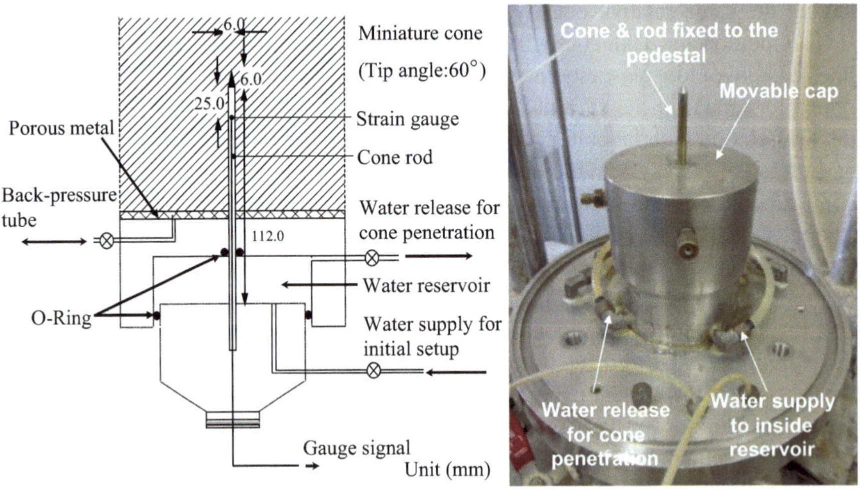

Fig. 3.6 Cross section (*right*) and photograph (*left*) of the modified pedestal in the lower part of triaxial apparatus

The penetration speed is about 2 mm per second, much slower than prototype CPT and is almost constant irrespective of the difference in relative density. The miniature cone is 6 mm diameter and 60° tip angle, about 1/6 smaller in dimension than the prototype cone used in the field. The strain gauges to measure the penetration resistance are glued at the inner wall of the rod tube, 25 mm lower than the foot of the cone.

Beach sand (Futtsu sand) consisting of sub-round particles of hard quality was used for reconstituted specimens. Fines mixed with the sand was silty and clayey soils with low plasticity index of $I_p = 6$ sieved from decomposed granite in reclaimed ground of the Kobe city, Japan. For the test series for cementation, a prescribed quantity of Portland cement was completely mixed with the Masa fine soil to make it different in chemical activity. The cement content C_c, the weight ratio of cement to soil (including fines), varied from 0 to 1.0 %, and the fines content F_c including the cement changed from 0 to 30 %. This means that the ratio C_c/F_c, a parameter representing chemical activity of fines, varied from 0 to 20 %. Then the sand specimen with given F_c was prepared by moist tamping to make a nominal relative density D_r, which was changed in three steps, about $D_r = 30$, 50 and 70 %. The specimen was then completely saturated with de-aired water, and consolidated under the isotropic effective stress of 98 kPa with the back-pressure of 196 kPa. If cement was added to fines, the consolidation time was controlled exactly 24 h after wetting, while for tests without cement it was about 2 h.

Then, the specimen was fully saturated by using de-aired water in a double negative pressure method so that the B-value larger than 0.95 was measured, and isotropically consolidated with the effective stress of 98 kPa with the back-pressure of 196 kPa. In the test sequence, the penetration test was first carried out under undrained condition after consolidating the specimen. Then, after releasing pore pressure and reconsolidating it under the same confining pressure again, although the volume change by this procedure was almost negligible, the liquefaction test was carried out on the same specimen. The sinusoidal cyclic axial load applied with the frequency of 0.1 Hz was measured with a load cell in the pressure chamber. The cell pressure and the pore-water pressure were measured with electric piezometers and the axial deformation was measured with LVDT of 50 mm maximum capacity. It may well be suspected that, in such a test sequence, the liquefaction strength is possibly influenced by the preceding cone test and subsequent reconsolidation. However, it was already confirmed in previous research (Kokusho et al. 2005) that the cyclic stress ratio for liquefaction was almost unaffected by the cone test and reconsolidation.

3.4 Results on Reconstituted Specimens without and with Cement

In the test, relative density D_r, fines content F_c and cement content C_c of the specimen were parametrically varied to investigate their effects on penetration resistance and undrained cyclic strength. Figure 3.7a–e show cone resistance q_t

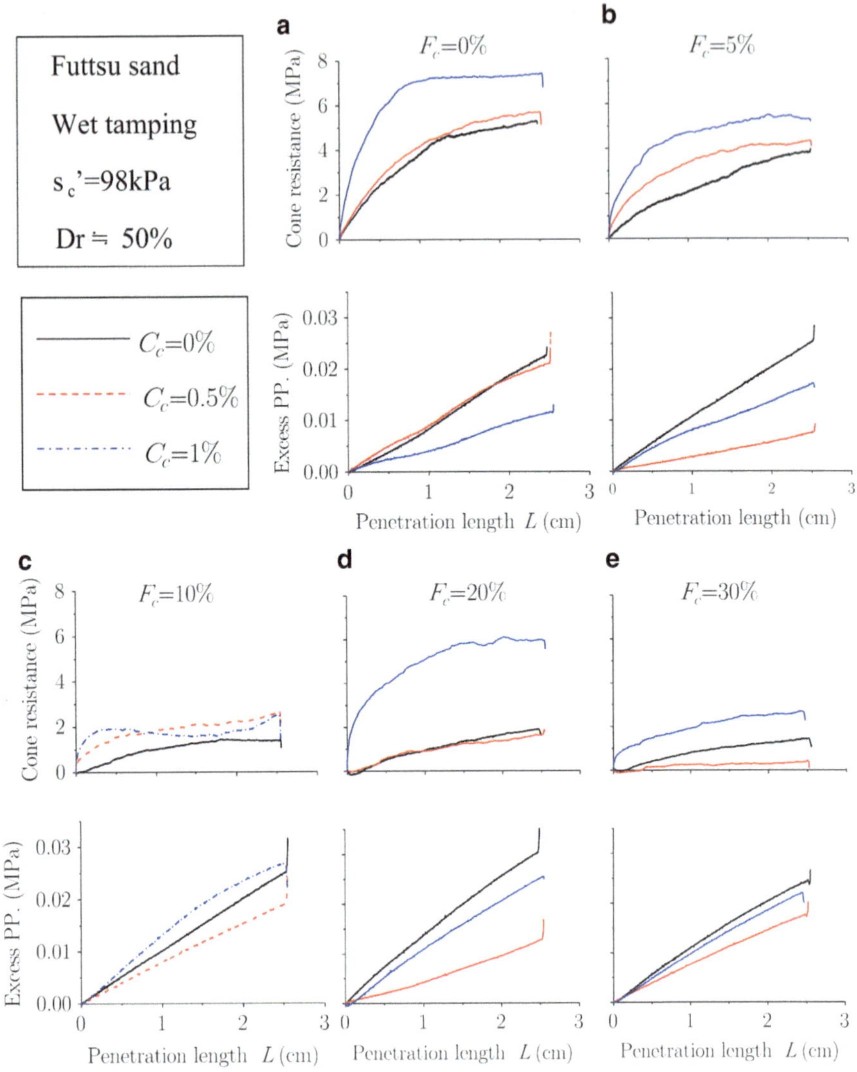

Fig. 3.7 Cone resistance or excess pore-pressure versus penetration length L for reconstituted specimens of $D_r \approx 50$ % with different fines content

and associated excess pore-pressure Δu plotted versus the penetration length L in the tests for $D_r \approx 50$ %, $F_c = 0, 5, 10, 20, 30$ % and $C_c = 0, 0.5, 1.0$ %. It is clearly observed that q_t for soils without cement (C_c =0) decreases with increasing fines up to $F_c = 20$ %. For individual F_c–values, q_t tends to increase more or less with C_c increasing from 0 to 1 %. The increase is particularly large for $F_c = 20$ % and $C_c = 1$ %. The excess pore pressure Δu building up almost linearly with the penetration length L tends to decrease with increasing C_c. In Fig. 3.8a, b, similar

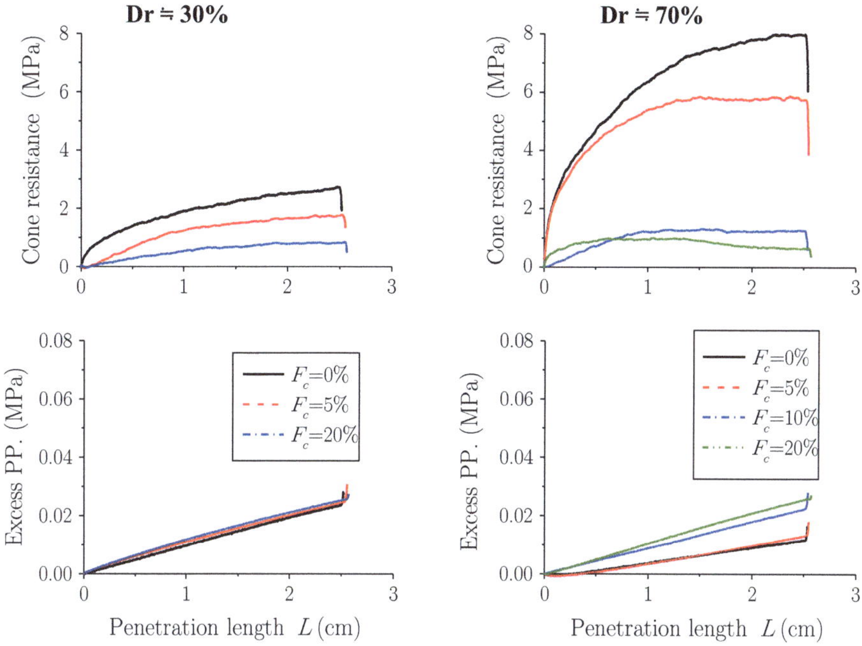

Fig. 3.8 Cone resistance or excess pore-pressure versus penetration length L for reconstituted specimens of $D_r \approx 30$ & 70 % with different fines content without cement

penetration test results are shown for sands of Dr ≈ 30 and 70 % with varying F_c but no cement. It is observed again that the increase in F_c tends to decrease cone resistance q_t.

Figure 3.9a–e show relationships between the stress ratios R_L ($\sigma_d/2\sigma_c'$: σ_d = cyclic stress amplitude , σ_c' = effective isotropic confining stress) versus the number of loading cycles N_c for the double amplitude axial strain $\varepsilon_{DA} = 5$ % obtained by undrained cyclic loading tests on sand specimens having $D_r \approx 50$ %, $F_c = 0$, 5, 10, 20, 30 %, and $C_c = 0$, 0.5, 1.0 %. Obviously, the increase in fines content tends to reduce liquefaction strength R_L in specimens without cement as already demonstrated in previous researches (e.g. Kokusho 2007). Also indicated in the figure is that the R_L-value increases with increasing cement content for all the fines content F_c and the increment is larger for $C_c = 1$ % and F_c = 10–30 %. In Fig. 3.10a, b, the similar liquefaction strength results are shown for specimens without cement and $D_r \approx 30$, 70 % with varying F_c. Again, the increase in F_c obviously decrease the strength R_L for dense sand of $D_r \approx 70$ % in particular.

In Fig. 3.11, the R_L-values are plotted versus q_t for different values of D_r, F_c and C_c with different symbols. R_L in the vertical axis is defined as the stress ratio $\sigma_d/2\sigma_c'$ for $\varepsilon_{DA} = 5$ % and $N_c = 20$. The penetration resistance q_t in the horizontal axis is determined as the maximum value of the cone resistance during penetration from the $q_t - L$ curves shown in Figs. 3.7 and 3.8. The open symbols in Fig. 3.11 which

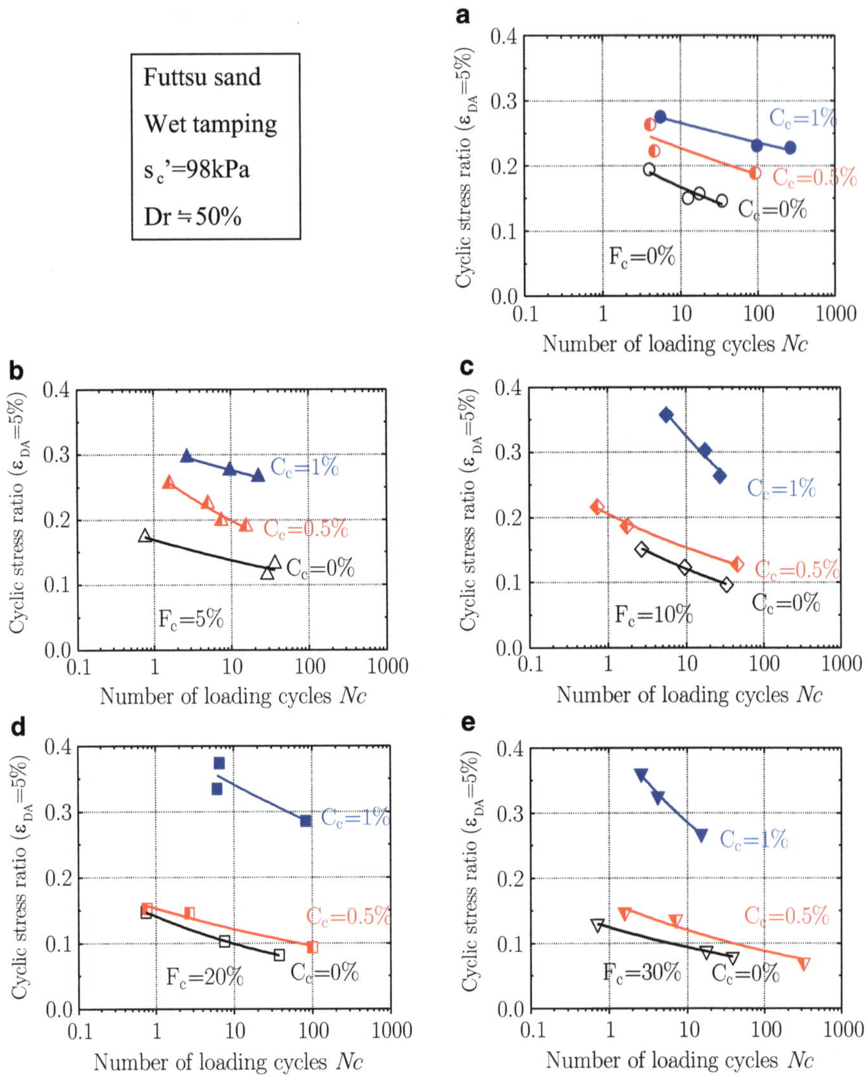

Fig. 3.9 Stress ratios R_L versus number of loading cycles N_c for reconstituted specimens of $D_r \approx 50$ % with different fines content

correspond to the specimens without cement are located along a thick dashed line in the chart despite wide differences in D_r (30–70 %) and F_c (0–30 %) as already pointed out in the previous research (Kokusho et al. 2005, 2009), indicating that the liquefaction strength R_L is uniquely correlated to q_t irrespective of D_r and F_c. This finding is contradictory with the current liquefaction potential evaluation practice, where R_L is to be raised according to F_c.

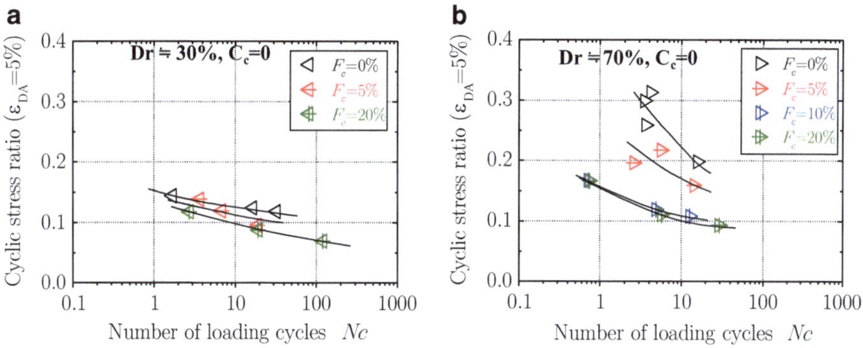

Fig. 3.10 Stress ratios R_L versus number of loading cycles N_c for reconstituted specimens of $D_r \approx 30$ & 70 % with different fines content without cement

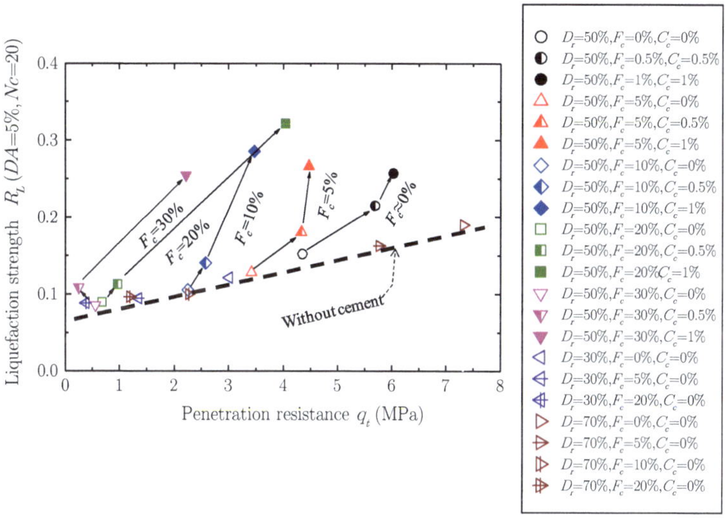

Fig. 3.11 Liquefaction strength R_L versus penetration resistance q_t diagram for different values of D_r, F_c and C_c

Half-close and close symbols in Fig. 3.11 are from the accelerated tests on specimens containing cement. Among them, upward triangles, for instance, represent the case $F_c = 5$ %, and they move up from the open ($C_c = 0$) to the half-close ($C_c = 0.5$ %) further to the full-close symbols ($C_c = 1.0$ %) as C_c/F_c changes from 0 to 20 % for the same value of $F_c = 5$ % as indicated by thin arrows in the diagram. In the similar manner, other symbols move up with increasing C_c or C_c/F_c for the same fines content of $F_c = 10$, 20, and 30 %. Specimens with higher value of C_c/F_c may be considered as of longer geological age because of stronger cementation effect in the same soil in the accelerated test. This effect seems to push up the data points on the $R_L - q_t$ diagram from the unique line of no cement and gives higher liquefaction strength under the same cone resistance.

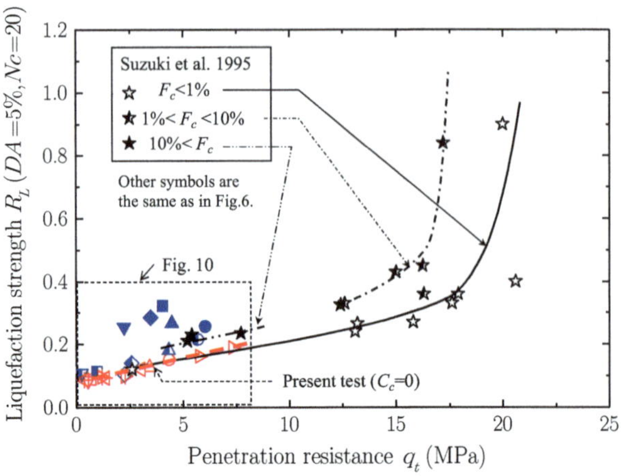

Fig. 3.12 Liquefaction strength R_L versus penetration resistance q_{tt} compared with field investigations combining prototype CPT and triaxial tests on intact samples

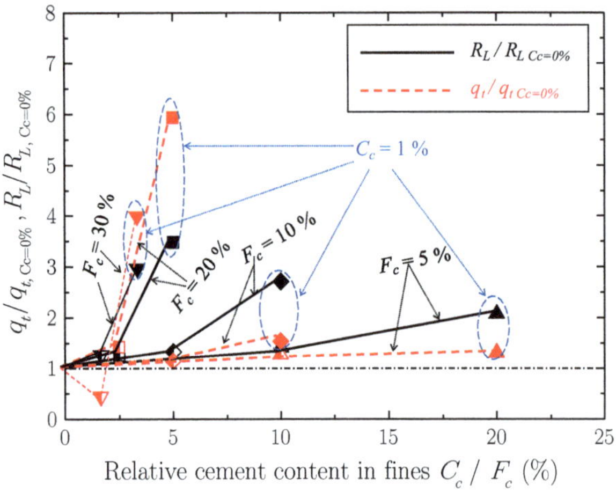

Fig. 3.13 $q_t/q_{t,\,Cc=0\,\%}$ or $R_L/R_{L,\,Cc=0\,\%}$ versus C_c/F_c

In Fig. 3.12, the same data points as in Fig. 3.11 are plotted again together with star symbols representing in situ soil data by Suzuki et al. (1995). In their research, prototype cone tests in situ and laboratory undrained cyclic triaxial tests on intact samples recovered from the same soil deposits by in situ freezing technique were combined. It demonstrates a clear difference in the $R_L - q_t$ relationships due to different fines contents of F_c <1.0 %, F_c = 1.0–10 % and F_c >10 %. Also noted is that the two research results are in a good coincidence not only qualitatively but also quantitatively, particularly in the case without cement, despite a large difference in the cone size and CPT test procedures.

Based on the data shown in Fig. 3.11, the ratio of increase in R_L or q_t with respect to those without cement is shown versus the C_c/F_c -value with the fines content F_c as a parameter in Fig. 3.13. With increasing chemical activity C_c/F_c, the R_L-value increases more than the q_t-value for $F_c \leq 10$ %, but it is reverse for $F_c \geq 20$ %.

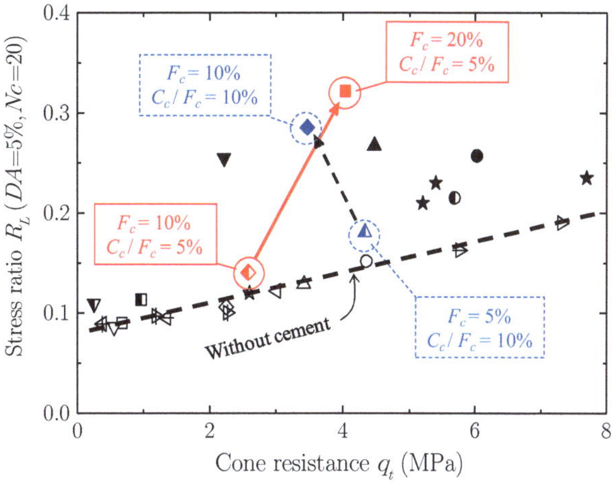

Fig. 3.14 Close-up of Fig. 3.6 showing the effect of fines content F_c for the same C_c/F_c

Also noted in Fig. 3.13 is that not only the C_c/F_c -value but also the fines content F_c increasing up to $F_c = 20$ % tends to considerably increase R_L and q_t even under the same cement content of $C_c = 1.0$ % as demonstrated by the encircled plots but the trend seems to change at $F_c = 30$ %. Thus, there seems to be a kind of transition between $F_c = 20$ and 30 %, where the increasing trend in R_L or q_t with increasing C_c/F_c -value changes. This may possibly have something to do with the change in soil fabric from grain-supporting to matrix-supporting structure.

In Fig. 3.14, where the same data as in Fig. 3.11 are plotted again, it is recognized that the plot for C_c/F_c –value, 5 or 10 % shifts upward as F_c increases from the unique line of no cement as indicated by the thick arrows in the graph. This indicates that, for the same C_c/F_c –value (simulating the same geological age), higher F_c-value results in higher liquefaction strength for the same cone resistance. This trend is compatible with the liquefaction potential evaluation practice currently employed. Thus, the present research results by the accelerated test clearly indicate that not the fines content itself but the cementation effect is responsible for the higher liquefaction strength for larger F_c under the same cone resistance.

3.5 Test Results on In Situ Intact Specimens

In order to know the direct $R_L - q_t$ relationship for natural soils and compare them with the accelerated test results mentioned above on reconstituted specimens using small amount of cement, intact soils have been sampled from in situ soil deposits. They have been tested in the same way to measure mini-cone penetration resistance and undrained cyclic strength of the same specimens. Pleistocene sands and Holocene sands have been recovered from two sites in Kanto area near Tokyo by block sampling using PVC tubes as shown in Fig. 3.15.

Fig. 3.15 Photographs of block sampling using a PVC tube of 100 mm diameter and 200 height (*Left*: Pleistocene sand, *Right*: Holocene sand)

Pleistocene soil was taken from two sand layers A and B interbedded in a river terrace near Narita-city in Chiba Prefecture. The age of the deposit was evaluated as about 100×10^3 and 150×10^3 years old, respectively (Mitani 2006). The overburden soil depth for the sample was about 5 and 11 m for layers A and B, respectively. The relative density and fines content of Layer A are $D_r = 101–107$ % and $F_c = 6.1–8.5$ %, and those of Layer B are $D_r = 75–81$ % and $F_c = 12.1–18.7$ %, respectively. Soils sampled in PVC tubes were frozen in the laboratory and then trimmed into the specimen of 100 mm in diameter and 200 mm in height in order to secure the intactness, because it was difficult to handle them without freezing. Holocene soils A with $D_r = 35–81$ %, $F_c = 13–25$ % and B with $D_r = 72–100$ %, $F_c = 17–31$ % were taken from a river terrace sand layer in Tateyama-city in Chiba Prefecture by using the same PVC tube. Its age was evaluated as $2–4 \times 10^3$ years old (Shishikura et al. 2005). The overburden depth for the samples was about 5 m for both A and B. For this soil, it was possible to trim the sample into the specimen size without freezing.

A hole for the mini cone was drilled in advance at the center of the specimen bottom by a drill machine, 6 mm in diameter and 43 mm in depth. It was then set on the pedestal of the modified triaxial apparatus. The specimen was fully saturated with de-aired water to secure the B-value larger than 95 %, consolidated by isotropic pressure of $p'_0 = 98$ kPa for all the intact samples. Then, the mini-cone penetration test was conducted in the undrained condition. After that, the specimen was reconsolidated by $p'_0 = 98$ kPa and undrained cyclic test was carried out in the same manner as in the reconstituted specimens mentioned before. Figure 3.16 shows cone resistance q_t and associated excess pore-pressure Δu plotted versus the penetration length for all intact specimens of Pleistocene layers A and B with thick solid curves. The corresponding data are also shown with dashed curves for reconstituted specimens of the same soils, which were once disturbed and compacted in a mold by the moist-tamping method to reproduce the original density. Despite some data dispersions, the cone resistance q_t for intact soils is generally larger than that for reconstituted soils. The similar data for Holocene soils are also superposed in Fig. 3.16 with thin curves. The difference in q_t between intact and reconstituted specimens are again obvious despite wide variation in D_r in situ.

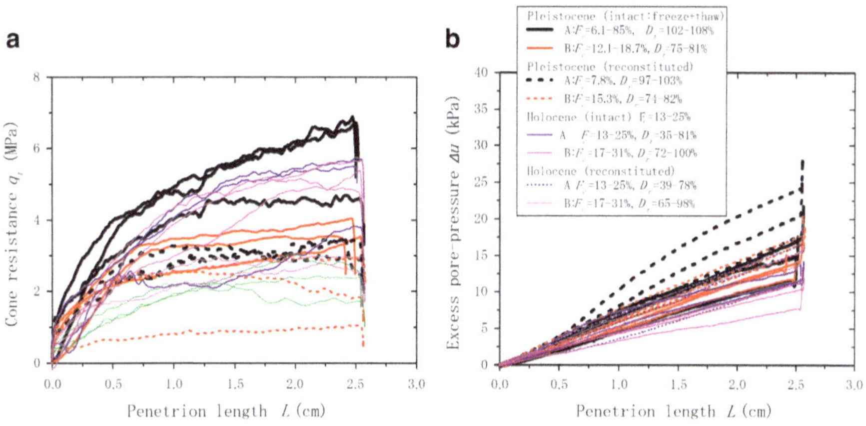

Fig. 3.16 Cone resistance (**a**) or excess pore- pressure (**b**) versus penetration length for intact and reconstituted Pleistocene/Holocene sands

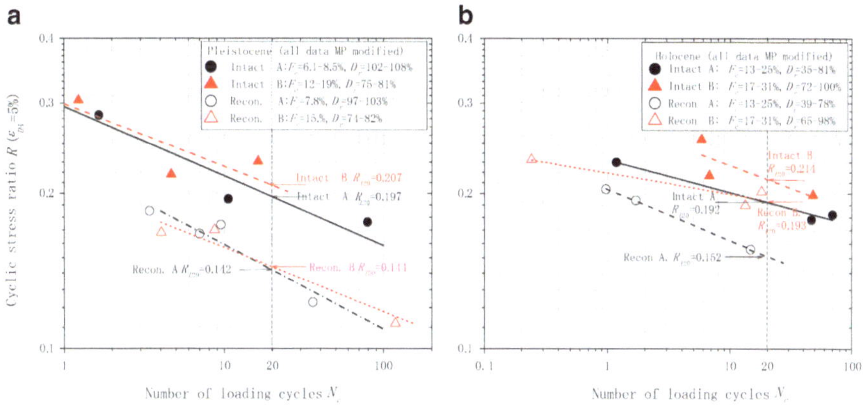

Fig. 3.17 Stress ratios R_L versus number of loading cycles N_c for intact specimens from Pleistocene sands

In Fig. 3.17, the cyclic stress ratios $R = \sigma_d/2\sigma_c^{'}$ are plotted versus the number of load cycles obtained from cyclic loading tests conducted for the same intact or reconstituted specimens of Pleistocene and Holocene soils after the mini-cone penetration tests. For all the test results both for intact and reconstituted, the R-values were modified considering the effect of rubber membrane penetration because of their rugged side face. For Pleistocene soils shown in (a), R-values of intact specimens are evidently larger than those of reconstituted soils for both Samples A and B in spite of the data dispersions. For Holocene intact soils (b), the data points are represented by unique curves on the $R - N_c$ chart for A and B despite the big difference in D_r, and evidently larger than those of reconstituted soils.

In Fig. 3.18, the liquefaction strength R_L and the penetration resistance q_t for Pleistocene and Holocene soils are directly compared on the $R_L - q_t$ diagram using

Fig. 3.18 Stress ratios R_L versus cone resistance q_t of intact Pleistocene and Holocene soils with reconstituted specimens

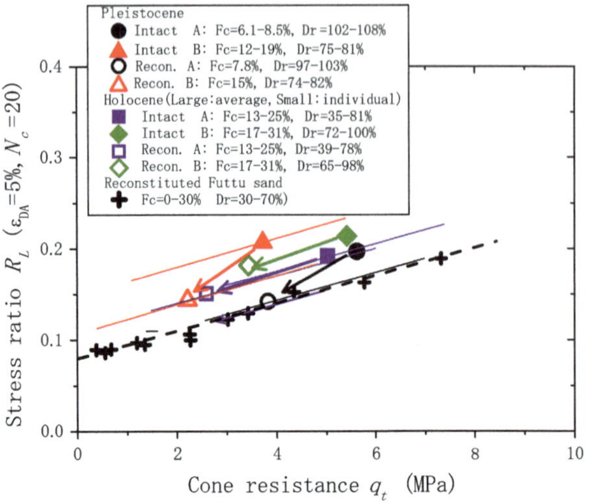

different symbols. The liquefaction strength R_L in the vertical axis is defined as the stress ratio R for $\varepsilon_{DA} = 5\ \%$ and $N_c = 20$ as indicated in Fig. 3.17. The penetration resistance q_t is determined as the maximum value of the cone resistance during penetration from the $q_t - L$ curves shown in Fig. 3.16. The q_t–value in the horizontal axis of Fig. 3.18 is the average for 3–4 penetration tests on the same soil.

The cross symbols in Fig. 3.18 indicate the $R_L - q_t$ plots obtained for Futtsu beach sand. As already explained, they are located along the thick dashed line irrespective of difference in D_r and F_c. The test results of Pleistocene soils indicate that the plots for reconstituted specimens (the open circle and triangle) is not completely coincident with the dashed line but slightly higher than that. However, the plots for intact soils (the solid circle and triangle) are found still higher than those of the reconstituted soils if the lines parallel with the dashed line passing through them are compared. For Holocene soils, the difference in R_L is not so large as that in q_t between intact (the close square and diamond symbols) and reconstituted (the open symbols), so that the lines connecting them are almost in parallel with the thick dashed line. It may reflect the difference in ages between the two soils from Pleistocene and Holocene. Although more test data on intact soils from in situ are needed to reliably quantify the aging effect of natural deposits considering various influences of soil sampling and testing methods, it has been shown that both R_L and q_t are measured higher in intact samples than reconstituted samples.

3.6 Conclusions

A series of experimental study by miniature cone penetration tests and subsequent cyclic loading tests were carried out in the same triaxial test specimen to investigate the aging effect on liquefaction strength by cementation with fines content F_c as a key parameter. Accelerated tests on the cementation effect using Futtsu beach sand without or with a small quantity of cement yielded the following major findings;

- For reconstituted sands without cement, the liquefaction strength R_L is uniquely related to the cone penetration resistance q_t, forming a single $R_L - q_t$ line, irrespective of relative soil density D_r and fines content F_c, quite contradictory to the current liquefaction potential evaluation practice.
- This laboratory test result coincides with in situ $R_L - q_t$ relation for clean sand by Suzuki et al. (1995) quantitatively despite big difference in the test procedures and cone size.
- Specimens with higher value of C_c/F_c (simulating longer geological age) results in higher liquefaction strength under the same cone resistance, indicating that the cementation effect tends to raise the $R_L - q_t$ line from that for soils without cement.
- For the same C_c/F_c-value (simulating the same geological age), higher fines content results in higher liquefaction strength under the same cone resistance, which is consistent with the trend found in the field investigations.

In order to examine that the above test results are compatible with natural deposits of long geological ages, intact specimens sampled from Pleistocene and Holocene deposits were tested in the same way to know the direct $R_L - q_t$ relationship of natural soils and compared with reconstituted soils, yielding the following;

- Differences are clear in both penetration resistance q_t and liquefaction strength R_L between intact and reconstituted specimens of Pleistocene or Holocene deposits.
- Though more tests are needed to conclude, a direct relationship between R_L and q_t seems to be compatible with that of the accelerated tests using cement, so that the $R_L - q_t$ line of intact specimens is located higher than that of reconstituted specimen. The difference between the lines tends to be larger with increasing soil age.

Consequently, the series of tests on reconstituted and intact natural sands revealed that the reason why higher fines content leads to higher liquefaction strength does not depend on fines content itself but cementation effect, which tends to be pronounced as fines increase. This further indicates that the current practice on liquefaction resistance modification by fines content may lead to dangerous results if it is applied to very young deposits.

Acknowledgments Geologists, Y. Mitani, S. Ishiwata and K. Momikura, experts in Pleistocene geology in Japan, who provided us with publications on local geology and guided us to outcrops of known ages for block sampling are gratefully acknowledged. Mr. Makoto Kamimura of Something Co. Ltd, in Japan, who provided the soil investigation data in Fig. 3.3 is gratefully appreciated.

References

Inada T (1960) On the use of Swedish weight sounding test results, *Tsuchi-to-Kiso*. Jpn Geotechnol Soc 8(1):13–18 (in Japanese)

Kokusho T (2007) Liquefaction strengths of poorly-graded and well-graded granular soils investigated by lab tests. In: Proceedings of the 4th international conference on earthquake geotechnical engineering, Thessaloniki. Springer, pp 159–184

Kokusho T, Murahata K, Hushikida T, Ito N (2003) Introduction of miniature cone in triaxial apparatus and correlation with liquefaction strength. In: Proceedings of the annual convention of JSCE, III-96, pp 191–192 (in Japanese)

Kokusho T, Hara T, Murahata K (2005) Liquefaction strength of fines-containing sands compared with cone-penetration resistance in triaxial specimens. In: Proceedings of the 2nd Japan-US workshop on geomechanics. ASCE Geo-Institute publication no 156, Tokyo, pp 356–373

Kokusho T, Ito F, Nagao Y (2009) Effects of fines and aging on liquefaction strength and cone resistance of sand investigated in triaxial apparatus. In: Proceedings of the earthquake geotechnical engineering satellite conference, 17th ICSMGE, Alexandria, Paper no I.16, CD-publication

Mitani Y (2006) Earthquake-induced liquefaction and fluidization in middle Pleistocene in the northern area of Chiba Prefecture, Japan. Earth Sci (Chikyu Kagaku) 60:241–252 (in Japanese)

Papadopoulou A, Tika T (2008) The effect of fines on critical state and liquefaction resistance characteristics of non-plastic silty sands. Soil Found 48(5):713–725

Seed HB, De Alba P (1984) Use of SPT and CPT tests for evaluating the liquefaction resistance of sands. In: Proceedings of the in-situ'86, ASCE geotechnical special publication no 156, pp 356–373

Shishikura M, Kamataki T, Takada K, Suzuki K, Okamura Y (2005) Survey report of emerged beach ridges in the southwestern part of Boso Peninsula -timing of the Taisho-type Kanto earthquake. Active Fault & Paleo-earthquake research, No. 5, pp 41–68 (in Japanese)

Suzuki Y, Tokimatsu K, Taya Y, Kubota Y (1995) Correlation between CPT data and dynamic properties of in situ frozen samples. In: Proceedings of the 3rd international conference on recent advances in geotechnical earthquake engineering and soil dynamics, St. Louis, vol 1, pp 249–252

Tokimatsu K, Yoshimi Y (1983) Empirical correlation of soil liquefaction based on SPT N-value and fines content. Soil Found 23(4):57–74

Chapter 4
Liquefaction in Tokyo Bay and Kanto Regions in the 2011 Great East Japan Earthquake

Kenji Ishihara, Kazuhiro Araki, and Kamata Toshiyuki

Abstract Severe damage to houses, roads and buried pipelines caused by liquefaction of the ground was the characteristic feature of destruction at the time of the 2011 Great East Japan in 2011. Widespread areas along the Tokyo Bay and in the downstream reaches of the Tone River suffered the liquefaction-associated damage, despite of the distance as long as 450–500 km from the epicenter of the quake. Typical examples of the damage are presented herein with reference to conditions of soil profiles. As a measure to gauge its destructiveness, the ground settlements resulting from liquefaction were calculated based on volume decrease characteristics of sandy soils and their outcome was compared with the settlements actually observed on the ground surface.

There were several accounts by eye-witnesses and video-pictures which are tacitly indicative of advent of surface waves or sloshing-like movements of the ground surface. Although conceptionally, some interpretation is given to these new features of motions which have not been hitherto addressed.

4.1 Introduction

An extremely large earthquake rocked the widespread area in the northern part of Japan Mainland at 14:46:33 (JST) on March 11, 2011. The magnitude of the quake was M = 9.0, an event accompanied with an unprecedentedly large amount of energy release at a centroid depth of 24 km where the Pacific tectonic plate subsides into the North America plate underneath the Japan Mainland. This quake was the largest ever recorded during the last 150 years since the start of seismic observation

K. Ishihara
Department of Civil Engineering, Chuo University, Tokyo, Japan
e-mail: kenji-ishihara@e-mail.jp

K. Araki • K. Toshiyuki (✉)
Design Department, Chemical Grouting Co. Ltd, Tokyo, Japan
e-mail: k-araki@chemicalgrout.co.jp; t-kamata@chemicalgrout.co.jp

M. Maugeri and C. Soccodato (eds.), *Earthquake Geotechnical Engineering Design*, Geotechnical, Geological and Earthquake Engineering 28,
DOI 10.1007/978-3-319-03182-8_4, © Springer International Publishing Switzerland 2014

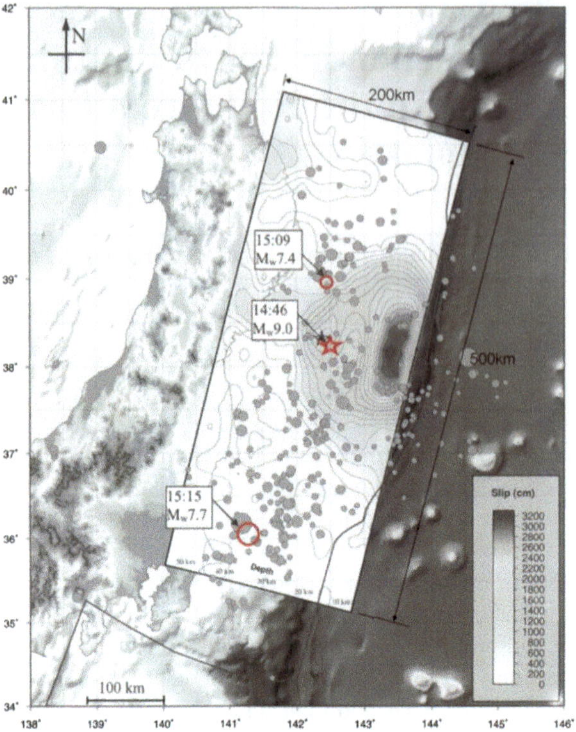

Fig. 4.1 Epicentral zone of the 2011 Great East Japan Earthquake (Hayes et al. 2011)

in Japan. Subsequently, the two big-scale aftershocks occurred, one at 15:09 with M = 7.4 in the north and the other at 15:15 with M = 7.7 in the south on the same day as displayed in the map of Fig. 4.1.

The fault zone covered a wide area about 500 km long and 200 km wide. It is noted that the biggest first quake with whopping M = 9.0 occurred at an epicentre under sea 175 km off the coast of Sendai, followed by the second largest event at an epicentre about 150 km east of northern coast. The fault ruptures moved southward, generating the third largest aftershock at a southern site 30 km off-shore from the coast of Ibaragi prefecture. Strong shaking was felt and recorded at a number of K-NET and Kick-NET stations over widespread areas from the northern prefectures, Iwate, Miyagi and Fukushima, down to the southern region such as Ibaragi, Chiba and Tokyo metropolis. Extensive liquefaction developed in the area of Tokyo Bay as well as over the flat lowlands surrounding lower reaches of the Tone River north of Tokyo. The features of liquefaction and associated damage have been reported by Kawabe et al. (2012), Towhata et al. (2011a, b), Tsukamoto et al. (2012a, b), and Urabe (2011). The features of the damage from different angles will be described in some details in the following pages of this paper focusing on those in Tokyo Bay area. Particularly noticeable was the phenomenal manifestation of the ground surface distorsion such as large settlements, offsets amongst various objects and push-over of the road pavements. Some conceptional interpretations are given to some types of seismic wave motions associated with such disastrous damage. There are many soil profile data in the area affected. With this information, some interpretation is given to the settlements of the ground resulting from liquefaction.

4.2 Ground Motion Characteristics in Tokyo Region

Shown in Fig. 4.2 are time histories of accelerations monitored at the K-NET station in Sendai, the area most severely shaken. It is noted that the peak acceleration monitored in the first M = 9.0 event was 1,000 gal in east-west direction (E-W component) and the duration of the main shaking was as long as 180 s. In the second quake that occurred 15 min later, the peak acceleration was 900 gal with the duration of about 120 s.

The trajectory of recorded accelerations projected on the plane of East-West and North-South axes is shown in Fig. 4.3 where it is noted that the largest shock had occurred in the N-S direction with the acceleration as high as 2,000 gal. The time histories of velocity obtained by integration of the recorded accelerations are shown

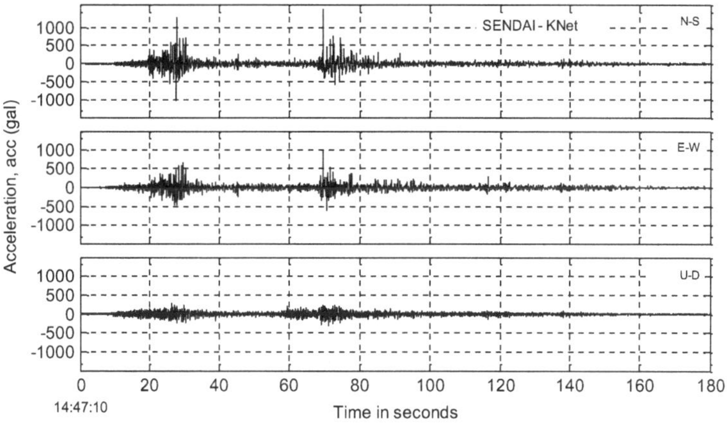

Fig. 4.2 Recorded motions at the Sendai K-NET station

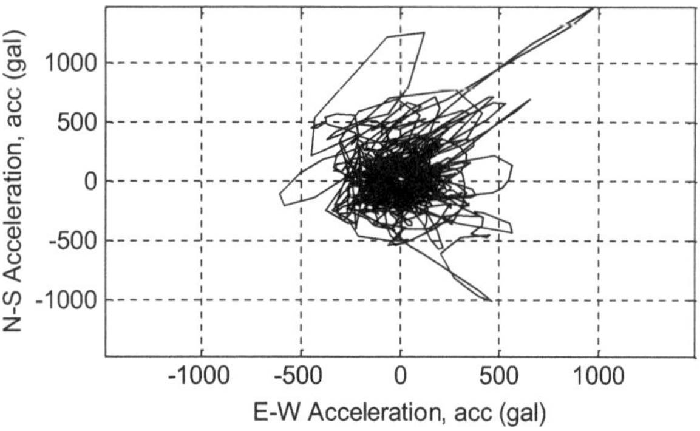

Fig. 4.3 Trajectory of recorded motion in Sendai projected on the E-W and N-Z plane (values less than 50 gal are removed)

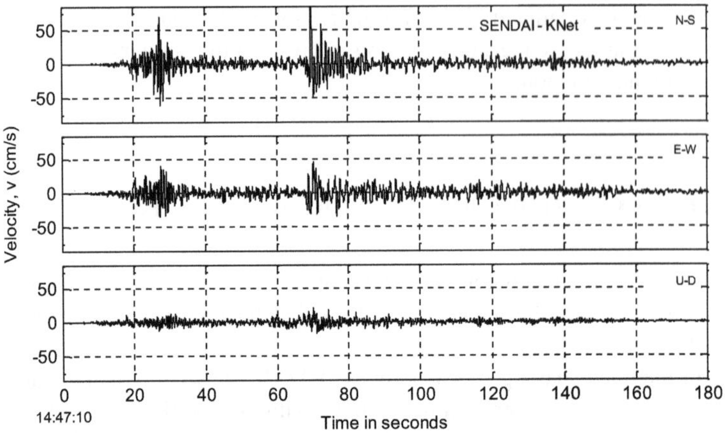

Fig. 4.4 Time histories of velocities obtained from recorded accelerations at the Sendai K-NET station

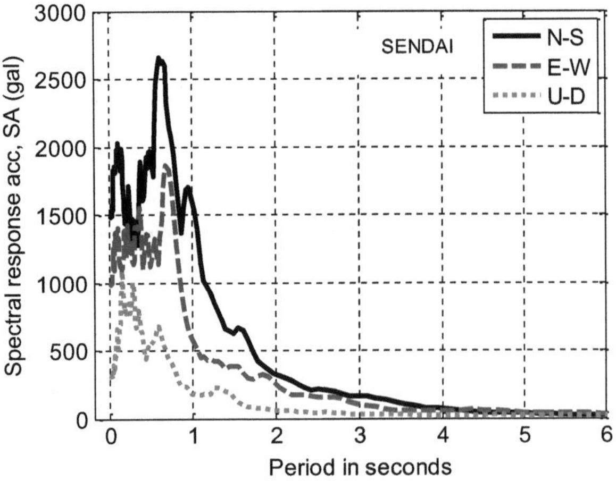

Fig. 4.5 Accelerations spectra at the Sendai K-NET station (for 5 % of critical damping)

in Fig. 4.4 where it is noted that the maximum velocities in the first event was as much as 80 cm/s. and that in the second quake was 70 cm/s.

The response spectra of recorded accelerations and computed velocities are displayed in Figs. 4.5 and 4.6, respectively, where it can be seen that the peak response acceleration and velocity occurred at the period of 0.7–1.0 s. It is also to be noted that the velocity at a period of 1.5 s is fairly predominant.

4.3 Ground Motion Characteristics in Tokyo Regions

There are several K-NET and KiK-net stations installed in the Kanto region where strong motion recorders were triggered. Locations of some of these stations are shown in Fig. 4.7. Typical motions obtained at Inage, Chiba K-NET station is

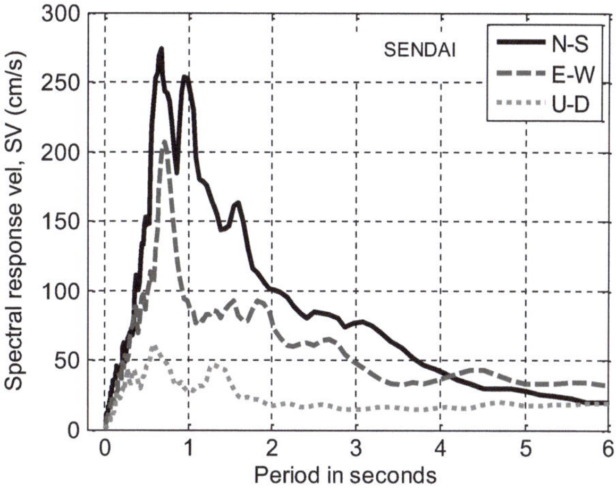

Fig. 4.6 Velocity response spectra at the Sendai K-NET station (for 5 % of critical damping)

Fig. 4.7 Liquefaction-affected areas in the Tokyo Bay

demonstrated in Fig. 4.8. It is noted here that the main shock by the largest quake was triggered at Inage, 108 s after the start of the motion at Sendai. This difference is considered approximately the time required for the tremor to travel through the distance of about 350 km from Sendai to Tokyo area. Thus, the velocity of propagation of the first wave front is estimated to have been 3.1 km/s. It can be seen in Fig. 4.8 that the peak horizontal acceleration (PGA) was 230 g in E-W direction. Note that this K-NET station is located in the section in Inage city where there was sign of liquefaction on the ground surface. Therefore, the peak acceleration is seen being followed by long period motions.

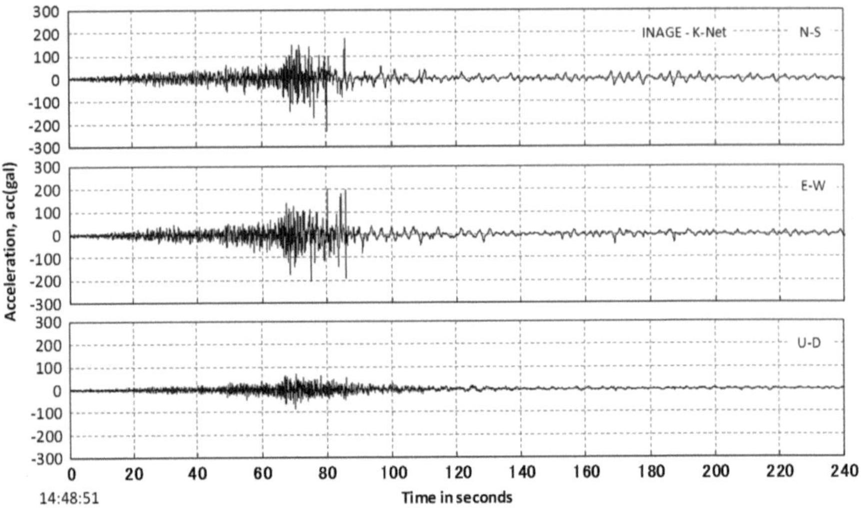

Fig. 4.8 Recorded acceleration at the K-NET station at Inage K-NET

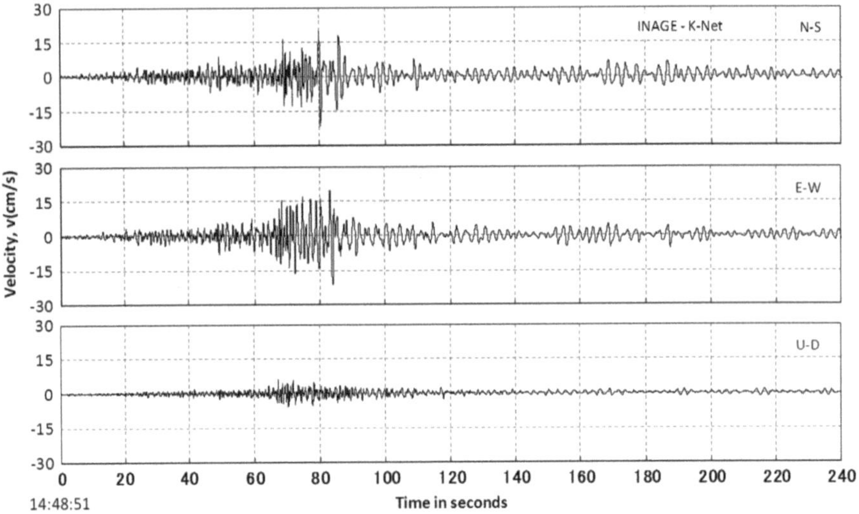

Fig. 4.9 Time history of velocities obtained from recorded accelerations at Inage K-NET recorded acceleration at the K-NET station at Inage K-NET

A majority of records obtained in Tokyo Bay area had a PGA in the range of 150–230 gals. The motion recorded at Urayasu was somewhat smaller being about 160 gal. The time histories of computed velocities in Inage are shown in Fig. 4.9 where it is noticed that the peak value is of the order of 20 cm/s. The trajectory of the recorded acceleration at Inage is displayed in Fig. 4.10 where it is noted that the predominant motion was in the East-West direction. It is to be noticed in Fig. 4.8

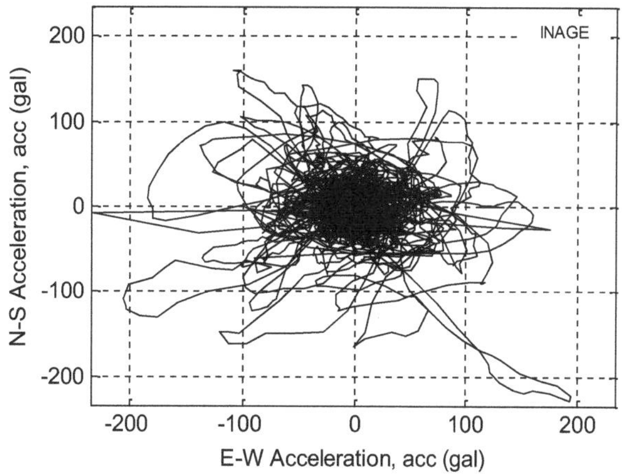

Fig. 4.10 Trajectory of recorded motion at Inage K-NET station plotted on the E-W and N-S plane

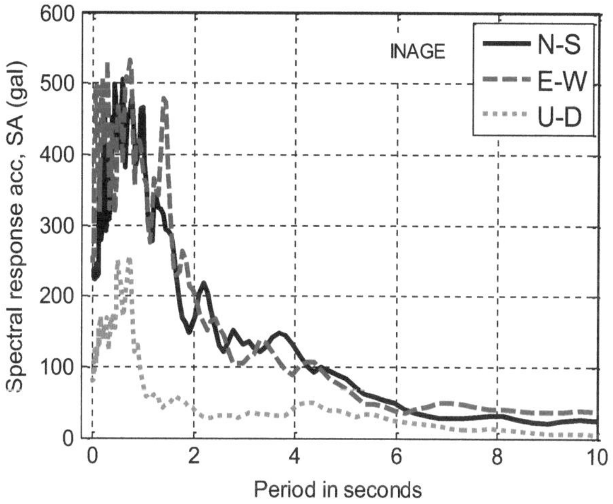

Fig. 4.11 Response of accelerations recorded at the Inage K-NET station (for 5 % of critical damping)

that the main shaking in Tokyo lasted as long as 150 s which are considered as being larger, perhaps, than any others ever recorded in the world. It should be mentioned herein that such a long duration of motions was the major cause of the liquefaction-related damage which were vastly disastrous as described below.

The response spectra of the recorded acceleration at the Inage K-NET station are displayed in Fig. 4.11. It is noted that the peak spectral acceleration occurred at the period of about 1.0 s. The spectral velocity at Inage is demonstrated in Fig. 4.12 where it may be seen that the large response occurred at the period of 1.5 and 4 s.

In summary, the motions in the area of Toyo Bay are characterized by the peak accelerations being of the order of 160–230 gal occurring primarily in the east-west direction with a predominant period of around 1.0–1.5 s.

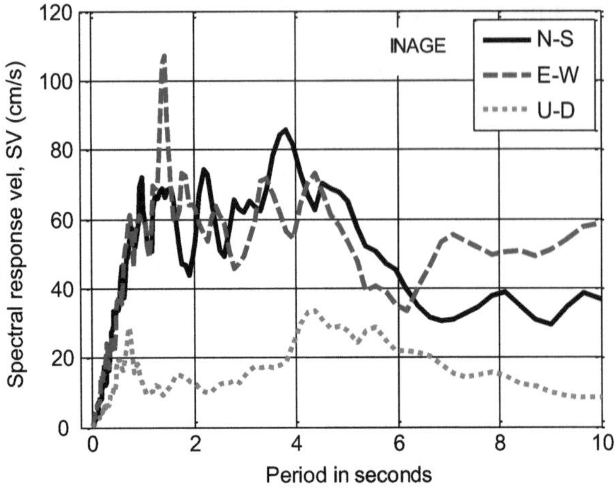

Fig. 4.12 Response spectra of velocities at the Inage K-NET station (for 5 % of critical damping)

4.4 Characteristic Features of the Ground Damage

4.4.1 The Damage in the North Tohoku Region

The eastern part of the north Tohoku region is generally mountainous with steep rocky slopes dipping sharply into the sea. There are dozens of small canyons and inlets formed by river channels along the coast from Hachinohe in the north down to Sendai in the south. This coastal geomorphology known as the "Rias" formation is characterized by the presence of steeply incised valleys which are filled with deposits of Pleistocene era. In such topography along the coast, the height of tsunami is amplified significantly in the course of its invasion into the inlet, resulting in the apocalyptically dreadful disaster. This has been indeed the case in the past with the tsunami-generated devastation in the north Tohoku region. There might have been liquefaction-associated damage by the seismic shaking, but because of significant erosion and scour of the subsurface soil deposits caused by repetition of inflow and retreat of the tsunami, traces of the ground damage, if any, seem to have been washed away, and could not be identified visually in the investigation after the event.

4.4.2 Liquefaction in Kanto Region in the South

Extensive liquefaction was induced in reclaimed deposits along rivers and bay areas in the plain of Kanto region including Ibaragi, Chiba and Tokyo which are located as far as 300–400 km south west of the epicentre of the main shock. The total land area in which signs of liquefaction were observed is purported to have been of the order of 70 km^2. In terms of the long distance from the epicentre and also in terms of

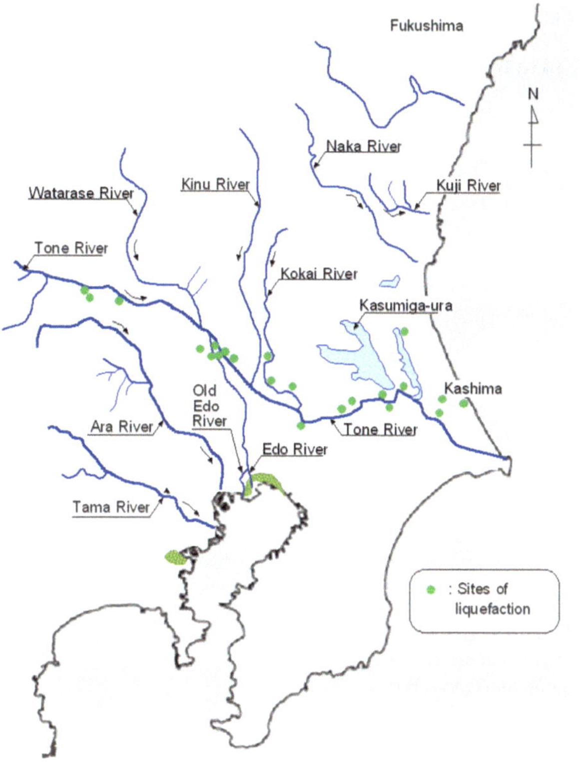

Fig. 4.13 Liquefaction – affected area in Kanto Region (S. Yasuda, 2011)

the large expanse net area, the liquefaction in the Kanto region by the 2011 earthquake was unprecedented and truly record-breaking. Shown in Fig. 4.13 is the distribution of places where apparent signs of liquefaction were observed such as sand oozing, boiling, ground cracking and associated ground settlements. These may be classified as (4.1) those which developed in the reclaimed waterfront area along the Tokyo Bay, (4.2) the liquefaction in the lakeshore district south of Kasumiga-ura in the lower reach of Tone River, and (4.3) the spotwise occurrence of liquefaction at many locales along the Tone River and its tributaries where landfills had been conducted in originally marshy flat lands. Somewhat detailed account on the features of the damage is given below, focusing on those in the area of Tokyo Bay.

Figure 4.14 shows roughly the coastal area along the Tokyo Bay affected by liquefaction, where it is noted that its effects extended south to the city of Kanazawa-bunko in Yokohama. Several cities in Chiba prefecture affected by liquefaction is shown in an enlarged map of Fig. 4.15.

One of the features specific to the present event was the occurrence of liquefaction over the area farther than any location ever recorded. The areas affected in Kanto area do span widely from the water-rich flood plain around the Tone River in

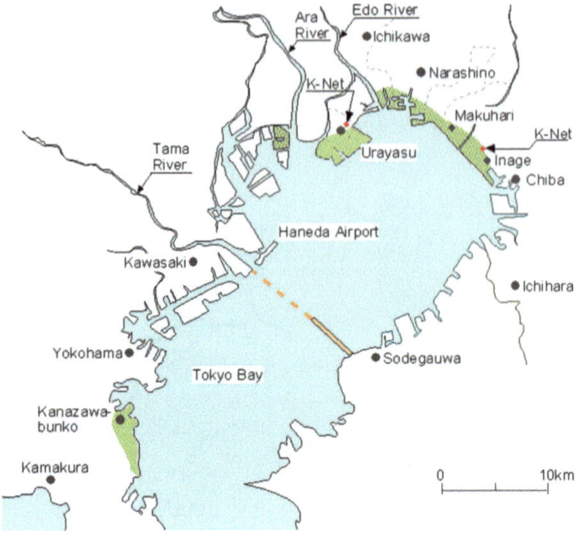

Fig. 4.14 Liquefaction-affected areas in the Tokyo Bay

Fig. 4.15 Liquefaction-affected areas in Chiba along the Tokyo Bay

the north to the alluvial or reclaimed areas in Tokyo Bay in the south. The distress to buildings and infrastructures caused by the liquefaction may be classified roughly into four types as described below.

4.4.2.1 Liquefaction in the Level Ground

There were a countless number of places where apparent signs of liquefaction were observed in flat ground. Sand boiling and oozing occurred, spurting a considerable amount of sand or silt accompanied by ground settlements as much as 10–100 cm all over the area nearby. These places included parks, play ground of schools, soccer fields, baseball fields and etc. One of the characteristic features was the fact that the sand spurting did occur preponderantly along long cracks or fissures which were opened in the level ground with a linear length of 5–20 m. This fact appears to indicate that not only horizontally polarized shear wave but also surface waves or reflected waves might have been induced in this area. It is also to be mentioned that the huge amount of silt or sand spurted produced deposits as thick as 50 cm on the ground surface. It is thus suspected that large voids of various sizes are left open under the ground.

4.4.2.2 Liquefaction Around Buried Lifelines

There are a number of complicatedly knit networks of pipelines such as sewage pipes, water supply pipes, and gas supply lines. These were constructed during various eras in near-surface deposits at depths 1–3 m, by excavating ditches and backfilling by sandy soils. Many times, these lifelines were laid down in sand deposits but in some cases in clayey or silty deposits containing gravels. No matter whether they lie in sandy or clayey soils, the backfilling was made by sands without compacting them sufficiently. Thus, it has been quite common in the past earthquakes as well to observe uplifting of manholes and sometimes pipes themselves particularly for sewage drainage lines. This type of damage occurred widely in the reclaimed areas in Tokyo Bay. It is due mainly to the fact that the majority of the pipelines was embedded not in the centre of roads under pavement but under sideways or walkways where there is no stiff mantle on the ground surface.

4.4.2.3 Effects of Liquefaction on Buildings and Houses

Many of light residential and commercial houses are founded on mat-type flat foundation or continuous reinforced concrete footings which are placed at depths 0.5–1.0 m directly on the underlying near-surface soil deposits. If the sandy soils underneath such structures had not been stabilized, liquefaction often developed readily causing deleterious effects on the foundations of houses involving settlements which were more or less accompanied by varying amount of tilt. Even in the case of a small tilt of the order of 0.005, houses became uninhabitable. The tilt of private residences accompanied by overall settlement of the order of 10–70 cm was observed over a number of residential sections. Buildings higher than three-stories, which are typically constructed on pile-supported foundations or on the flat foundations sitting on improved near-surface deposits, remained by and large intact and usable. High-rise buildings such as hotels and offices remained free from any damage.

4.4.2.4 Flow Slides in Gentle Slopes

Near the waterfront area and sloping ground as well where there is difference in elevation, lateral displacements or slides occurred at many places with horizontal displacement on the order of 0.5–3.0 m. In a long mileage of coastal line along the Tokyo Bay, the waterfront is protected by different types of seawalls retained by mounds of stones or stacks of wave-breaking concrete blocks. There was little damage to these seawall lines except for some sections.

4.5 Liquefaction-Affected Tokyo Bay Area

4.5.1 Damage in Urayasu City

Amongst the cities in Chiba Prefecture, Urayasu was the one most seriously affected by liquefaction. The major areas in Urayasu are indicated in Fig. 4.16 where apparent signs of liquefaction were observed at the time of the first strongest shaking on March 11, 2011 as well as in the aftershocks. Most of the areas affected were those reclaimed over the period of 1960–1980 and developed to provide landspace for residential houses, parks, schools and warehouses for industries. There was a sharp contrast between the areas of liquefaction or no liquefaction which is divided by the old shore-line before the reclamation started each in 1960s. This boundary line is indicated on the map in Fig. 4.16.

Most severely affected areas in Urayasu were districts where land expansion was undertaken towards the shallow sea waters starting from early in 1960s. The progress of land reclamation is indicated in Fig. 4.17 where it can be seen that the filling in the reclaimed area had been finished by the end of 1970s. The landfilling was conducted by dredging shallow-depth seabed near the shores. The sandy soils were transported hydraulically by pipes and sedimented under water. The photos in Fig. 4.18 show the reclamation works then underway. After filling hydraulically, dry sand was transported onland and spread by bulldozers to a thickness of 1–2 m. No means was taken further to compact the soils. Soil improvements were left to the hands of owners who purchased the land. Roughly speaking, the city of Urayasu is divided into three zones, as indicated by dashed lines in Fig. 4.16. These are (1) the old town (Moto-machi) north of the old beach line, (4.2) the middle town (Naka-machi) located on the man-made land southeast of the old town and (4.3) the new town (Shin-machi) on the man-made land due southeast of the middle town.

The old town part of Urayasu was practically free from any distress by the quake due probably to relatively stiff nature of the surface soil crust and also of the aged alluvium. The middle town area which is residential and commercial zone suffered serious damage due to liquefaction, because of the newly filled loose sandy soil

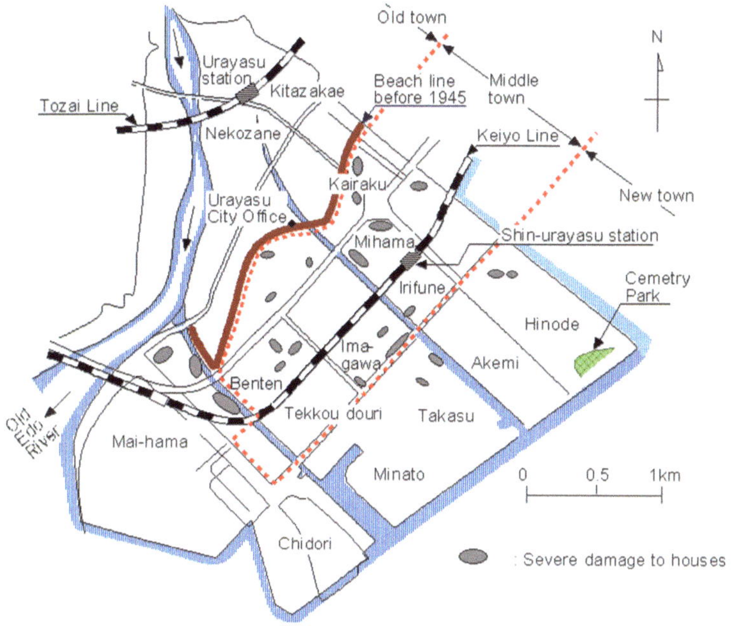

Fig. 4.16 Places of serious damage to private houses in Urayasu

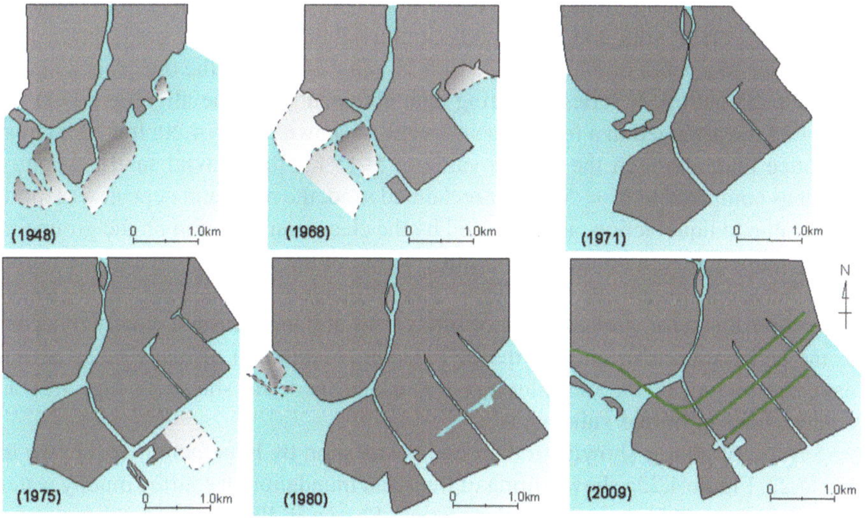

Fig. 4.17 Progress of reclamation work in Urayasu

deposit existing near the ground surface and also because of the high ground water table shallower than 1.0 m. In the new town area, the ground had been raised to an elevation of 4–5 m above the sea level by placing additional fills transported on land. Thus, the ground water table was rather lower. The majority of structures were hotels,

Fig. 4.18 Photos showing the reclamation work by the hydraulic method (Offered by Urayasu city office). (**a**) Distant view. (**b**) Pipes for hydraulic transmission

office building etc. which were constructed on sound foundations supported by improved soils and long piles. Because of these favourable conditions, the damage was minor, except for several buildings and facilities with poor foundation works. The soil profile at Urayasu consists generally of loose sand layers of reclaimed and alluvial origin underlain by a thick soft clay-silty layer to a depth of 40–70 m.

A typical soil profile in the middle town at a site at Irifune 3-chome is shown in Fig. 4.19, with the exact location indicated in the inset. It can be seen that a soft clayey silt deposit of alluvial origin exists to a depth of 45 m at this site overlaid by alternate layers of silty and sandy soils of alluvial era. The loose sandy layer near the surface to a depth of 5 m having a SPT N-value of 3–10 is the one placed newly by the reclamation. A series of boring data obtained along the alignment D-D' in Fig. 4.21 are arranged in a form of a side view as shown in Fig. 4.20. It is of interest to notice that, although the SPT N-value of 5–15 for the alluvial sand is slightly larger as compared to $N = 5–10$ for reclaimed sand, the new sand deposit was more vulnerable to liquefaction, as evidenced by the clear manifestation on the ground in the reclaimed area in contrast to no liquefaction in the old town area in the north. This fact seems to indicate that effects of solidification of fines due to aging might have contributed for strengthening of sandy soils although it is not clearly reflected in the penetration resistance of the SPT. Upon compilation of many other boring data, the bottom of the old alluvium was established and shown in Fig. 4.21 as contour lines of buried valleys.

Several of photos showing the ground devastation by liquefaction are shown in Fig. 4.22. Figure 4.22a shows tilting of trees and inundation of spurted muddy water over the road in Mihama residential section. Shown in Fig. 4.22b is uplifting of a sewage manhole on the sideway about 1.5 m above the ground. The features of the ground damage in general are shown in Fig. 4.22c. It should be noted that the ejected sand was mostly laden with silt and clay, making it easy to flow and spread over the ground surface. Figure 4.23a shows a photo indicating settlement and tilt of

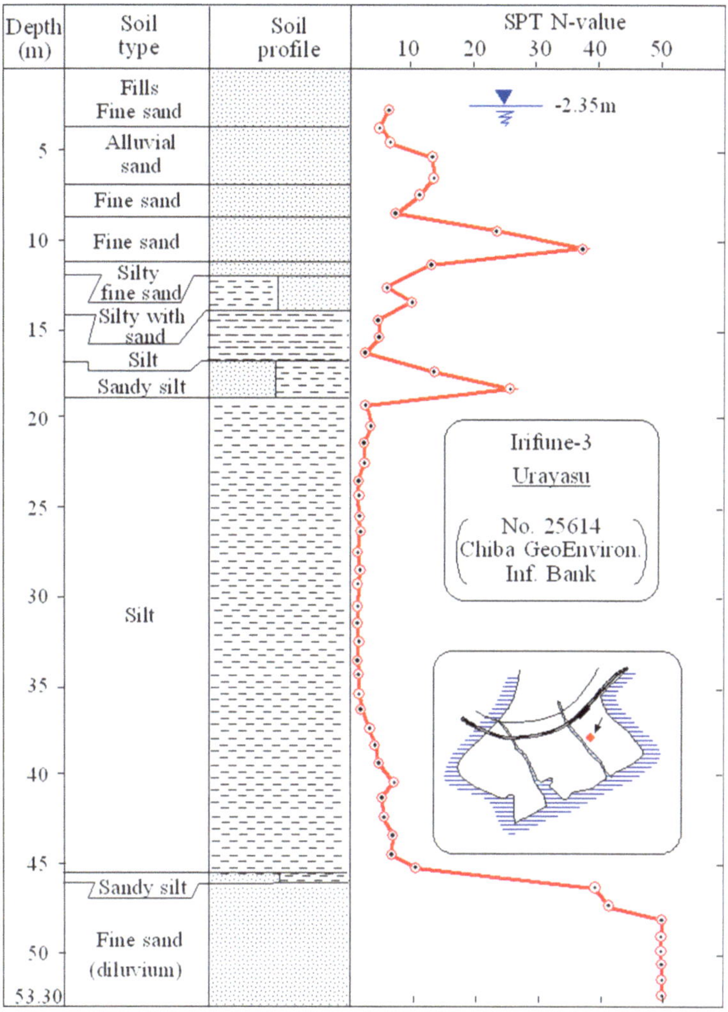

Fig. 4.19 Soil profile at Irifune 3-chome

a house. Figure 4.23b shows a house tilted and settled as a result of liquefaction. In the building supported by firm foundations such as piles, considerable vertical offset developed relative to the surrounding ground surface, as typically displayed in Fig. 4.23c. The gap about 50 cm between the building foundation and the subsided ground in its vicinity did result in breakage or tear-off of several pipelines and cables entering the buildings such as water and gas supply, sewage pipe, and electricity and telecommunication cables. The breakage of the lifelines as above was fatal for continued maintenance of operational conditions for the buildings.

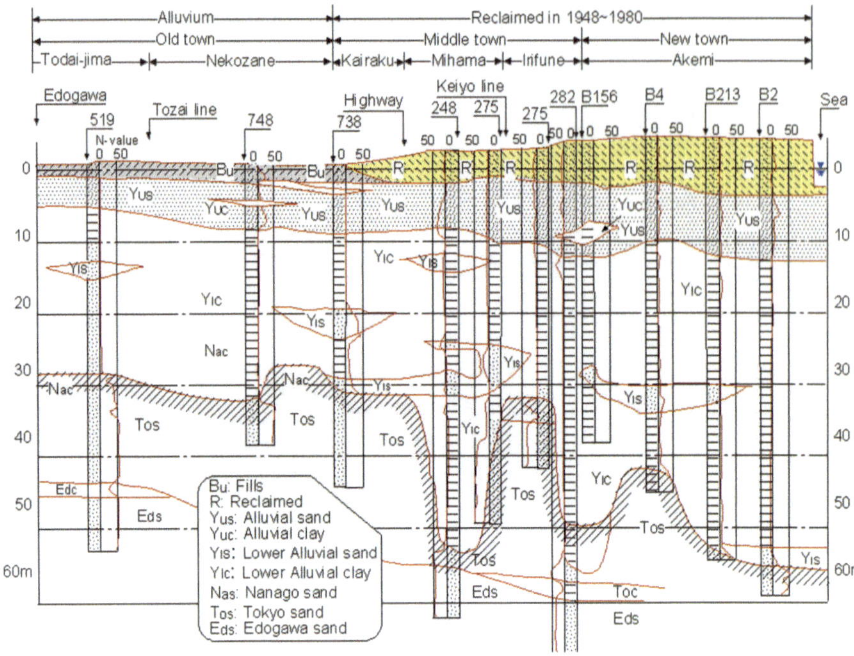

Fig. 4.20 Cross section D-D' in Fig. 4.25, Urayasu (From the Urayasu city office)

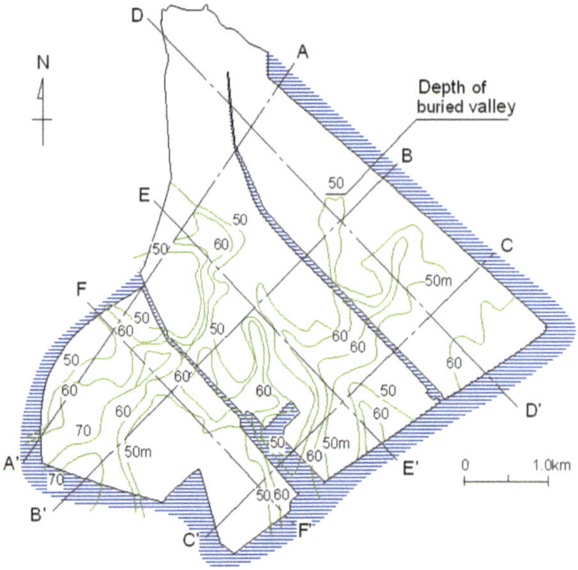

Fig. 4.21 Counter lines of valleys buried by alluvial deposits at Urayasu

Fig. 4.22 Destruction of the ground due to liquefaction (Offered by Urayasu city office). (**a**) Ground devastation in a residential section. (**b**) Floating of sewage manhole. (**c**) Spreading of sand over the road

4.5.2 Flow Slides in Waterfront Area in Urayasu

There were some places in Urayasu City where large lateral displacements took place in the man-made fills behind the seawalls. Most spectacular of this type did take place in a gentle seaside slope behind the cemetery park located in the east-south edge of Hinode-8. Exact location is shown in Fig. 4.16. Figure 4.24a shows an opening of cracks and holes about 2.0 m wide on the landside. A cross section in the middle of the slide area is roughly depicted as shown in Fig. 4.25 where it is noted that there is a concrete-made seawall. In front of the wall, there is a paved apron about 30 m wide which is retained by the sheetpile-supported reinforced concrete wall and wave-breaking concrete blocks stacked on the seaside.

It appears likely that the man-made fills underneath the apron as well as those on the landside of the concrete seawall were all placed by lightly compacting sandy soils by means of bulldozers which remained still loose enough to trigger liquefaction. The landslide in the cemetery park shown in Fig. 4.24 seems to indicate that the overall liquefaction through the distance of about 110 m in the cross section is accompanied by the lateral flow of the liquefied silt-laden sand. In fact, the seaward edge of the asphalt-paved apron was seen having moved laterally through

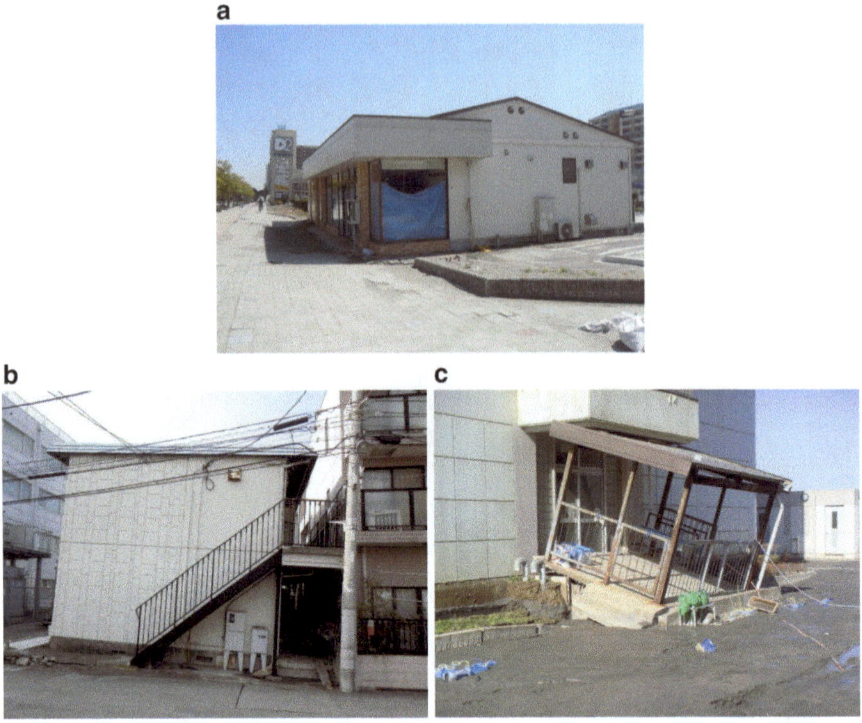

Fig. 4.23 Settlements and tilts of houses due to liquefaction (Offered by Urayasu city office). (**a**) Settlement and tilt of a house. (**b**) Sinking of an apartment house. (**c**) Offset between an undamaged building and settlement of the surrounding ground

1.0–1.5 m, thereby breaking off its edge and falling down into the wave-breaking concrete blocks as shown in Fig. 4.24b. As viewed in a photo in Fig. 4.24c, traces of sand ejection were observed through open seams of pavements in the apron. The concrete seawall in the landside was also seen having deformed seawards by about 30 cm as indicated in Fig. 4.26, which is indicative of the mass of liquefied sand deposit having slid along the likely sliding plane indicated by a dashed line in figure.

4.5.3 Ichikawa City

The city is located immediately east of Urayasu city as shown in Fig. 4.15. The major damage was induced by liquefaction in the area of man-made coastal zones in the south as indicated in Fig. 4.26. Shown in the photo of Fig. 4.26a are sand boils and resulting ground settlements observed in the belt zone about 3 m wide just behind the wave protection wall along Shiohama. The robust seawall unit consists

Fig. 4.24 Lateral spreading towards the sea behind the cemetery park at Hinode-8, Urayasu. (**a**) A depression of the ground inland of the seawall – a view to the east. (**b**) Damage at the edge of the apron – a view to the west. (**c**) Sand gushing from a seam of pavement over the apron in front of the seawall – a view to the west

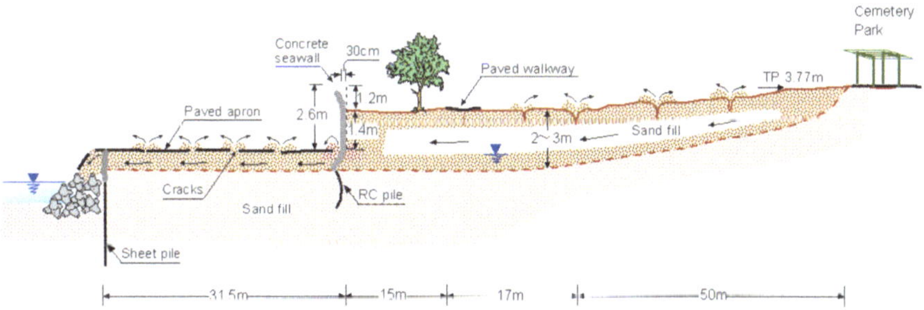

Fig. 4.25 A conceptual picture of the lateral spreading at the seaside zone in the south of the Cemetery Park, Urayasu

of stacks of stones (about 1.0–1.5 m in size) in front of the sheet-pile supported concrete wall. Although there are some variations in cross section, the coastal line along the Tokyo Bay in this area had been firmly protected by this type of compound structure and, hence performed well without damage. Although there

Fig. 4.26 Ground damage at Shiohama, Ichikawa, Chiba-city. (**a**) Damaged steps in the fishery port with the lateral displacement of about 50 cm – a view to the east. (**b**) A view of the undamaged seawall (center) with extensive liquefaction at inland walkway – a view to the east

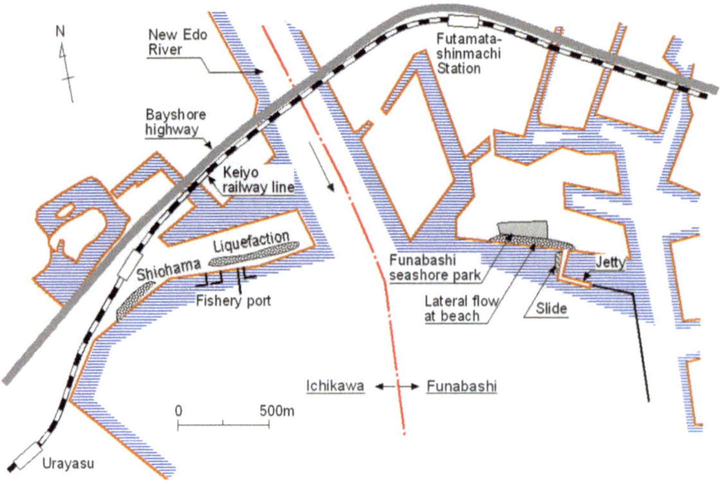

Fig. 4.27 Ichikawa and Funabashi City

was strong shaking of the order of 150–200 gal plus the following rise of about 1.0 m in the sea level due to tsunami, the seashore protection structures as above remain almost intact and fulfilled its intended purpose.

Shown in the photo of Fig. 4.26b is the damage to fishery port in Ichikawa involving lateral displacement of the order of 30–50 cm at the ground surface level. Considerable evidences of liquefaction were observed also along the belt zone along the seashore road as shown in the map of Fig. 4.27.

a b

Fig. 4.28 Damage to the ground by liquefaction in the Seashore Park at Funabashi. (**a**) Sinking of a small house with overall spreading of liquefied sand – a view to the coast. (**b**) Lateral spreading of the beach in front of the Seashore Park – a view to the west

4.5.4 Funabashi City

Most of the shoreline structures are warehouses, industrial buildings, quaywalls of private industries which were difficult to make access for site investigation, but these appear to be protected by the pile or H-beam supported concrete seawalls. Most pronounced damage was to houses and facilities in Funabashi Seashore Park located behind the sand dune of the beach line as shown in Fig. 4.28. Overall ground subsidence of the order of 50 cm occurred due to extensive liquefaction. Outdoor small shopping houses supposedly supported by flat foundations sunk by 50 cm due to inundation of sand as shown in the photo of Fig. 4.28a. Along the beach line at the park, cracks or fissures of the ground were observed, as displayed in Fig. 4.28b, resulting from lateral flow of thc liquefied sand.

4.5.5 Makuhari City

The city of Makuhari had been developed in 1970s over reclaimed lands, based on well-conceived city planning and all the buildings, parks, infrastructures etc. are arranged in an orderly manner. These facilities themselves were constructcd on stabilized soils and supported, in addition, by different types of embedded to stiff strata, and did not experience any damage. However the lands such as roads, squares, parks, schools, play grounds and parking places had not been improved and did experience extensive damage due to the ground liquefaction involving cracking, sand ejection and overall settlements of the order of 30–50 cm. Particularly noticeable was the vertical offsets between the periphery of pile-supported buildings and the paved aprons or sideways in their vicinity.

There is a river about 20 m wide called Hanami River south of Makuhari (see Fig. 4.15). On Makuhari side near the mouth of the river, the area including the road about 100 m wide and 200 m long, as indicated on the map of Fig. 4.29, suffered a

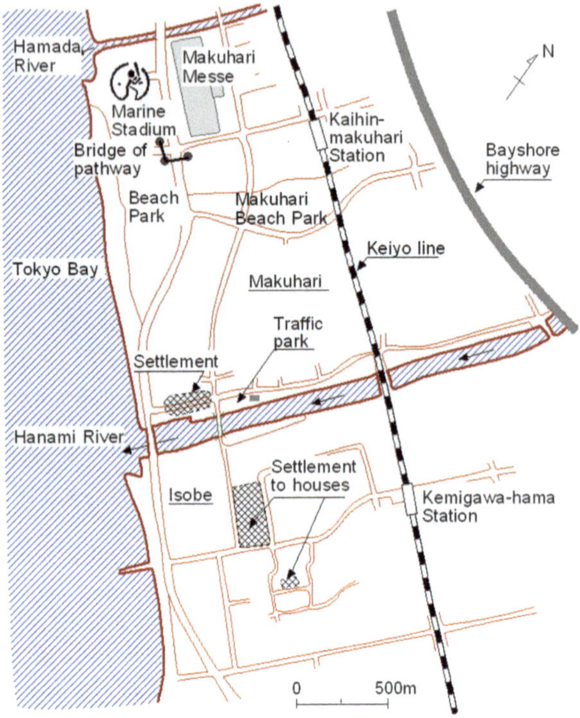

Fig. 4.29 Makuhari-Isobe, Mihama-ku, Chiba

Fig. 4.30 Damage features associated with large settlements over the area at the north bank of the Hanami River near its mouth. (**a**) Vertical offset in the park on the other side of the road. (**b**) Small facilities at the portal of the building

subsidence of about 50–100 cm due to liquefaction. The photo in Fig. 4.30a shows a vertical offset of about 50 cm that had extended into the park through a distance of about 200 m. The Fig. 4.30b showed the vertical offset in front of the 10 storey building on the north side of the road. It may be seen that the vertical gap of the order of 50–70 cm in front of the building produced damage to facilities such as

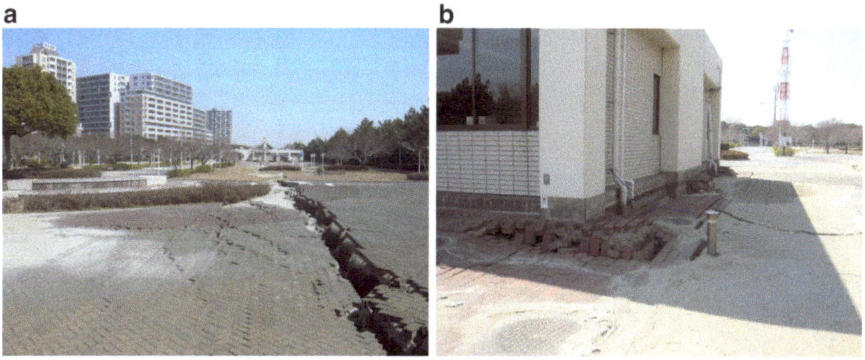

Fig. 4.31 Open cracks through the premise of the Traffic Park and the punching settlement of the office building. (a) Open cracks through the park. (b) Punching settlement along the periphery of the office building

portal steps and fences. In the Traffic Park, east of the above mentioned area, traces of considerable liquefaction was observed on the ground surface, including development of widely open cracks and sand boiling. The photo in Fig. 4.31a, viewed from the west shows open cracks, as wide as 0.7–1.0 m, running through the park in parallel to the line of the Hanami River. Shown in Fig. 4.31b is the end of the cracks, viewed from the north, blocked by a one-storey office building. It is of interest to note the large push-in of the ground that has developed along the periphery of the building due probably to punching interaction between the concrete foundation and the surrounding soil deposits that had liquefied. The depth of the push-in was 50–70 cm relative to the flat level of the surrounding ground.

The Traffic Park is bordered by a line of bush fence along the walkway about 1.0 m lower in elevation which is located along the Hanami River immediately on the south. The concrete wall along the river was turned down as viewed in the photo of Fig. 4.32a. There was a considerable amount of sand that had spread over the apron one step down the walkway as shown in Fig. 4.32b. It is highly likely that the liquefied sand over the park premise about 2 m higher in elevation had flowed underground towards the Hanami River and pushed out the walkway wall, resulting in the collapse of the wall and spreading of the sand over the apron. The scenario as such is displayed in the cross sectional sketch shown in Fig. 4.33.

In Makuhari district there are a number of high-rise buildings and modern infrastructures which had been laid out based on well-conceived grand design of the city. Since the area was reclaimed over the period of clay layers of alluvial or Holocene origin having a thickness of about 10–40 m. All of these facilities had been built up on the foundations supported by various types of piles embedded to the stiff deposits about 20–40 m deep.

In addition, the near-surface sand deposits had generally been compacted by various methods in order to provide sufficient resistance of the foundation pile

Fig. 4.32 Failure of the walkway wall on the north of the Hanami River. (a) Collapse of the wall north of The Hanami River – viewed from the west. (b) Collapse of the wall north of the Hanami River – viewed from the east to the sea

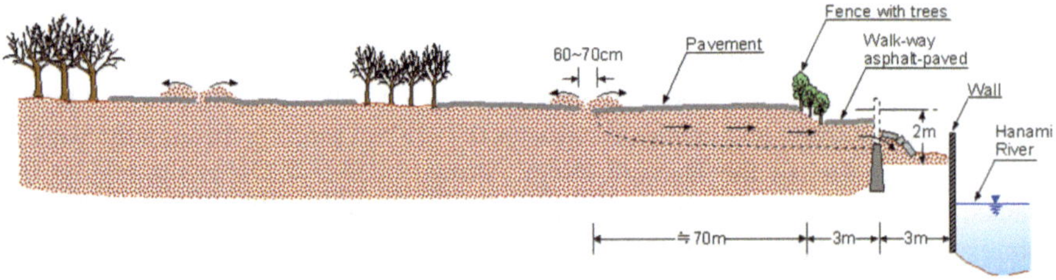

Fig. 4.33 Conceptual sketch across the Traffic Park for the lateral flow of the liquefied sand towards the Hanami River

system against the lateral force induced by strong earthquakes. In contrast, a large portion of the reclaimed lands used for roads, parks, playgrounds, tree-planted sideways, green belts etc. had been left unstabilized.

Thus, while there was practically no damage to the well-engineered structures, extensive damage developed due to liquefaction over the wide-spread areas which had been left intact without implementing stabilization of sandy deposit. The overall settlements of the order of 30–50 cm were observed here and there with fissures in parallel alignments. These injuries were always accompanied by sand spurting or boiling, indicating apparently that cracking or offsets in the pavements and corner stones or sometimes with wavy distortion of road the occurrence of liquefaction was the root cause of the distress. Shown in Fig. 4.34a is a photo showing annular-shaped sag of the liquefied sand surrounding a 1.5 m in-diameter column supporting the pathway bridge pier over the road near the entrance to the parking area in Marina Baseball stadium behind the coastal line (see Fig. 4.29). A distant view of the pathway bridge is shown in a photo of Fig. 4.34b.

a b

Fig. 4.34 Settlement of the ground near the cross-road walkway bridge at the entrance to the parking area of the Marine Stadium at Mihama, Chiba. (**a**) Annular sag in the liquefied ground surrounding the column. (**b**) Settlements of the ground around the walkway bridge

4.6 Ground Motion Characteristics Related with Liquefaction

It has been widely known and accepted generally that the repetitive shear stresses on the horizontal soil element produced by upward propagation of shear wave are the predominant component of external stresses producing pore water build-up and consequent liquefaction. The evaluation of the liquefaction potential for the level ground has been made exclusively on this assumption. However, in the case of the present earthquake there are many testimonies by eye-witnesses saying that they observed wavy movement of flat grounds following the shaking of the main shocks and aftershocks.

In fact, it should be mentioned that the most devastating features of liquefaction-induced damage were associated not only with the conventionally observed sand boiling, but also with large settlements, and sideward pushing up of roads and walkways. In flat ground such as parks and playgrounds, fissures or cracks were open in linear manner through a length of 5–30 m. An example of such fissures developed over the flat playground in Sports park in Maihama, Urayasu, is displayed in Fig. 4.35a. Figure 4.35b shows another example of linear fissures over the dry elevated flood plain on the stream-side of the Tone River. The characteristic features as above appear to suggest that, in addition to the spatially uniform propagation of shear wave, there must have been other modes of waves generated in the course of travel from the hypocenter as far as 400–500 km away. The surface waves, reflected or refracted waves or a forced oscillation within the liquefied sand basin might be the types of waves that are conceivable to be generated. It is to be noted that the latter type of wave is similar to what is called "seiche" in hydraulic or coastal engineering.

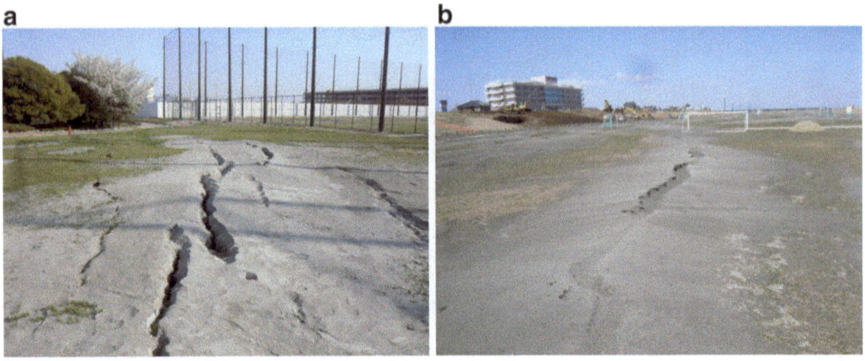

Fig. 4.35 Linear fissures in wide flat ground. (a) Cracks at Sports Park at Urayasu. (b) Cracks on the elevated bed of the Tone River

4.6.1 Motions by Surface Waves

As wellknown, there are two types of surface waves, that is, Rayleigh wave and Love wave. Both waves are known to be generated over flat plains located far away from the hypocenter as a result of underground reflection or refraction of the incident body waves. It is also known that the surface waves are generally felt after arrival of the body wave and propagate with a relatively long period of motions of the order of 2–10 s.

One of the significant characteristics of the surface waves is that the shear stress or shear strain components is not equal to zero in the soil element near the ground surface. In the case of Love wave, the shear stress acting on the vertical plane perpendicular to the direction of its propagation is non-zero near the surface. In the Rayleigh wave the deviator stresses induced in the horizontal plane and in the vertical plane perpendicular to the direction of propagation is non-zero near the ground surface. It is also to be remembered that behaviour of cohesionless soils is governed, not by the shear stress itself, but by the shear stress ratio. Since the confining stress due to overburden is very small near the surface, the shear stress ratio induced by the surface waves could be very large making it easy for soil elements to fail or to develop liquefaction. It is also important to notice that non-zero components of stresses near the surface could easily induce cracking or offsetting on the ground surface, as described below, leading to the near-surface disruption.

(a) One of the evidence suggesting existence of the Rayleigh wave is displayed in a photo in Fig. 4.36a which was taken in the vast plain of Hachiro-gata, Akita prefecture, at the time of the Nippon-kai-chubu earthquake in 1983. This is the agricultural land created by the drain-off of the large expanse of a lake. The deposits below its bottom consist of loose sand and silty clay. The wave length as inferred from the photo is about 100 m and the crest-trough amplitude is about 50 cm.

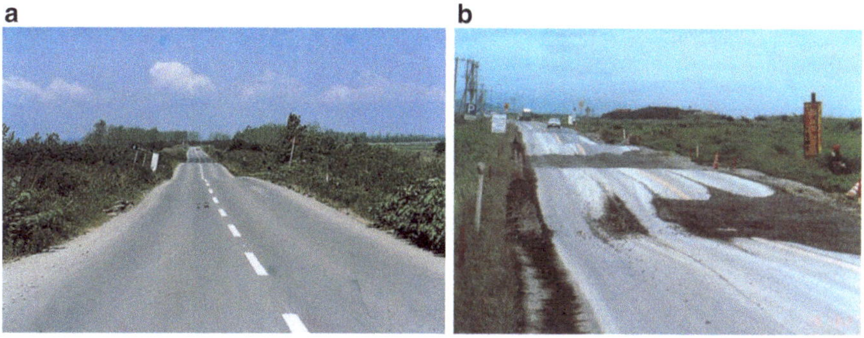

Fig. 4.36 Photos showing wavy deformation on the ground surface. (**a**) A photo showing wavy distorsion of a road in Hachiro-gata at the time of the 1983 Nipponkai-chubu earthquake. (**b**) A photograph taken at Toyotsu in the No. 5 National Highway at Noshiro at the time of Hokkaido-Nansei-oki Earthquake

(b) Another example of surface disruption indicative of advent of the surface wave is shown in a photograph of Fig. 4.36b which was taken at Toyotsu in the No. 5 National Highway at the time of the Hokkaido-nansei-oki earthquake ($M = 7.8$) in 1993 (Kazaoka 2003). This is the place where iron-containing sand had been excavated to a depth of about 4 m and filled back later on by sands. Somewhat detailed measurements in the field showed a wave length of about 30–50 m and the crest- trough amplitude of about 50 cm. The resistance by Swedish cone tests performed at a place 200 m apart showed $Wsw \cong 100$ kg in static phase of penetration and $Nsw = 0$ in rotational phase of penetration, indicating that the corresponding SPT N-value is about 3–5. It is not known whether this is the Rayleigh wave or not. More detailed studies may need to be made before this is identified as the Rayleigh wave. There are several other examples like this indicating the wavy motions on the ground surface.

(c) Still other example showing effects of the surface wave is reported by Kazaoka (2011). At the time of the earthquake on March 11, 2011, he happened to be in the field performing the portable-type penetration tests on the lawns in Inage Near-Shore Park along the coast of Chiba. Its location is shown in Fig. 4.15. He took motion pictures from the start of the shaking. The sequence of events described by him may be summarized as follows.

1. The shaking began from around 14:47 on March 11, 2011. About 2 min later, with increased ground motions, the movement became visible clearly by eyes.
2. At the same time, there appeared fissures open around the place where sand boiling is known to have occurred in the previous Chiba-Toho-oki earthquake in 1987. Then, cycles of sand boiling and sucking were observed in the fissures shown in a photo of Fig. 4.37a. This cycle was repeated with a period of about 5 s and continued for a few minutes. The crest-to-trough wave height of the motion was about several ten of centimetres.

Fig. 4.37 (**a**) Around 14:50: Sand boiling began at the place of precious liquefaction from around 14:48 (Kazaoka 2011). (**b**) Around 14:51: After the main shock with increased discharge of mud water (Kazaoka 2011). (**c**) Around 14:52: Increased water discharge with enlarged cracking (Kazaoka 2011). (**d**) After the maximum aftershock, widespread zone of the park was inundated with sand-water (Kazaoka 2011)

3. Around 14:49, there occurred a large wavy motion with a wave length of about 100 m. Then with increased wave height that followed, the ground fissures were enlarged in length and a large amount of sand-laden water began to spring out and spread over the lawns. The situation around 14:51 is displayed in Fig. 4.37b.
4. Around 14:52, the fissures were further stretched and discharge of sand and water continued until around 15:30. The ground condition around 14:52 is displayed in Fig. 4.37c.
5. At the same time, the ground settlement began to take place gradually and by the time of the largest aftershock at 15:15, the settlement had increased to about 30 cm, creating a depression for water pooling, as shown in Fig. 4.37d.
6. At the time of the aftershock at 15:15, the ground wave having a wave length of 50–100 m and a crest-to-trough amplitude of about 50 cm was observed in the water-pooled pond. The situation around 15:10 prior to the aftershock is shown in Fig. 4.37d. It indicates that nearly entire zone of the Inage-shore Park were filled with water as if it were a lake.

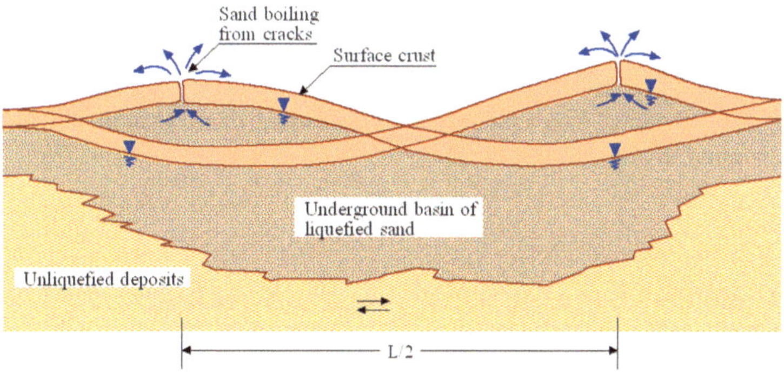

Fig. 4.38 Conceptional picture of the motion in a underground basin created by liquefaction

There are several other motion pictures taken by local residents showing the cyclic opening and closing of surface cracks accompanied by spurting and sucking of sand and water. The period of these repetitive motions is also about 5 s. Summarizing several observations as above, two conceivable interpretations may be drawn, as follows, regarding the nature of ground motions following the main shock.

4.6.1.1 Surface Wave

Judging from the traces of the motions such as those obtained at Hachiro-gata in 1983 and at Toyotsu in 1993, it is conceivable that the Rayleigh wave was generated after the arrival of the shear wave. This wave could be a source of disturbance inducing liquefaction but considering the fact that the surface wave is generated at a later time of seismic shaking, the main component of the motion triggering liquefaction would be the cyclic shear stress emerging from the shear wave propagation. After soil deposits have been softened as a result of liquefaction, the surface wave is generated in conformity to the spatial extent of softened soil deposits near the surface. It is to be noticed, however, that, the level of the damage would become worse and more disastrous, with stronger destructive forces by the surface wave.

4.6.1.2 Entrapped Ground Wave

It is likely that liquefaction does not occur uniformly over wide areas because of complex nature of subsurface stratification varying spatially depthwise and laterally particularly in reclaimed area. Thus, it is conceivable that a kind of underground basin of liquefied soils is created by the strong shaking in the mainshock. This is depicted conceptionally in Fig. 4.38. Upon continued arrival of the body waves that follow, this underground basin would be excited, thereby generating a kind of forced oscillation having its own wave length and period. In the field of hydraulics and coastal engineering, the free oscillation which occurs within enclosed waters is referred to as "seiche". When the oscillation occurs in storage tanks or any small-sized containers, it is called "sloshing". From the observation by Kazaoka (2011) and others as described above, it is likely that a kind of oscillation similar to sloshing or seiche must have taken place in the underground basin filled with viscous fluid generated by liquefaction.

4.7 Considerations for the Waves and Cracking of the Ground Surface

4.7.1 *Identification of Wave*

From the considerations as above, it is envisaged that there would be a type of wave or oscillation generated near the ground surface which is different from the body wave coming directly from the bottom of deposits. The movements of the ground as above are characterized by the three features, that is, (4.1) the movement appeared after the main shaking, (4.2) the ground moved periodically at a long period of about 5 s with a wave length of about 50–100 m, and (4.3) cracks developed soon afterwards and opened and closed periodically during the phase of this motion.

If it is the Rayleigh wave, the velocity of propagation would be $V = L/T \cong 50/5 \sim 100/5 \cong 10 \sim 20$m/sec., which is a value likely to be the case. There is a possibility that the motion could be a kind of trapped wave or seiches. If it is assumed as the seiche in enclosed waters, the longest period, T, of the motion is calculated by a simple equation shown in Fig. 4.39.

In that formula, h is the depth to which liquefaction penetrates and L is a representative distance in the direction of motions in an area within which the seiches take place. It is also to be noticed that the seiche is a free oscillation in non-viscous fluid with fixed boundary on the periphery. The liquefied sand may exhibit strong viscosity and therefore the oscillation may need to be considered as a forced vibration within a basin-like deposit with fixed boundary on the periphery excited at its base by continuously incoming body waves, which would be the case in the long duration motions such as the Great Tohoku Earthquake in 2011.

Sloshing would be an alternative interpretation. The period of sloshing motion within a cylinder with a radius, L, filled with non-viscous fluid to a height, h, is estimated by a formula shown in Fig. 4.39. In this formula, $\varepsilon_1 = 1.841$ and g is specific gravity. It is to be noticed that the sloshing is generally referred to as the non-viscous liquid motion having fixed boundary conditions at the periphery of containers.

Anyhow, estimate of the period of motion for the case of liquefied basin with a dimension of the order of $h = 15$ m and $L = 50$–100 m would require additional scrutiny. Thus, identification of the wave type would remain a matter yet to be resolved. In the following, the wave will be referred to as entrapped oscillation.

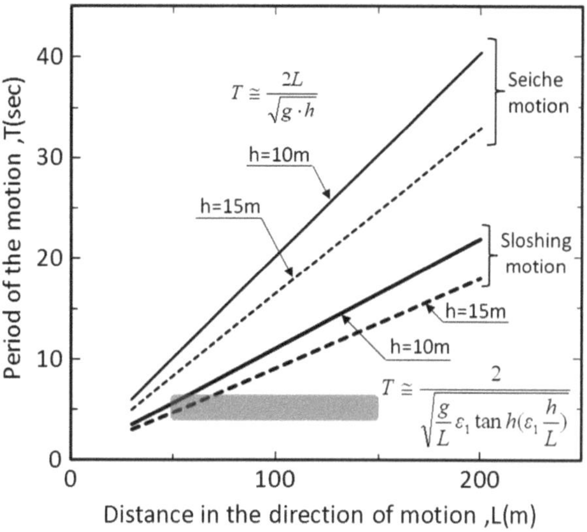

Fig. 4.39 Wave motion characteristics of seiche and sloshing

Shown in Fig. 4.39 are plots of the two formulae, one for the seiche and the other for the sloshing motion. In the light of obscure ground movements by eye-witness and motion pictures indicating that $L = 50–100$ m, and $T \cong 5$ seconds, the sloshing motion is more likely scenario for the motion. The conceptional picture for the motion in the liquefied sand basin is shown in Fig. 4.38.

4.7.2 Developments of Surface Cracks

The conditions for liquefied sand whether it can gush or not up on the ground would also depend upon the depth of the crack opening. If it is shallow, the liquefied sand can not go up on the surface, but if the crack penetrates deep enough, it is easy for the sand-laden water to gush and spread area the surface. Depth of the crack penetration, z, may be estimated roughly as follows. Let the uniaxial strength of a soil element at the bottom be devoted by q_u, as illustrated in Fig. 4.40. If the vertical stress, $\sigma_v = \gamma_t z$ is greater than q_u the crack will be closed because of the large lateral displacement at failure. Thus, a simple criterion as expressed by

$$\gamma_t . z \leq q_u \tag{4.1}$$

would define the crack penetration. Assuming the unit weight of soil to be γ_1 =1.8 t/m2, and q_u =2–5 t/m^2, the crack depth is given as

$$z \leq q_u/\gamma_t = 1.0 \sim 2.8m. \tag{4.2}$$

There are several in-situ studies by pit excavations to observe the paths of sand-water flow through cracks towards the ground surface. According to visual

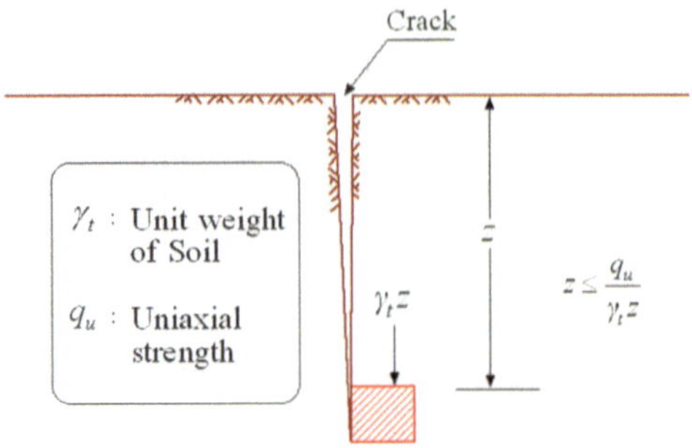

Fig. 4.40 Possible depth of crack penetration

observation of side-walls of the pits excavated to a depth of 1.5–2.0 m, remains of pipe-shaped vertical holes with variable diameters have been detected indicating clear presence of paths or vents for liquefied sands towards the surface. It is to be noted, on the other hand, that the crack can not penetrate into the water-saturated soil deposits. Thus, the crack penetration is limited to the depth above the ground water table. In the light of the fact that the ground water table is located less than 3 m below the surface in the majority of liquefied sites in the present earthquake, it may as well be mentioned that the cracks must have developed to the depth of the ground water tables due to the wavy motions of the ground surface having the crest-to-trough amplitude of about 50 cm and wave length of about 50–100 m.

4.8 A Sequence of Scenarios in Liquefaction and Consequent Damage

On the basis of the several observations as described above, it becomes now possible to envisage a sequence of scenarios regarding the onset of liquefaction and consequent occurrence of ground damage at the time of the 2011 earthquake. The conceived scenario is summed up in a form of a chart shown in Fig. 4.41. Generally speaking, the whole event in the mainshock or aftershock may be divided into two phases, that is, the initial part leading to the triggering of liquefaction and the subsequent events related to the ground destruction such as crack opening, sand boiling and surface distorsions.

1. Onset of liquefaction
   The shaking in the main shock lasting for 60–90s with the peak acceleration 170–200 gals is considered to have occurred as result of upward propagation of the shear wave. Loose deposits of saturated sands below the ground table was subjected to repetitive shear stress under undrained conditions thereby accompanied by gradual buildup of excess pore water pressure. Around 80–90 s after the start of shaking, the sand deposits became extremely softened as a result of onset of liquefaction. At this stage, the liquefied sand deposits are covered by the crust of unliquefied soils near the ground surface.

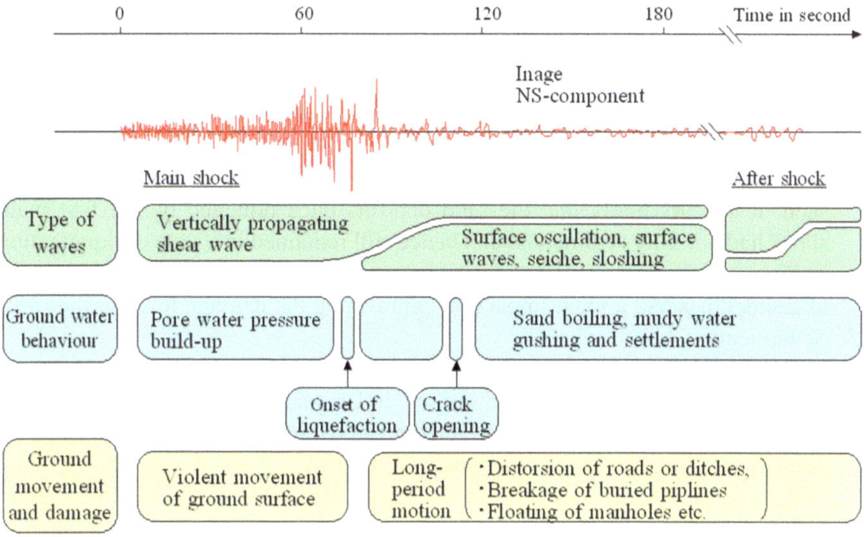

Fig. 4.41 Scenario of events from liquefaction to the consequent damage of the ground

2. Long-period ground oscillation
   In the reclaimed areas such as those in Urayasu and Inage in the Tokyo Bay or in Itako and Kamisu as well, the liquefiable sand deposits do not exist in the form of uniformly stratified horizontal layers. In some areas, liquefiable deposits extend as deep as 15 m, but only down to a few meters in its neighbourhood. Thus, it may be with reasons to assume that a basin-like liquefied zone was created under the ground covered by the surface crust about 1.0–3.0m thick. Thus, the surface crust appears to behave as if it were a huge raft floating on the hidden liquefaction-pond or lake. With this conception in mind, it would be easy to understand the occurrence of ground motions with a period of about 5.0 m seconds in the later part of the seismic motion as accordingly indicated in Fig. 4.41. It is not clear whether this kind of long-period motions is to be identified as seiche or sloshing. It is literally referred to as "ground wave" by a specialist group involved in the geoenvironmental study in Japan (Nirei 2011). It may be rather called as "entrapped wave".
3. Crack opening, sand boiling and ground distorsion
   Following several cycles of the long-period oscillation, some cracks started to open on the ground surface or along lines of junctions or discontinuities with objects such as side stones or pavements on the roads. Through the fissure opening or spotwise vents, the liquefied muddy water started to ooze or gush on to the ground surface. In many such sites, oozing and sucking were repeated with a period of about 5 s accompanied by the opening and closing of the cracks. Due to the prolonged shaking, the destruction in this phase continued for about 5–10 min. Settlements of the ground are said to have kept on going 20–30 min afterwards.

4. Effects of the aftershock
   At 15:15 pm, a large aftershock with M = 7.7 hit the area of liquefaction in the Tokyo Bay as well as the downstream region of the Tone River. The peak acceleration was of the order of 150–200 gal. A sequence of events as illustrated above must have occurred similarly in the aftershock. In the area devastated by the main shock, sand boiling occurred again thereby ejecting the sand-laden sand. It appears likely that the sand deposits once liquefied by the first main shock had not fully sedimented and hence still remained in a state of liquefaction leading to the continued sand boiling and spurting. It is obvious that the aftershock has had a great impact on aggravating the damage level further on.
5. Settlements
   In the area devastated by liquefaction, a huge amount of sands was expelled from underground and spread over the surface. In Urayasu city, for example, the sand was collected from all over the affected area covering about 5–6 km^2. The total amount of sands collected was about 75,000 m^3. Then, the average ground settlement due to the loss of sand from underground is estimated to be about 1–2 cm. In the light of the overall average of the ground settlement roughly estimated to be of the order of 5–10 cm, the settlement due to the loss of sand would be about 10 % of the total settlement. Thus, the remaining 90 % of the settlements is considered to have resulted from the volume reduction of sand deposits following liquefaction. Similar situations are deemed to have prevailed in other regions affected by liquefaction.

4.9 Estimate of Ground Settlements

4.9.1 Procedures for Settlements Estimate

The ground settlements following liquefaction may be taken as a measure to evaluate a level of destructiveness of the ground damage. The settlement can be assessed by summing up vertical displacements at each layer of sandy soils which could occur as a result of dissipating excess pore water pressures. The method employed here is the one proposed by Ishihara-Yoshimine (1992) with some additional data compilation modified by Tsukamoto et al. (2004). A series of steps for this aim are summarized as follows.

4.9.1.1 Step 1

First of all, the factor of safety against liquefaction is estimated for each of stratified layers, based on the SPT N-value, using some empirical formulae stipulated in the codes. While there are several codes or charts, these are known to yield more or less the same results. In this paper, the formulae in the Japanese design code for the bridge foundation will be adopted. Since the SPT N-value tends to decrease with

Table 4.1 Mean grain size and fines content stipulated in the Japanese design of bridge foundation

Soil type	Mean grain seize D_{50} (mm)	Fines content F_C (%)
Surface soil	0.02	80
Silt	0.025	75
Sandy silt	0.04	65
Silty fine sand	0.07	50
Very fine sand	0.1	40
Fine sand	0.15	30
Medium sand	0.35	10
Coarse sand	0.6	0
Gravel	2.0	0

increasing fines content while the resistance to liquefaction does not change so significantly, the formulae for estimating equivalent N-value are put forward in the code so that effects of fines can be taken into consideration. In the records of boring logs, soil types in each layer are described by the words such as fine sand, silty sand, sandy silt, etc. It is, thus, necessary to assess the fines contents, based on the descriptions as above. In this paper, the fines contents, Fc, are estimated in accordance with the rules stipulated in the Japanese design of bridge foundations as shown in Table 4.1.

4.9.1.2 Step 2

It is known that the fines contents exert significant influence on the post-liquefaction volume change of sandy soils as well. This aspect was addressed and examined to some extent by Cubrinovski-Ishihara (2002) and Tsukamoto et al. (2004). It was proposed that, instead of using the fines contents, the void ratio range, $e_{\max} - e_{\min}$, be a better parameter that can be recommended for taking into account overall effects of the fines contents. Thus, the chart shown in Fig. 4.42 will be utilized to determine the value of $e_{\max} - e_{\min}$. The volumetric strain following liquefaction is closely related to the possible potential volume change, ε_{vr}, that can be achieved when a sand is compacted from its loosest to densest state. The plots made up by collecting a large number of data in Japan in which ε_{vr} is plotted versus $e_{\max} - e_{\min}$ are shown in Fig. 4.43. However, if the values of $e_{\max}$ and $e_{\min}$ are known, the potential volume change ε_{vr} can be calculated by the formulac shown in Fig. 4.43.

4.9.1.3 Step 3

The post-liquefaction volumetric strains that can occur in the field deposits are largely governed by the looseness or denseness of in-situ sandy soils. It is, therefore, necessary to assess the in-situ relatively density, D_r, based on from the SPT

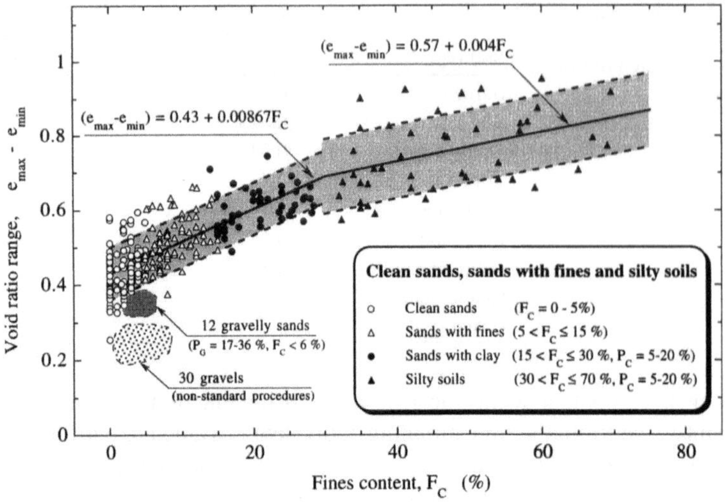

Fig. 4.42 Relationship between void ration range and fines content of sandy soils (Cubrinovski-Ishihara 2002)

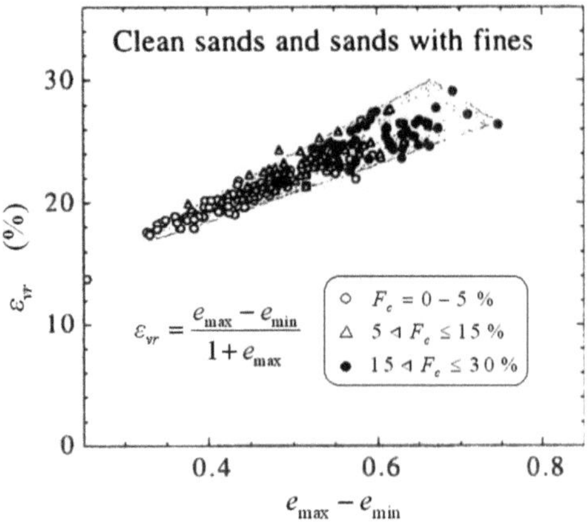

Fig. 4.43 Volumetric strain potential as a function of ($e_{\max} - e_{\min}$) of sands (Chubrinovski-Ishihara 2002)

N-values. In view of the fact that the N-value depends on the fines contents in sandy soils, the following empirical formulae were proposed by Cubrinovski and Ishihara (2002).

$$D_r = \left[\frac{\left(e_{\max} - e_{\min}\right)^{1.7}}{9} N_1\right]^{1/2} \tag{4.3}$$

Where N_1 denotes the SPT N-value normalized to that mobilized at a overburden pressure of 1.0 kg/cm^2. In this paper, the relative density will be estimated using (4.3).

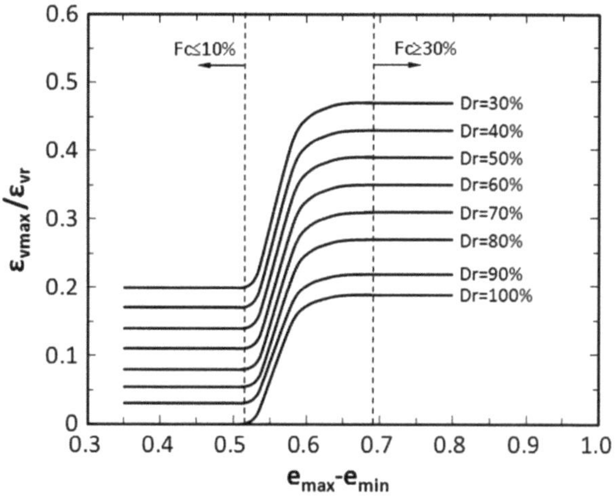

Fig. 4.44 Chart for determining the potential in-situ value of volumetric charge, $\varepsilon_{v\max}$

Note that the relative density is an important factor as well for determining the cyclic resistance of in-situ sandy soils, and therefore, to evaluate the factor of safety, F_{ℓ}. The effect in this respect is incorporated separately in a different manner in the formulae in the code.

4.9.1.4 Step 4

It is to be remembered that the dissipation of excess pore water pressure following liquefaction takes place under the gravitational environment. Therefore, in deposits in the field, in-situ void ratio, e, will never reach its minimum value, $e_{\min}$. It is thus necessary to know how much volume change or volumetric strain, $\varepsilon_{v\max}$, can indeed take place potentially in the gravitational environment in the field, after the sandy soils have been disturbed vastly due to liquefaction with development of double-amplitude strains in excess of 5–10 %. The only way to explore this kind of information would be to have recourse to the data from laboratory tests on undisturbed samples where water is drained out of the samples after they develop liquefaction. The data in this context from the laboratory triaxial cyclic loading tests on undisturbed samples were collected and summed up in a form of a diagram shown in Fig. 4.44 where the laboratory-measured maximum values, $\varepsilon_{v\max}$, normalized to ε_{vr} is indicated as a function of $e_{\max} - e_{\min}$ for various relative densities. Note again that ε_{vr} implies the volumetric strain by the formulae in Fig. 4.43, and $\varepsilon_{v\max}$ is the maximum possible volumetric strains obtained on undisturbed samples in the laboratory-tests that are envisioned to occur in the in-situ gravitational field. Not that the chart in Fig. 4.44 is based on a limited number of laboratory tests now available and thus far from complete. This chart will be used, however for the settlement calculation presented hereafter.

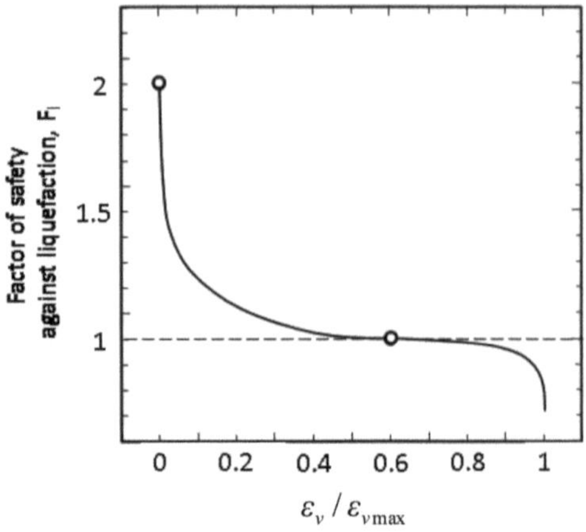

Fig. 4.45 Chart correlating factor of safety F_ℓ and $\varepsilon_v/\varepsilon_{v\ max}$

4.9.1.5 Step 5

The determination of the post-liquefaction in-situ volumetric strain, ε_v, will be made using the chart shown in Fig. 4.45, where the factor of safety, F_ℓ, is plotted versus the value of $\varepsilon_v/\varepsilon_{vmax}$. The curve in the figure is a summary of similar plots from various sands containing some percentage of fines. Actual procedures to determine, ε_v, is first to calculate the factor of safety, F_ℓ, based on the method as described in Step 1. Then, with reference to the chart in Fig. 4.45, the ratio, $\varepsilon_v/\varepsilon_{vmax}$, is determined. On the other hands, the value of $e_{max} - e_{min}$ has been estimated using the relation shown in Fig. 4.42. With the known value of $e_{max} - e_{min}$, the potentially feasible value of the volumetric strain, ε_{vr}, is evaluated through the relation shown in Fig. 4.43. Then, entering into the chart in Fig. 4.44, with the known values of $e_{max} - e_{min}$ and relative density, D_r, the ratio, $\varepsilon_{vmax}/\varepsilon_{vr}$ is determined.

4.9.1.6 Step 6

With the known values of $\varepsilon_v/\varepsilon_{vmax}$, ε_{vr} and $\varepsilon_{v\ max}/\varepsilon_{vr}$, the volumetric strain, ε_v, at each depth of a deposit due to dissipation of pore water pressure is finally determined.

4.9.2 Settlements at Sites of Liquefaction in Urayasu and Isobe

Based on the procedures described above, the settlements were calculated for several sites in Urayasu and other districts in Chiba. Examples of such calculation will be given below. In this analysis scheme, one of the important assumptions that

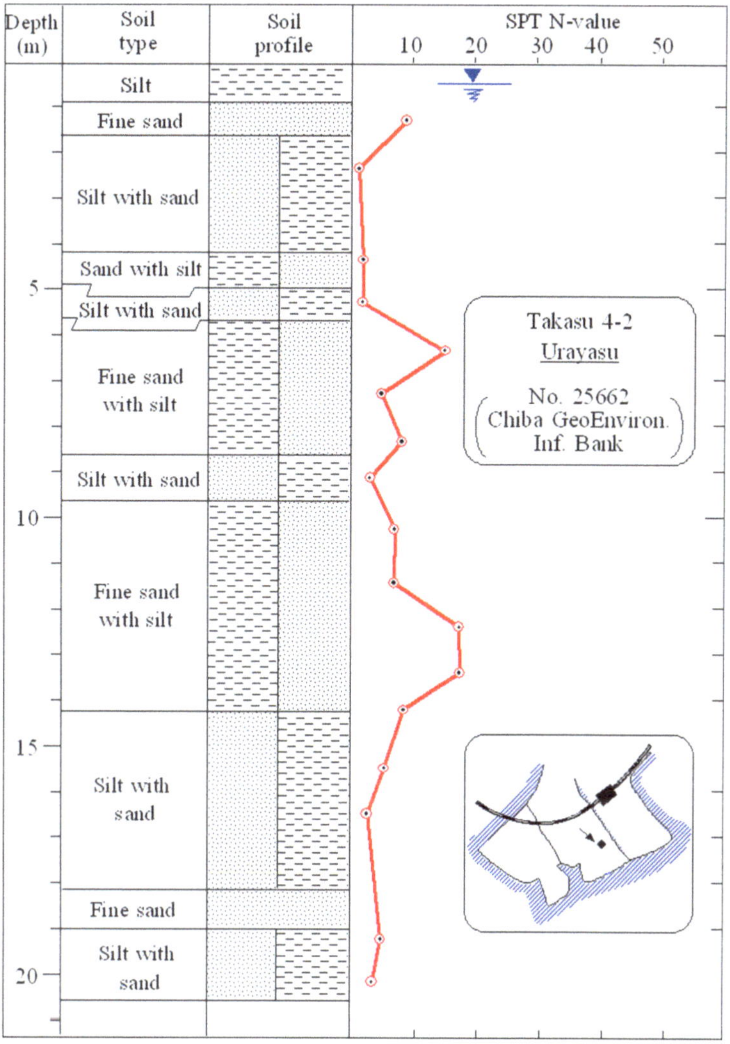

Fig. 4.46 Soil profile at Takasu 4-2, Urayasu

is to be made is to specify the depth of the deepest key layer below which the volumetric strain could be practically neglected. In the present paper, this key layer was set as an unliquefiable layer with a factor of safety greater than $F_{\ell} = 1.2$ which is encountered first depthwise in the alluvial layer. The depth from which the volumetric strains were integrated upwards to the surface is indicated in each of the figures that follow.

A site at Takasu 4-chome was chosen as a place for liquefaction analysis. The soil profile and its location are shown in Fig. 4.46. The results of the liquefaction analysis are shown in Fig. 4.47. This is the site where signs of liquefaction such as

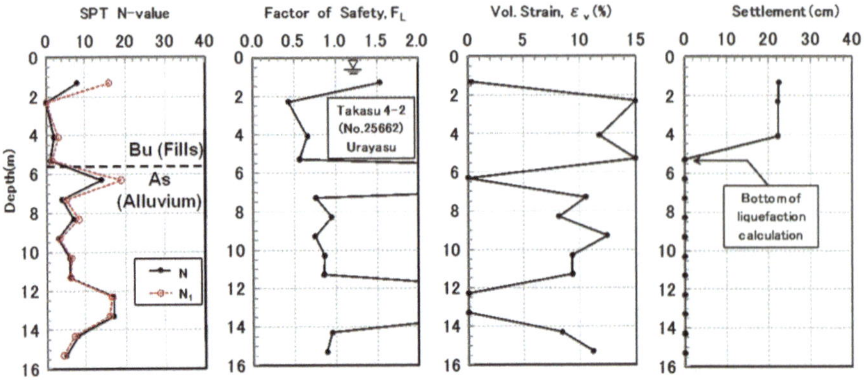

Fig. 4.47 Analysis of liquefaction for a site at Takasu 4-2, Urayasu (No. 25662)

sand boiling and settlements were observed at the time of the earthquake. It is to be noticed that the deposits below the depth of 5.8 m are identified as the natural alluvium. The computed factor of safety with the 0.16 g acceleration becomes greater than 1.2 at this depth as shown in Fig. 4.47. Then, the volumetric strains below 5.8 m were not counted. The settlement of the ground surface was obtained by integrating the volumetric strain at each depth, coming up with a value of 23 cm. The actual settlement exactly at this place was estimated to be 20–50 cm judging from the post-quake ground elevation relative to nearby buildings supposedly supported by piles. The value of the estimated settlements is displayed in Fig. 4.56 versus the range of the settlement judged from observation after the quake.

Another comparison between the estimated and observed values was made for the liquefied site at Mihama, Chiba of which the soil profile is shown in Fig. 4.48 together with its location indicated in the inset. Note that this site suffered the ground settlement as obviously seen by the photos in Fig. 4.34. Results of analysis are shown in Fig. 4.49 where estimated settlement is 61 cm. The actual settlement was inferred to be 40–60 cm by comparing the post-earthquake ground level to the trace of the pre-earthquake ground surface marked on a pile-supported column of a nearby overpass walkway bridge. Thus, the estimated statement indicates good coincidence with the observed value, as indicated in Fig. 4.56.

Sill other example of liquefaction analysis was made for the site at Isobe 8-chome, Mihama-ku, Chiba, where serious damage due to liquefaction occurred to private houses. As seen in the soil profile in Fig. 4.50, silty sand or sandy silt deposits extend down to the depth of 20 m, apparently indicating a high possibility of liquefaction. The results of the analysis is demonstrated in Fig. 4.51, where it is noted that the settlements at each depth tend to increase almost in proportion to increasing elevation of the deposits, ending up with a value of 81 cm at the ground water level. At this location, observed settlement was as large as 60–80 cm. The values of settlement estimated and observed are shown plotted in Fig. 4.56.

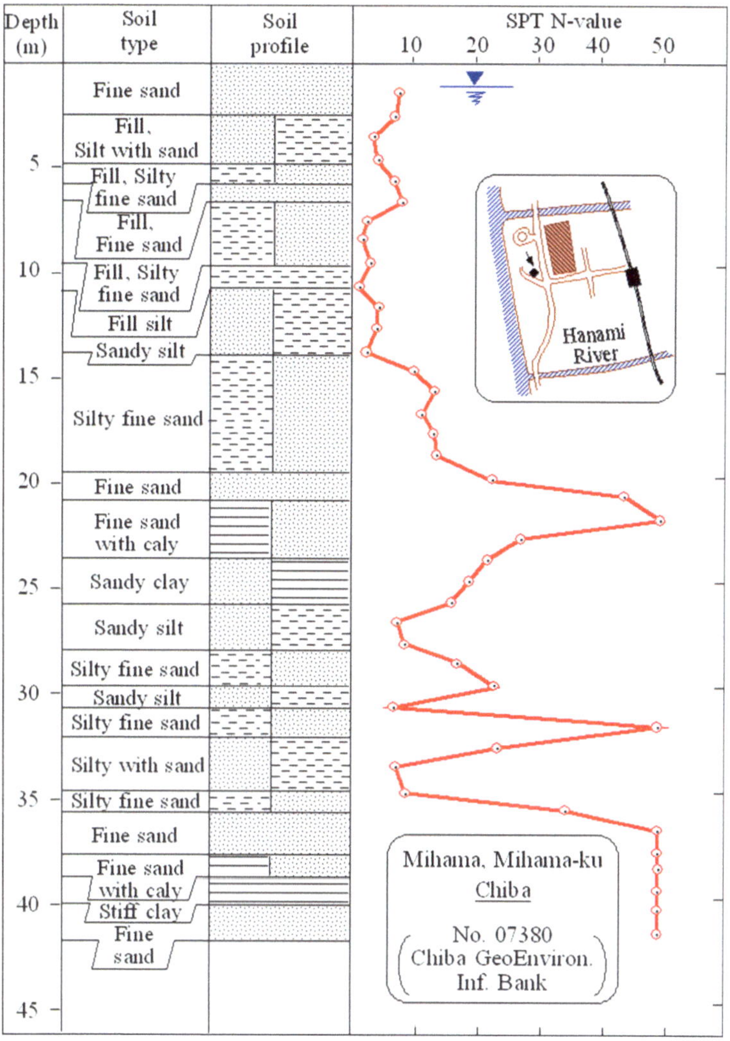

Fig. 4.48 Soil profile at Isobe-8, Mihama-ku, Chiba

4.9.3 Settlements at Sites of No-Liquefaction in Urayasu

In the old city section of north Urayasu, no visible sign of liquefaction was observed at the time of the 2011 earthquake. The settlement analyses were performed for two sites in this area with the ground acceleration of 0.16 g. One of the sites in Fujimi 3-chome has a soil profile as shown in Fig. 4.52. The result of liquefaction analyses yielded no settlement which is plotted in Fig. 4.53. Similar analysis was performed for another site of no-liquefaction, viz., at Kairaku 2-chome, where the soil profile

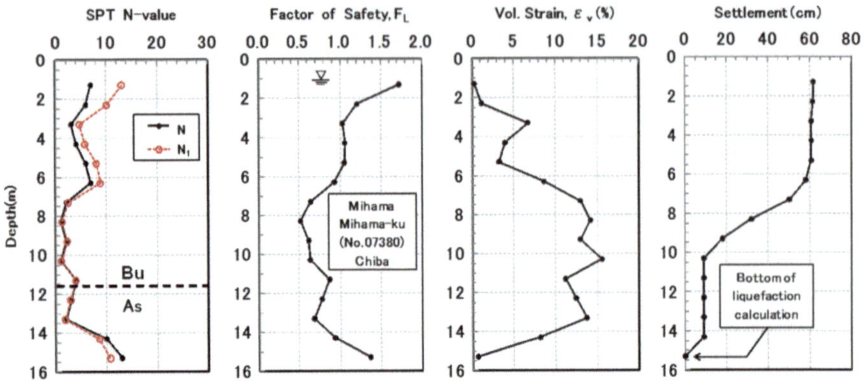

Fig. 4.49 Analysis of liquefaction for a site at Mihama, Mihama-ku, Chiba (No. 07380)

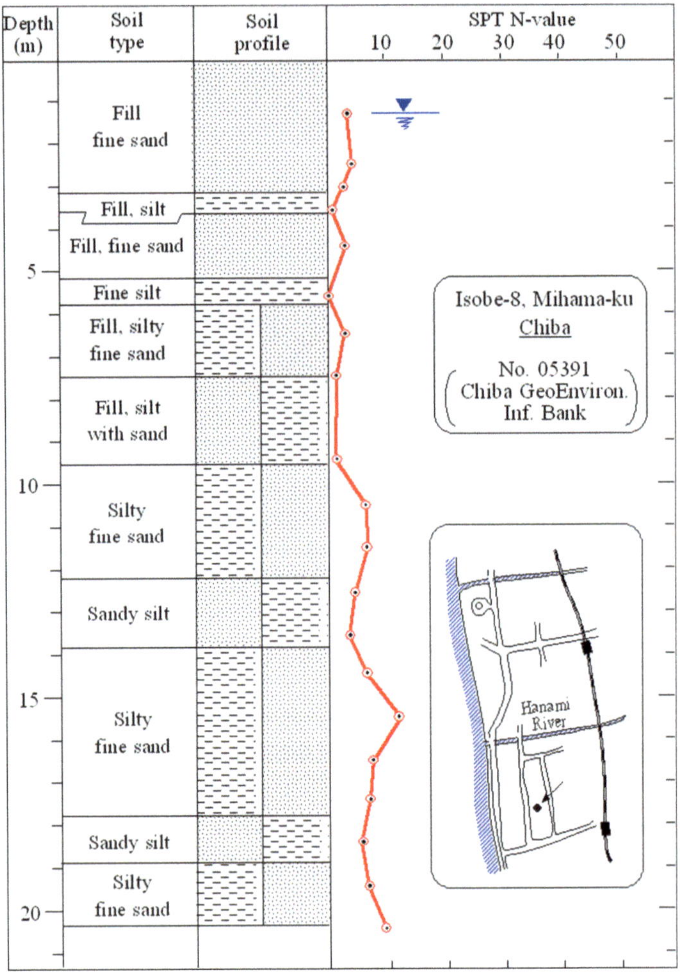

Fig. 4.50 Soil profile at Isobe-8, Mihama-ku, Chiba

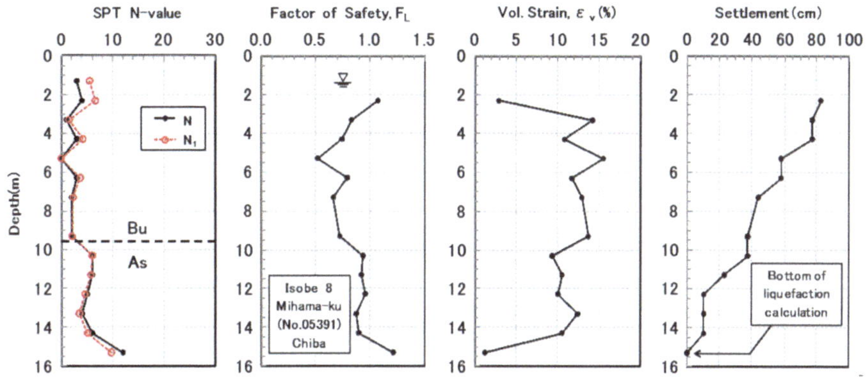

Fig. 4.51 Analysis of liquefaction for a site at Isobe-8, Mihama-ku, Chiba (No. 05391)

consists of about 1.2 m thick filled materials as shown in Fig. 4.54. The ground water table was 2.4 m. The calculated settlement is shown in Fig. 4.55 being of the order of 30 cm. This site is located in the area of no liquefaction in Urayasu, as accordingly indicated in the inset of Fig. 4.54. The analysis results at this site does not coincide with the observed ground behaviour after the quake (Fig. 4.56).

4.9.4 Comparison of Settlements Between the Calculated and Observed Values

There are several other sites in the city of Urayasu for which estimate was made by using the simple method as described above. All the data obtained by calculation are plotted together in Fig. 4.55 versus actually observed values. As there were several sites where actual values of settlements are difficult to determine uniquely, they are demonstrated in the form of ranges of variation in Fig. 4.56. Looking over the plots in Fig. 4.56, one can remark main points as follows.

1. Estimated values of settlements do coincide, by and large, with the settlements investigated a few months after the earthquake. However, there are some scatters in the comparison. This seems to emerge, for one thing, from the complexity of man-made or man-reformed nature of soil deposits prevailing over the areas affected by the liquefaction during this time earthquake.
2. For some of the sites with no evident sign of liquefaction, the computed values of settlement are larger than those observed actually after the earthquake. One of the probable reasons for such disagreement would be existence of surface mantle of soil deposits that are immune to liquefaction. For instance, when the ground water table is sufficiently deep, deleterious effects of liquefaction in underlying layer would not be manifest on the ground surface. This aspect was addressed by Ishihara (1985) in terms of the thickness, H_1, of un-liquefiable near-surface soils deposits which was plotted against the thickness, H_2, of underlying liquefiable soils.

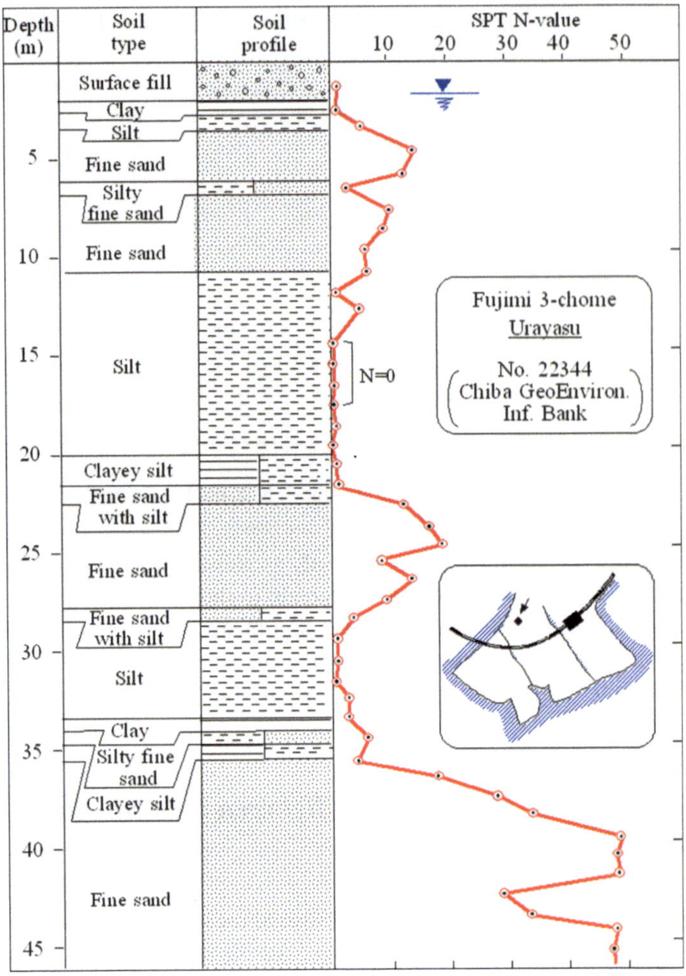

Fig. 4.52 Soil profile at Fujimi 3-chome, Urayasu

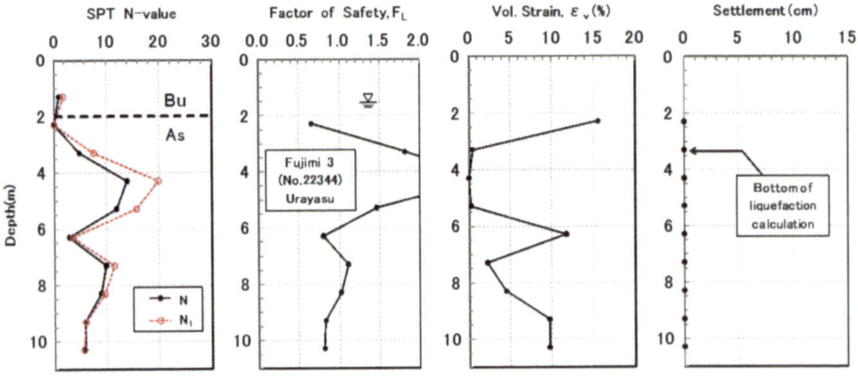

Fig. 4.53 Analysis of liquefaction for a site at Fujimi-3 Urayasu, Chiba (No. 22344)

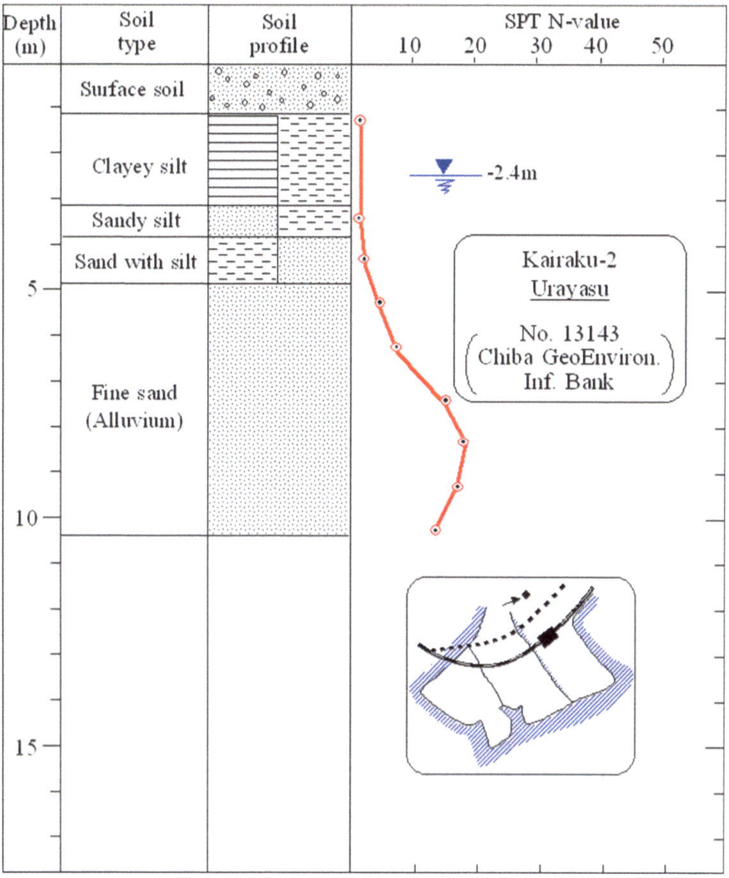

Fig. 4.54 Soil profile at Kairaku-2 chome, Urayasu

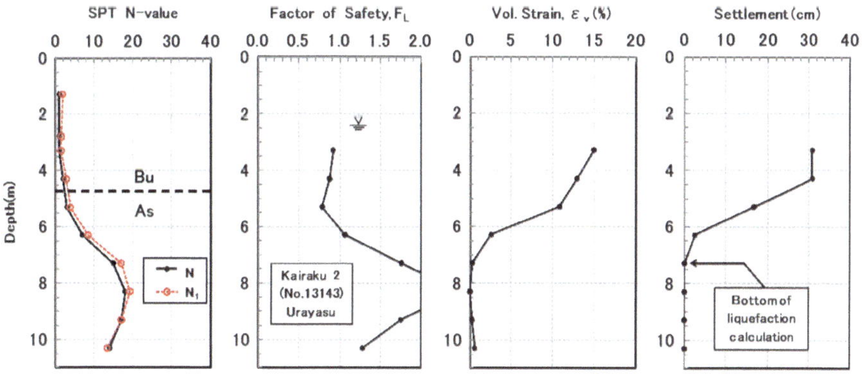

Fig. 4.55 Analysis of liquefaction for a site at Kairaku-2, Urayasu (No. 13143)

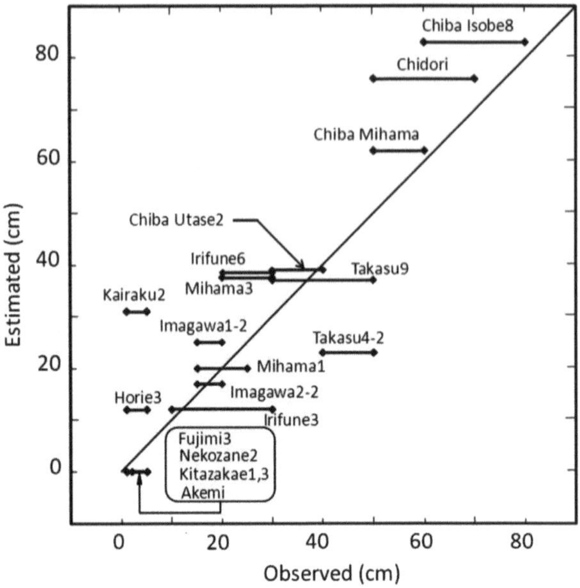

Fig. 4.56 Comparison of estimated and observed settlements in Urayasu and Mihama

4.10 Concluding Remarks

There were several characteristics in the occurrence of liquefaction and consequent damage in the 2011 event which are different from a number of cases experienced in the past earthquakes. These are summed up as follows.

1. It was probably the first case to the knowledge of today's geotechnical engineers to observe such a wide-spread occurrence of liquefaction in the area more than 350 km afar from the epicentre. This may be attributed to the rarely encountered huge magnitude-scale earthquake as great as M = 9.0. Corresponding to the large-scale event, the duration of shaking was as long as 2–3min. In the light of the 10–20 s duration hitherto encountered in the magnitude 7–8 events, the shaking duration in the East-Japan Great Earthquake can be cited as extraordinarily longlasting. This fact appears to have made the damage level worst, as compared to any other cases of liquefaction ever encountered.
2. In Urayasu city there was a section where practically no sign of liquefaction was observed. This was the old city part, while there was severe damage in the newly reclaimed area. In old deposits comprised of the alluvium, it is conceivable that effects of aging tended to act towards solidifying deposits more pronouncedly particularly for the fines containing sands. This could perhaps be a reason why there was a sharp contrast regarding the liquefaction occurrence between areas of alluvial deposits and newly reclaimed sections in Urayasu city.
3. Many observations by witness and movie pictures showed advents of wavy motions with a period of about 5 s and a wave length of 50–100 m. These

observations appear to show appearance of secondary surface waves such as sloshing or Rayleigh wave which were mobilized over what might be called underground basin created by liquefied soils.

4. Post liquefaction settlements nay be considered as a gage to evaluate the level of destructiveness of the ground due to liquefaction. In this context, estimate was made for the settlements based on the procedures proposed previously. The outcome of the estimated settlements was compared with those observed in-situ following the earthquake with the results that coincidence is relatively good except for the area where no surface of liquefaction was observed.

Acknowledgements In preparing this paper, the information on the progress of reclamation in Urayasu city and general features of the damage was given by the courtesy of Mr. Ichiro Ishii of the Urayasu city office. The settlement analysis was made by Mr. Keita Okamura of Chemical Grout Co. Ltd. The authors wish to express his deep thanks to the office and persons as mentioned above.

References

Chiba Geo-Environmental Data Bank. http://wwwp.pref.chiba.lg.jp/pbgeogis/servlet/infobank.index

Chubrinovski M, Ishihara K (2002) Maximum and minimum void ratio characteristics of sands. Soils Found Jpn Geotech Soc 42(6):65–68

Hayes GP, Earle PS, Benz HM, Wald DJ, Briggs RW, The USGS/NEIC Earthquake Response Team (2011) 88 hours: The U.S. Geological Survey National Earthquake Information Center response to the 11 March 2011 Mw 9.0 Tohoku earthquake. Seismol Res Lett 82:481–493

Ishihara K (1985) Stability of natural deposits during earthquakes. In: Proceedings of the 11th international conference on soil mechanics and foundation engineering, San Francisco, vol 2, pp 32576

Ishihara K, Yoshimine M (1992) Evaluation of settlements in sand deposits following liquefaction during earthquakes. J Jpn Geotech Soc Soil Found 32:173–188

Kawabe S, Tsukamoto T, Takahashi R (2012) Soil liquefaction observed at Katori city located along the lower stream of Tonegawa River during 2011 Great East Japan earthquake (Submitted to the present conference)

Kazaoka O (2011) Basic story on liquefaction, flow and wave phenomenon. In: Symposium of socio-geology society on liquefaction, flow and wave phenomena in the middle reaches of the Tone River, pp 1–18

Kazaoka O (2003) Major earthquakes in Japan inducing damage by liquefaction and flow (1964–2001). Urban Kubota (In Japanese)

Nirei H (2011) Composite ground hazard by liquefaction, flow, ground wave and tsunami in the lower reaches of the Tone River. In: Symposium of Socio-Geology Society on wave phenomena in the middle reaches of the Tone River, pp 71–78

Towhata I, Goto H, Kazama M, Kiyota T, Wakamatsu K, Wakai A, Yasuda S, Yoshida N (2011a) On gigantic Tohoku Pacific earthquake in Japan. Earthq News ISSMGE Bull 5(2):46–66

Towhata I, Goto H, Kazama M, Kiyota T, Nakamura S, Wakamatsu K, Wakai A, Yasuda S, Yoshida N (2011b) On gigantic Tohoku Pacific earthquake in Japan. Earthq News ISSMGE Bull 5(2):46–66

Tsukamoto Y, Ishihara K, Sawada S (2004) Correction between penetration resistance of Swedish weight sounding tests and SPT blow counts in sandy soils. Soils Found Jpn Geotech Soc 44(3):13–24

Tsukamoto Y, Kawabe S, Kokusho T, Araki K (2012a) Observed performance and estimated soil profiles of reclaimed and natural deposits at Kamisu city during 2011 Great East Japan Earthquake. In: Proceedings of the 9th international conference on urban earthquake engineering, Tokyo Institute of Technology

Tsukamoto Y, Kawabe S, Kokusho T, Araki K (2012b) Soil liquefaction observed at areas located along the lower stream of Tonegawa River during 2011 Great East Japan Earthquake. In: Proceedings of the present conference

Urabe A (2011) Liquefaction damage at the residential land in the inland area of Kanto district-Itako. Hinode and Kuki. Minami-Kurihashi. In: Proceedings of the symposium on liquefaction in the plain of Tone River, Socio-Geotechnical Society, pp 22–25 (in Japanese)

Urayasu City Official Site. http://www.city.urayasu.chiba.jp

Yasuda S (2011) Liquefaction-induced damage in Kanto Region. Report of investigation for the Tohoku Off-Ocean Earthquake, Japanese Geotechnical Society (in Japanese)

Chapter 5
Allowable Settlement and Inclination of Houses Defined After the 2011 Tohoku: Pacific Ocean Earthquake in Japan

Susumu Yasuda

Abstract The 2011 Tohoku-Pacific Ocean earthquake (Great East Japan earthquake) caused severe liquefaction not only in the Tohoku district of northeastern Japan, but also in the Kanto district. About 27,000 timber houses, a lot of buried sewage pipes and roads were damaged due to liquefaction. In Tokyo Bay area, the very long duration of the main shock and an aftershock 29 min later probably caused serious settlement and inclination of houses. The maximum inclination in Urayasu City was about 60/1,000. In the heavily tilted houses, inhabitants feel giddy and nausea and difficult to live in their houses after the earthquake though no damage to walls and windows were observed. Then, on May 2011, Japanese Cabinet announced new evaluation standard for the damage of houses by the two factors, settlement and inclination. Tilted houses more than 1/20, 1/20 to 1/60, 1/60 to 1/100 are judged as completely destroyed, large-scale partially destroyed, and partially destroyed houses, respectively in the standard. Tilting of houses is derived from non-uniform settlement. Several factors affect to the non-uniform settlement. Among them effect of adjacent houses was dominant. If two houses are closely constructed these houses tilt inward, and if four houses are close these houses tilt toward the center. Many inhabitants along Tokyo Bay are facing to the serious problem how to restore the damaged houses. Complicated problem is the re-liquefaction during aftershocks or future earthquakes.

5.1 Introduction

Many timber houses have settled due to liquefaction during past earthquakes in Japan. However detailed measurements of settlement have not been carried out up to the 2000 Tottoriken-seibu earthquake which caused the settlement of numerous timber houses. Of them, many houses settled and tilted at Abehikona housing lot in

S. Yasuda (✉)
Department of Civil and Environmental Engineering,
Tokyo Denki University, Saitama, Japan
e-mail: yasuda@g.dendai.ac.jp

M. Maugeri and C. Soccodato (eds.), *Earthquake Geotechnical Engineering Design*, Geotechnical, Geological and Earthquake Engineering 28,
DOI 10.1007/978-3-319-03182-8_5, © Springer International Publishing Switzerland 2014

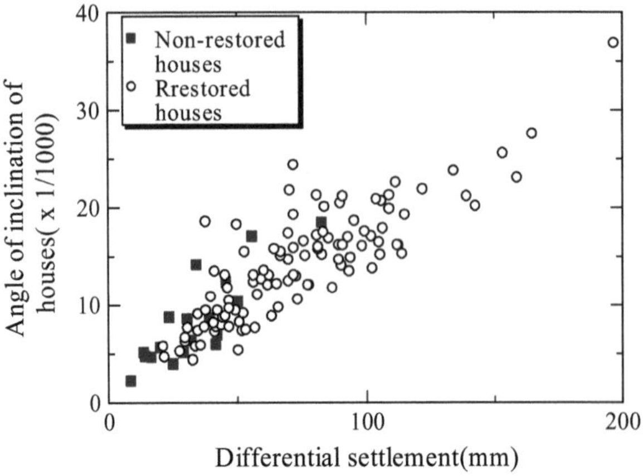

Fig. 5.1 Angle of inclination of restored and non-restored houses at Abehikona housing lot during the 2000 Tottoriken-seibu earthquake (Yasuda et al. 2004)

Yonago City (Yasuda et al. 2004; Yasuda and Ariyama 2008). The area of the housing development is almost a square of 360 m × 360 m. 169 timber houses with mainly two stories had been constructed before the 2000 Tottoriken-seibu earthquake. Of them 116 houses settled and tilted though no severe cracks were observed on the walls of the settled houses. As more than 100 houses settled and tilted due to liquefaction, angles of inclination of the settled houses were measured after the earthquake. The maximum angle of inclination and the maximum differential settlements in Abehikona housing development were 37.5/1,000 and 33 cm, respectively. Angle of inclination was classified into four rankings, (a) more than 15/1,000, (b) 10/1,000 to 15/1,000, (c) 5/1,000 to 10/1,000 and (d) less than 5/1,000. The numbers of houses within each of the four ranks were 47, 30, 39 and 53. In the heavily tilted houses, inhabitants felt giddy and nausea, and could not live in their houses after the earthquake, though walls, pillars and windows of the houses had no damage. Heavily tilted houses had to be restored to horizontal. In the restoration work, super-structures were uplifted by jacks, footings were repaired or reconstructed to become horizontal, and then the super-structures were replaced on the footings. The cost of the restoration work for one house was about 3–4million Yen (about US$25,000–35,000). By contrast, slightly tilted houses did not need to be restored. The author and his colleague compared critical angles of inclination whether the restoration work was necessary or not. Figure 5.1 shows the relationship between differential settlement and angle of inclination of houses. In this figure, non-restored houses and restored houses are indicated by different symbols, closed square and open circle. It can be seen that the critical angle of inclination requiring restoration was about 5/1000 to 15/1000.

This experience from the 2000 Tottoriken-seibu earthquake revealed that inclination of houses is more severe than average settlement of houses for inhabitants (Yasuda 2009). However, no official definition of the damage of houses based on the inclination has been proposed up to the 2011 Tohoku-Pacific Ocean earthquake. Liquefaction occurred in very wide area and about 27,000 houses were seriously settled tilted due to the liquefaction. Then, in Japan, governments, engineers and inhabitants have changed their minds to pay attention to the inclination.

5.2 Numbers of Damaged Houses Due to Liquefaction During the 2011 Tohoku-Pacific Ocean Earthquake

The 2011 Tohoku-Pacific Ocean earthquake (Great East Japan Earthquake), with a magnitude of Mw = 9.0 occurred in the Pacific Ocean about 130 km off the northeast coast of Japan's main island on March 11, 2011. The hypocentral region of this quake was about 500 km in length and 200 km in width. The quake was followed by a huge tsunami that destroyed many cities and killed about 20,000 people along the Pacific coast. In the geotechnical field, many houses and lifelines were damaged by soil liquefaction, landslides occurred, dams failed, and river dikes settled not only in the Tohoku district of northeastern Japan, but also in the Kanto district, surrounding Tokyo. According to the result of totaling by the Ministry of Land, Infrastructure, Transport and Tourism, about 27,000 houses were damaged due to liquefaction. Figure 5.2 shows numbers of the damaged houses in each prefecture. About a half of the damaged houses are located in Tokyo Bay area.

Total area of liquefied zones in the Tokyo Bay area, from Shinkiba in Tokyo through Urayasu, Ichikawa and Narashino cities to Chiba City was about 41 km^2. The reclaimed lands that liquefied were constructed after around 1966 from soils dredged from the bottom of the bay. The dredged and filled soils must have been liquefied by the earthquake (Yasuda and Harada 2011). Seismic intensities in the liquefied zones were not high as about 160–230 cm/s^2 by K-NET (NIED 2011), though the liquefied ground was covered by boiled sands. The very long duration of the main shock and an aftershock 29 min later probably induced the severe liquefaction and associated damage to houses. Sidewalks and alleys buckled at several sites, probably due to a kind of sloshing of liquefied ground. Moreover, much sand boiled from the ground

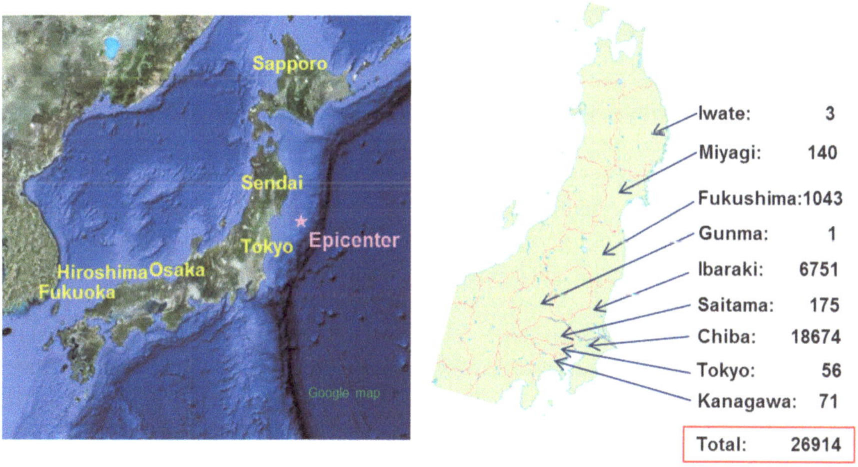

Fig. 5.2 Numbers of the damaged houses due to liquefaction in each prefecture during the 2011 Tohoku-Pacific Ocean earthquake

and the ground subsided significantly because the liquefied soils were very fine. Sewage pipes meandered or were broken, their joints were extruded from the ground, and many manholes were sheared horizontally. This remarkable damage might have occurred due to a kind of sloshing of liquefied ground (Yasuda and Ishikawa 2012).

5.3 Process on the Occurrence of Liquefaction in Urayasu City

As the earthquake occurred at 14:46 in the afternoon on Friday, many important photos and movies were taken at many sites along Tokyo Bay to study the process and mechanism of liquefaction. Among them a series of photos taken by Mr. Katsunori Ogawa just after the earthquake at Maihama in Urayasu City is introduced in Fig. 5.3. The followings are his comments on the photos:

<<<Shake due to main shock started at 14:47 in Urayasu>>>

<14:56>: Spew out of muddy water started in northeast direction. It took several minutes to start the boiling after the settle of the main shock.

<15:00>: Same direction as 14:56

<15:03>: Much muddy water was spouting from side ditches and covered a road. Same direction as 14:56

<15:10>: Muddy water was spouting with a height of about 20 cm from the holes of a lid of a sewage manhole.

<15:16>: Much muddy water spewed out before the strong aftershock at along Miake River.

<<<A strong aftershock hit Urayasu at 15:16>>>

<15:19>: A footway heaved and house settled along Miake River.

<15:21>: The southwest road covered completely with muddy water. Water pipes look like to break.

<15:22>: Many houses settled and tilted and cars submerged in the boiled muddy water.

The author sent out questionnaires to about 30 inhabitants in Irifune district which is also located in Urayasu City but about 3 km northeast from Maihama to ask the timing of boiling and height of boiled muddy water (Yasuda and Ishikawa 2012). Answers are summarized in Fig. 5.4. About 1/3 persons observed the boiling of muddy water immediately after the main shock, however another 1/3 persons recognized the spout of muddy water 5–9 min after the main shock. Other persons found the muddy water at different timing. Height of the muddy water was not high as mainly less than 9 cm after the main shock. About 2/3 persons mentioned that the boiling of muddy water continued up to the aftershock, and about 3/4 persons watched covered water until aftershock. On the contrary, about 3/4 persons observed spew out of muddy water just after the aftershock and the height of the water was apparently greater than the height after the main shock. This means the boiling accelerated due to the aftershock and more severe liquefaction occurred during aftershock at some sites. Question on the timing of the settlement of houses must be difficult to answer for inhabitants, however, a 1/3 persons and another 1/3 persons answered that the settlement of their houses was

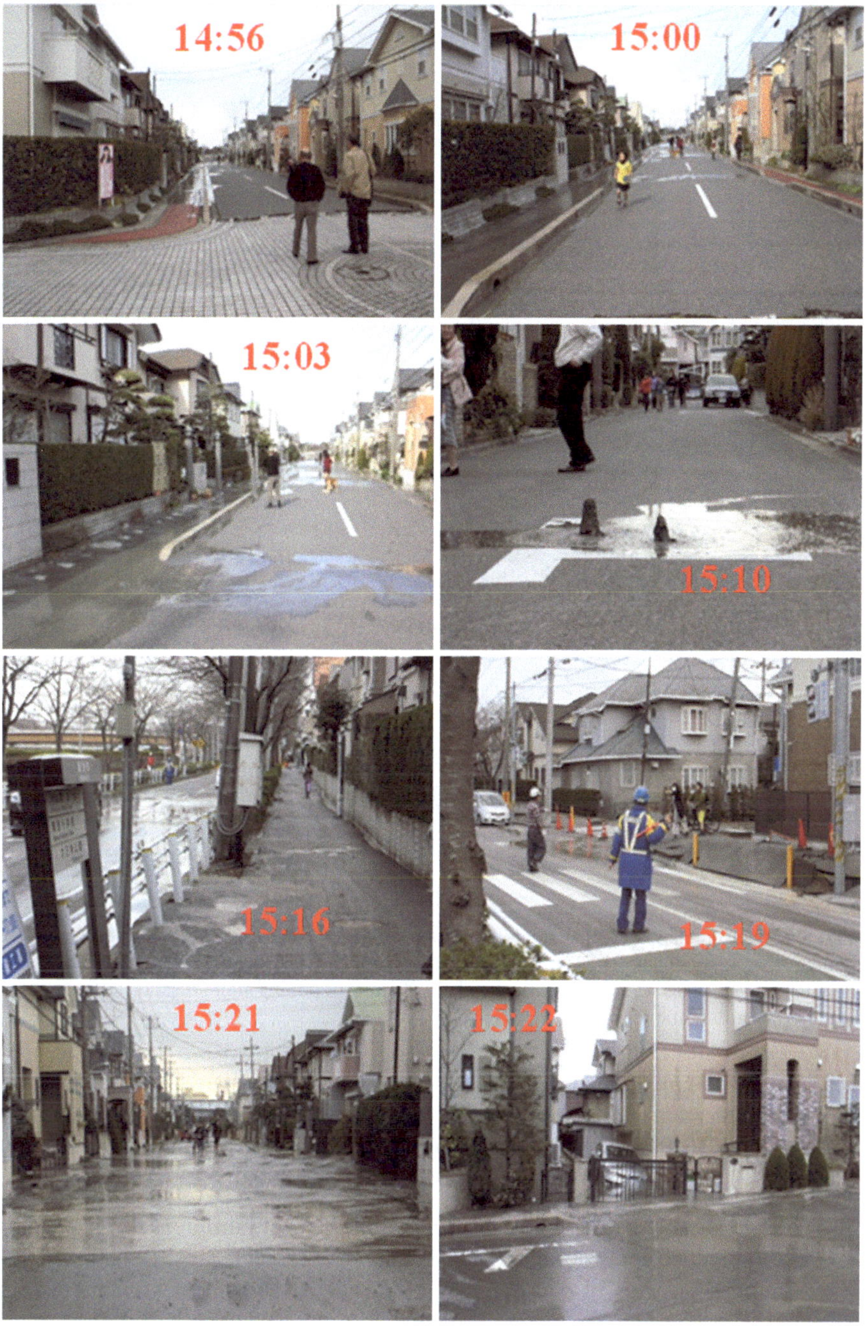

Fig. 5.3 Sequential photos taken by Mr. K. Ogawa after main shock in Urayasu

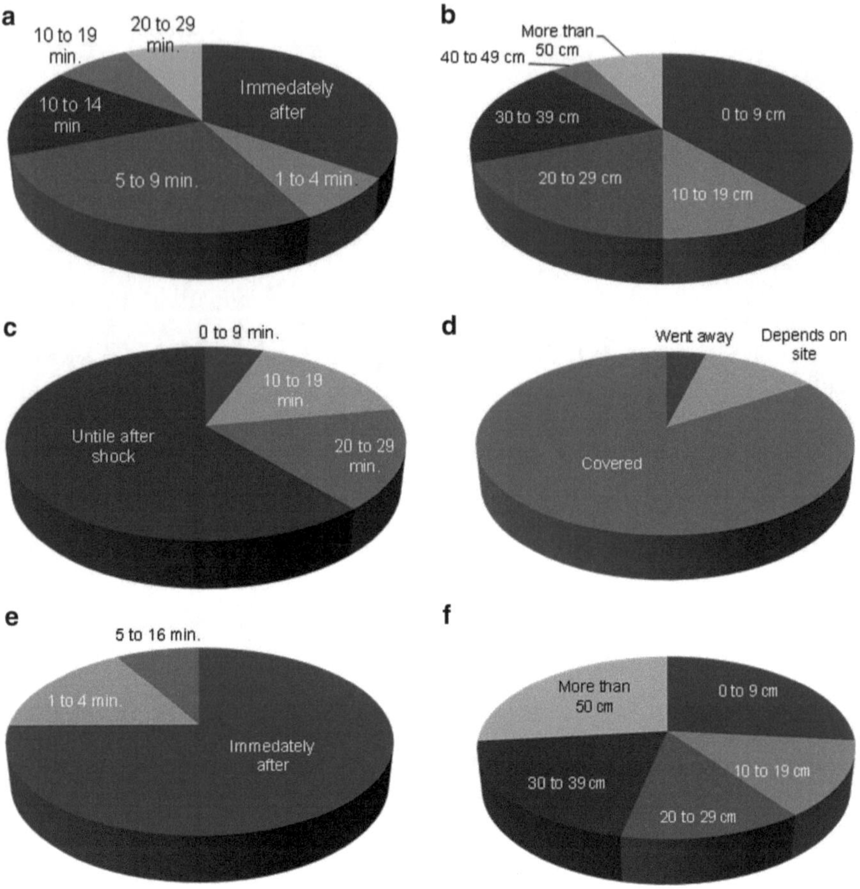

Fig. 5.4 Questionnaires to inhabitants in Irifune (Yasuda and Ishikawa 2012). (**a**) Timing of boiling after main shock. (**b**) Height of muddy water after main shock. (**c**) Time of the stop of boiling. (**d**) Surface water before aftershock. (**e**) Timing of boiling after aftershock. (**f**) Height of muddy water after aftershock

zero and 10–19 cm, respectively after the main shock. And, many inhabitants recognized the settlement on the next day.

According to the series of photos and questionnaires to inhabitants, several interesting process of liquefaction and boiling can be found:

1. Starting time of the boiling of muddy water are quite different at place. This must imply the depths of liquefied layer and/or water table are different at place. Though the answers of questionnaires from the inhabitants at Mihama district are not introduced in this paper, some inhabitants testified boiling did not occur during main shock but occurred during aftershock. Some excess pore water might increase but did not liquefy during main shock at this site.
2. Boiling of muddy water was not intense after the main shock like an ooze out, however boiling was accelerated during the aftershock. Increased water table

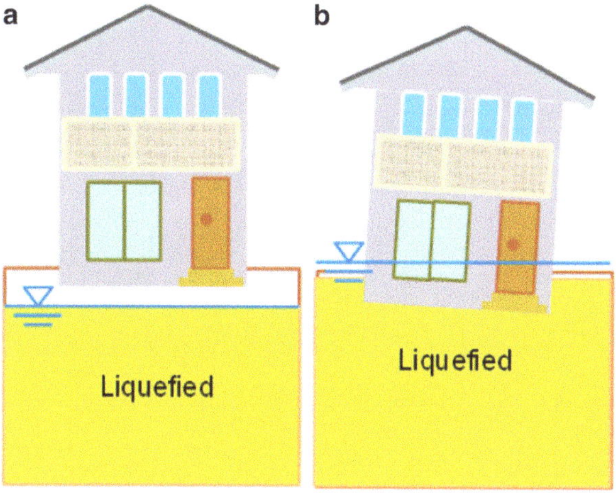

Fig. 5.5 Possible effect of aftershock on the settlement of house. (**a**) During main shock. (**b**) During aftershock

after the main shock might accelerate the boiling during the aftershock. Or, some fissure induced in the ground during main shock might accelerate the boiling during aftershock.

3. A part of houses settled during or immediately after the main shock though other houses settled after the aftershock. The houses settled during main shock might settle more during after shock. As mentioned above many persons watched covered water until aftershock. Then, settlement of houses is excitable during the aftershock though shaking amplitude was less than that during the main shock as schematically shown in Fig. 5.5.

5.4 New Evaluation Standard for the Damage of Houses Introduced After the Earthquake

Figure 5.6a, b show a damaged house in Urayasu taken from outside and inside, respectively. Though no damage to walls and windows were observed, the house settled and tilted about 40/1,000. Large settlement which means the penetration of houses to the ground, such as 30 cm caused damage to water, sewage water and gas pipe by tearing or bending. Moreover, in the heavily tilted houses, inhabitants feel giddy and nausea, and difficult to live in their houses after the earthquake as recognized at Abehikona housing lot during the 2000 Tottoriken-seibu earthquake mentioned above. Then, on May, Japanese Cabinet announced new evaluation standard for the damage of houses by the two factors, settlement and inclination, as shown in Table 5.1. New class "large-scale partially destroyed house" was also introduced, and tilted houses more than 1/20, 1/20 to 1/60, 1/60 to 1/100 are judged as completely destroyed, large-scale partially destroyed, and partially destroyed houses, respectively in the standard.

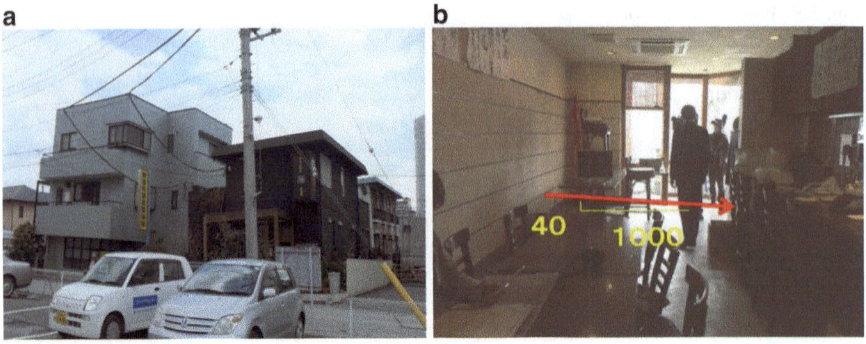

Fig. 5.6 Severely settled and tilted house in Urayasu City. (**a**) Outside of the tilted house. (**b**) Inside of the tilted house

Table 5.1 New evaluation standard for the damage of houses

	Evaluation method	
Grade of damage	Inclination	Settlement
Completely destroyed	More than 1/20	Floor to 1 m upper than floor
Large-scale partially destroyed	1/60 to 1/20	25 cm upper than footing to floor
Partially destroyed	1/100 to 1/60	Up to 25 cm upper than footing

Numbers of damaged houses in Urayasu City by the new standard are listed in Table 5.2 together with the numbers counted by old judging method (Urayasu City 2011). Numbers of completely and partially destroyed houses increased drastically, and the numbers of damaged houses more severe than partially destroyed enlarged to 3,680. Frequency distribution of the numbers of the damaged houses are counted from Table 5.1 and plotted on Fig. 5.7 together with the damage in Abehikona housing lot during the 2000 Tottoriken-seibu earthquake. The maximum inclinations in Urayasu City and Abehikona are about 60/1,000 and 37/1,000, respectively. Moreover though average inclination in Abehikona is 10/1,000 to 15/1,000, many houses tilted more than 15/1,000 in Urayasu City. Therefore it can be said that houses in Urayasu severely tilted than the houses in Abehikona. Settlements of the damaged houses are not clear because the measurement of the settlement is not easy. However, it is inferred that more large settlement occurred in Urayasu City than Abehikona.

One more interesting thing is the effect of depth of ground water table on the damage to houses. Figure 5.8 shows contour lines of the depth of ground water table in Urayasu City (Urayasu City 2011). Frequency distributions of the depth of water table at different level of damage are summarized in Fig. 5.9. Average depth of water table for completely destroyed, large-scale partially destroyed, and partially destroyed, partially damaged houses were GL-1.38 m, GL-1.78 m,GL-1.84 m, GL-2.10 m, respectively (Urayasu City 2011). Then the author has an image of the effect of the depth of water table as illustrated in Fig. 5.10.

Table 5.2 Numbers of damaged houses in Urayasu City by old and new standard (Urayasu City)

	Number of houses	
Grade of damage	Old standard	New standard
Completely destroyed	8	18
Large-scale partially destroyed	0	1,541
Partially destroyed	33	2,121
Partially injured	7,930	5,096
No damage	1,028	1,105
Total	8,999	9,981

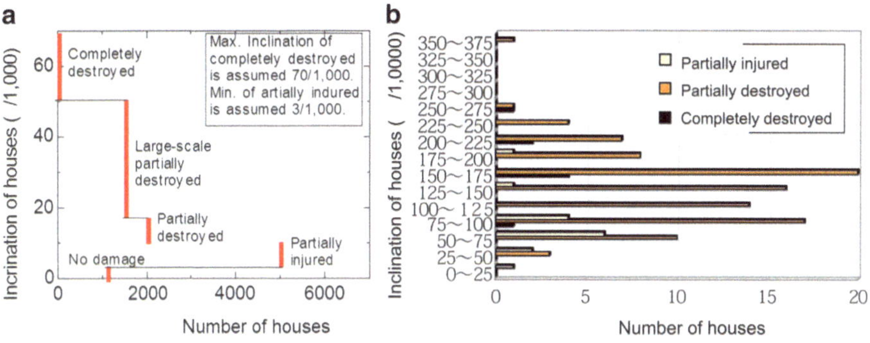

Fig. 5.7 Frequency distributions of the numbers of the damaged houses. (**a**) Urayasu City. (**b**) Abehikona in Yonago City (Yasuda et al. 2004)

Fig. 5.8 Estimated depth of water table (Urayasu City 2011)

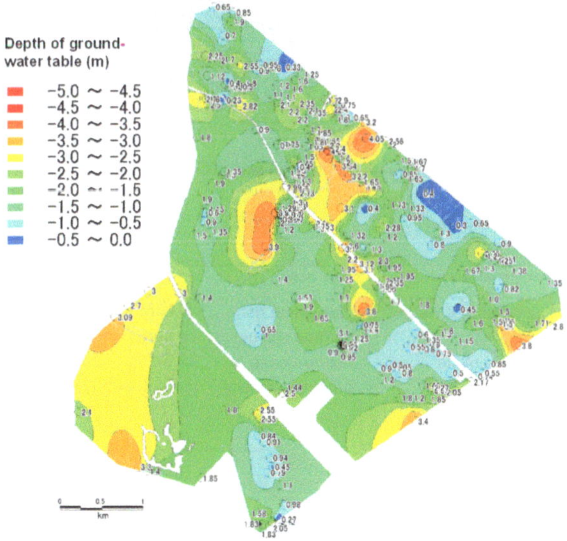

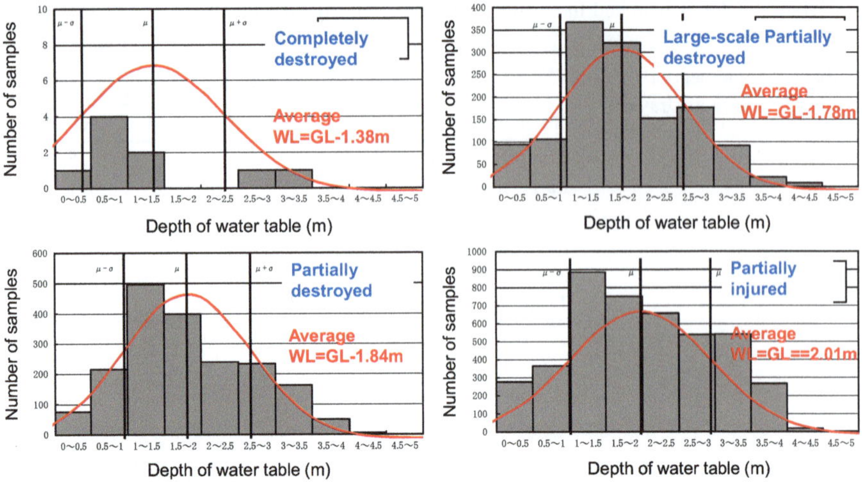

Fig. 5.9 Frequency distributions of the depth of water table at different level of damage (Urayasu City 2011)

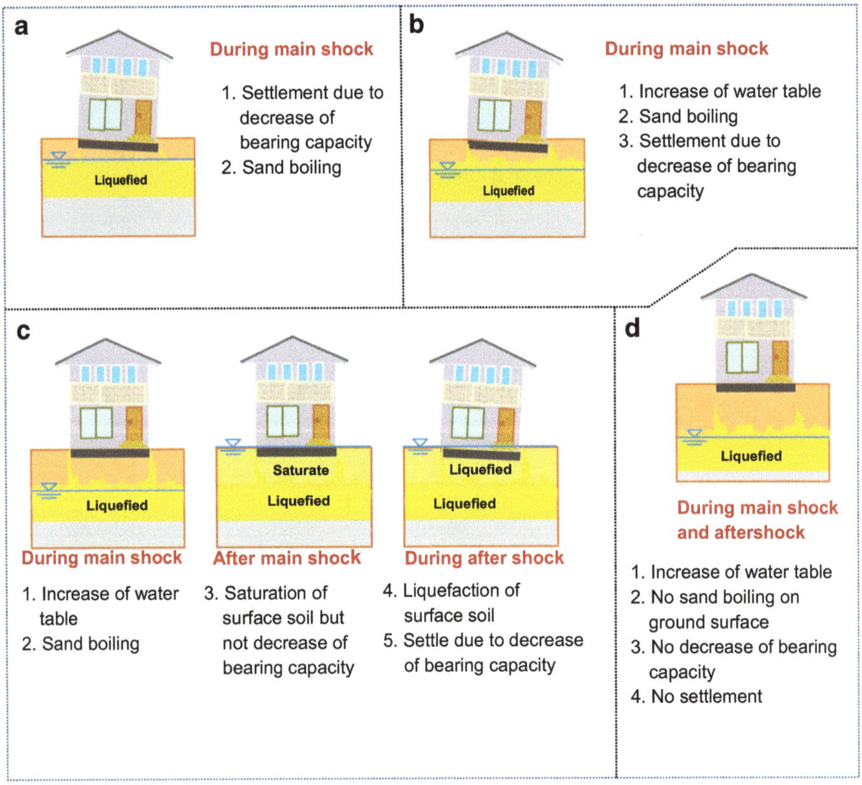

Fig. 5.10 An image of the effect of water-table depth on the damage of houses; (**a**) shallower than GL-1 m, (**b**) about WL=GL-1.5 m, (**c**) about WL=GL-2 m, (**d**) deeper than GL-3 m

Now many inhabitants along Tokyo Bay are facing to the serious problem how to restore the damaged houses. Complicated problem is the re-liquefaction during aftershocks or future earthquakes. Not only the restoration but also some countermeasure against re-liquefaction must be applied, but the problem is the cost and the technique to treat the ground under existing structures. Then early development on effective and economic measures against liquefaction for existing timber houses are facilitating by several organizations. Moreover, applicability of special measures to improve an area by decreasing ground water table is studying.

5.5 Factors Affect to Non-uniform Settlement

Tilting of houses is derived from non-uniform settlement. According to author's previous study on the non-uniform settlement of houses, several factors affect to the non-uniform settlement (Yasuda 2010). Among them effect of adjacent houses was dominant in Abehikona housing lot (Yasuda and Ariyama 2008). Similar tendency was observed in Tokyo Bay area as schematically shown in Fig. 5.11. If two houses are closely constructed these houses tilt inward, and if four houses are close these houses tilt toward the center. Figure 5.12 shows actual tilted directions of adjacent houses at Irifune in Urayasu City. Not only two lows of houses tilted toward each other but also distance of adjacent houses influenced the inclination to right and left directions.

Figure 5.13 shows two houses tilted toward each other at Abehikona during the 2000 Tottoriken-seibu earthquake. On the contrary, two buildings shown in Fig. 5.14 tilted away from each other in Adapazari City during the 1999 Kocaeli earthquake. The author then conducted detailed studies on the damages of houses and buildings during the Tottoriken-seibu and the Niigata earthquakes (Yasuda and Ariyama 2008). In the investigation, the relative angle of inclination of adjacent houses, $\theta_r = \theta_1 - \theta_2$ is defined as shown in Fig. 5.15. If two houses inclined inwards, θ_r was defined as positive. Figure 5.16 shows relationship between relative

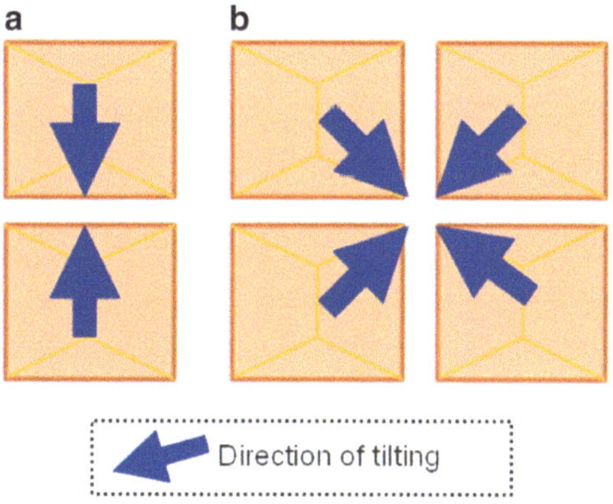

Fig. 5.11 Effect of adjacent houses on the inclination of houses; (**a**) two houses are close, (**b**) four houses are close

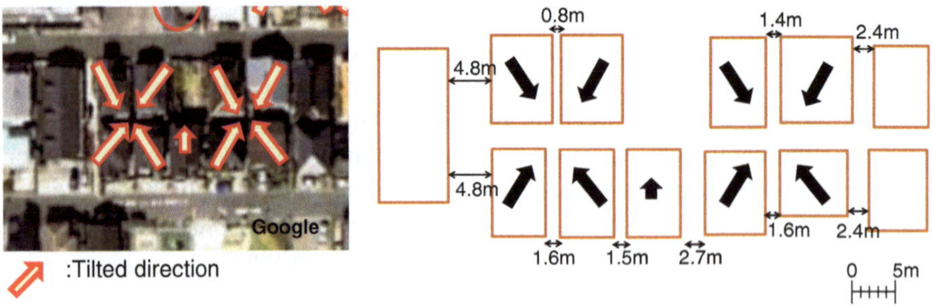

Fig. 5.12 Actual tilted directions of adjacent houses at Irifune in Urayasu City

Fig. 5.13 Two houses tilted toward each other at Abehikona during the 2000 Tottoriken-seibu earthquake

Fig. 5.14 Two buildings tilted away from each other in Adapazari City during the 1999 Kocaeli earthquake

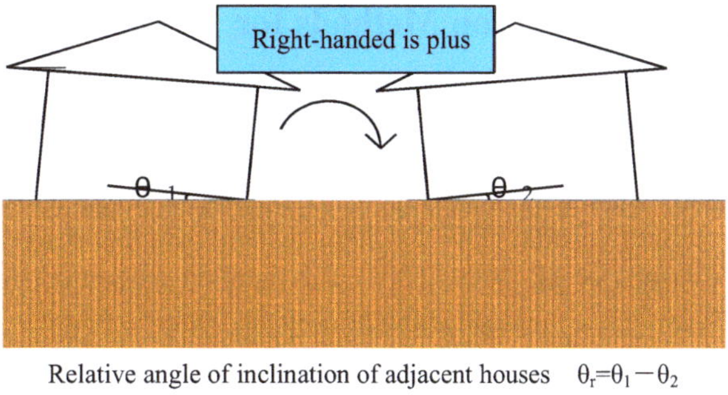

Fig. 5.15 Definition of relative angle of inclination of adjacent houses

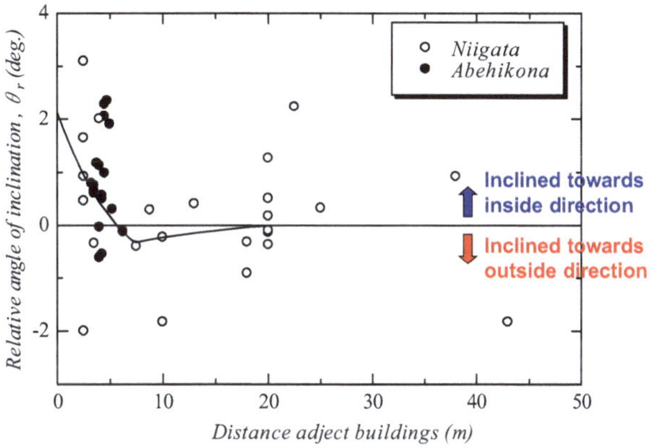

Fig. 5.16 Measured relationship between relative angle of inclination and distance between adjacent houses (Yasuda and Ariyama 2008)

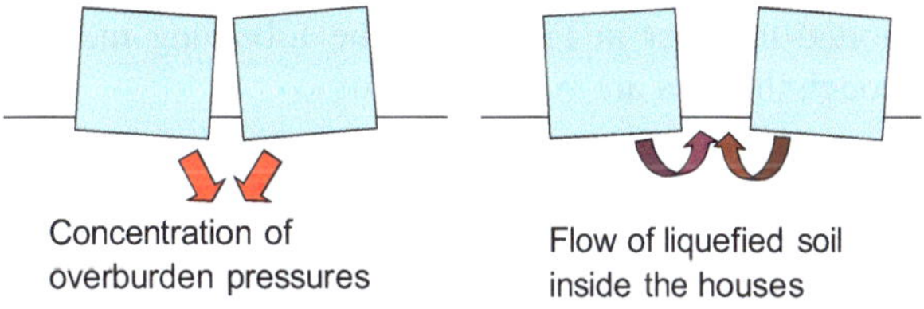

Fig. 5.17 One idea of differential settlement of adjacent houses

angle of inclination and distance between adjacent houses. As shown in the figure, relative angle of inclination is about 1–3° if the distance is as narrow as 1–5 m. And the relative angle deceased with distance up to almost 7 m. At this distance, the relative angle was negative. Then the relative angle increased with distance to become zero. Effect of the distance on the inclination may be explained by the mechanism shown in Fig. 5.17.

5.6 Estimation Methods for Non-uniform Settlement

Estimation methods for liquefaction-induced deformation of structures and ground are classified into three grades: empirical methods, static (residual deformation) analyses and dynamic (seismic response) analyses. Accuracy and cost are different among these grades (Yasuda 2009). Of them, a static (residual deformation) method named ALID (Yasuda et al. 1999, 2003) was applied to demonstrate the factors which affect liquefaction-induced differential settlement of houses (Yasuda and Ariyama 2008). Soil condition, seismic condition and dimension of houses for the Basic Model Case (Case 2-0) were assumed to be the same conditions as at Abehikona housing development. Several factors were then changed to demonstrate their effect on the differential settlement. Figure 5.18 shows the deformation of the Basic Model Case (Case 2-0). As show in the figure, two houses settled and inclined inwards. Average settlement of the houses was about 150 mm. And relative angle of inclination was about 1.2°. These values are similar to the actual settlement and inclination. Figures 5.19 and 5.20 show relationships between relative angle of inclination and distance of adjacent houses, and safety factor against liquefaction F_L, respectively. If the distance between adjacent houses is narrow, less than 3 m, two adjacent houses tilt inwards, towards one another. By contrast, two adjacent houses tilt outwards, away from one another, if the distance is 3–15 m. This relationship is similar to the actual relationship shown in Fig. 5.16. Relative angle of inclination increased with the decrease of F_L. The author and his colleagues conducted shaking table tests also to demonstrate the effect of the distance of adjacent houses on relative settlement. Though test results are not introduced in this paper, similar relationships to those shown in Figs. 5.19 and 5.20 were obtained. However, new analyses and shaking table tests are necessary for the soil condition at Tokyo Bay area because liquefied soils have much fines.

It seems the angle of inclination increases with average settlement of houses. The author investigated the relationship between average settlement and inclination for the buildings damaged by the 1964 Niigata, the 1990 Luzon and the 1999 Kocaeli earthquakes (Yasuda et al. 2001; Yasuda 2010). Three data sets were plotted together in Fig. 5.21. The following mean relationship may be derived though the data are very scattered:

$$\theta = 0.05 S_{av} \tag{5.1}$$

Where, θ is the angle of inclination (°) and S_{av} is average settlement (cm). Similar relationship for timber houses must be investigated base on the data during the 2011 Tohoku-Pacific Ocean earthquake.

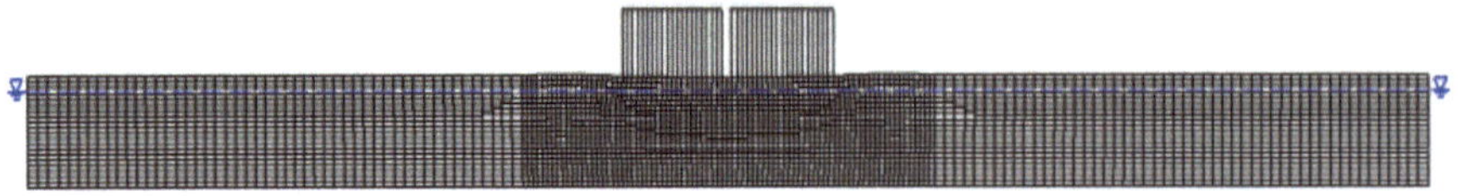

Fig. 5.18 Deformation of the Basic Model Case (Case 2-0) (Yasuda and Ariyama 2008)

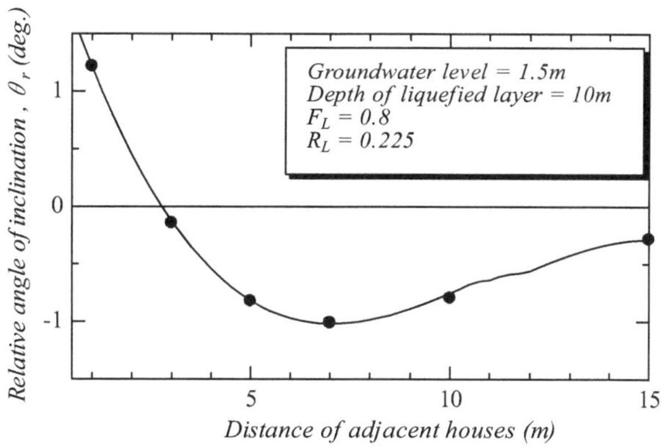

Fig. 5.19 Relationships between relative angle of inclination and distance of adjacent houses (Yasuda and Ariyama 2008)

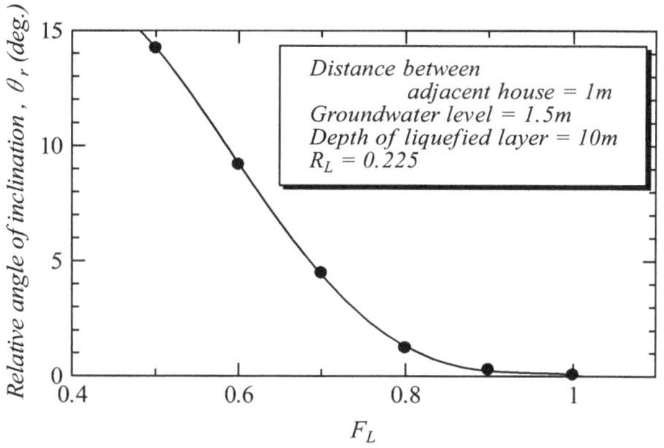

Fig. 5.20 Relationships between relative angle of inclination and F_L (Yasuda and Ariyama 2008)

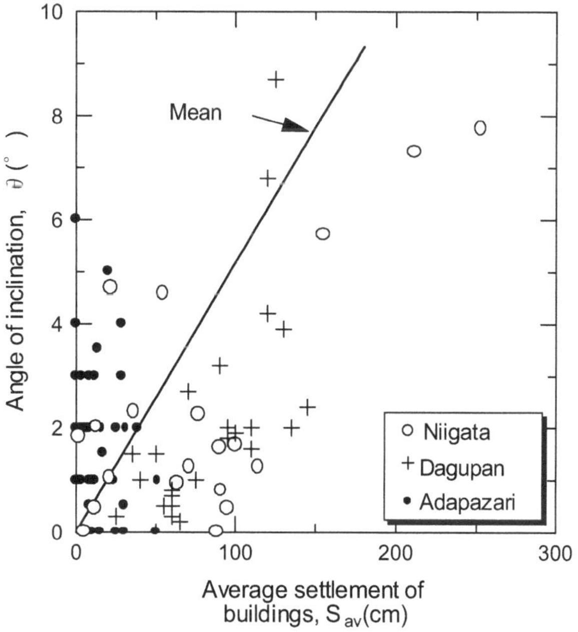

Fig. 5.21 Relationship between average settlement and angle of inclination (Yasuda et al. 2001)

5.7 Conclusions

About 27,000 houses were damaged due to liquefaction in Tohoku and Kanto districts of Japan during the 2011 Tohoku-Pacific Ocean earthquake. Especially, severe liquefaction occurred in reclaimed lands along Tokyo Bay. The very long duration of the main shock and an aftershock 29 min later should have induced the severe liquefaction and consequent damage to houses. Though no damage to walls and windows were observed, inhabitants feel giddy and nausea, and difficult to live in the heavily tilted houses after the earthquake. Then, Japanese Cabinet announced new evaluation standard for the damage of houses by the two factors, settlement and inclination. Many inhabitants along Tokyo Bay are facing to the serious problem how to restore the damaged houses. Early development on effective and economic measures against liquefaction for existing timber houses is necessary.

Acknowledgements Mr. K. Ogawa provided me important photos. This work was supported by MEXT Grant-in-Aid for Special Purpose (22900002). The author would like to express his sincere appreciation to them.

References

National Research Institute for Earth Science and Disaster Prevention (NIED) (2011). K-NET WWW service, Japan. http://www.k-net.bosai.go.jp

Urayasu City (2011) Report by the technical committee on the measures/against liquefaction (in Japanese)

Yasuda S (2009) Relevant soil investigations and laboratory tests to estimate liquefaction-induced deformation of structures. In: Proceedings of the international conference on performance-based design in earthquake geotechnical engineering- from case history to practice, pp 259–263

Yasuda S (2010) Damage to structures due to soil liquefaction. In: Williams et al (eds) Proceedings of the 11th IAEG congress, geologically active. Taylor & Francis Group, pp 15–52

Yasuda S, Ariyama Y (2008) Study on the mechanism of the liquefaction-induced differential settlement of timber houses occurred during the 2000 Totoriken-seibu earthquake. In: Proceedings of 14th world conference on earthquake engineering, Beijing, Paper no. S26-021

Yasuda S, Harada K (2011) Liquefaction-induced damage in the reclaimed lands along Tokyo Bay. Geotech Eng Mag JGS 59(7):38–41 (in Japanese)

Yasuda S, Ishikawa K (2012) Several features of liquefaction-induced damage to houses and buried lifelines during the 2011 Great East Japan earthquake. In: Proceedings of the international symposium on engineering lessons learned from the giant earthquake, Paper no 46

Yasuda S, Yoshida N, Adachi K, Kiku H, Gose S (1999) A simplified analysis of liquefaction-induced residual deformation. In: Proceedings of the 2nd international conference on earthquake geotechnical engineering, pp 555–560

Yasuda S, Irisawa T, Kazami K (2001) Liquefaction-induced settlements of buildings and damages in coastal areas during the Kocaeli and other earthquakes. In: Proceedings of the satellite conference on lessons learned from recent strong earthquakes, 15th ICSMGE, pp 33–42

Yasuda S, Ideno T, Sakurai Y, Yoshida N, Kiku H (2003) Analyses for liquefaction-induced settlement of river levees by ALID. In: Proceedings of the 12th Asian regional conference on SMGE, pp 347–350

Yasuda S, Hitomi T, Hashimoto T (2004) A detailed study on the liquefaction-induced settlement of timber houses during the 2000 Tottoriken-seibu earthquake. In: Proceedings of the 5th international conference on case histories in geotechnical engineering, Paper no 3.33

Part III
River Levees and Dams

Chapter 6
Seismic Performance of River Levees; Experience and Prediction

Ikuo Towhata

Abstract After the 2011 gigantic earthquake in Japan, the author has been investigating damage extents, damage mechanisms, restoration policies, and proposal of future technological development for river levees. Those activities are summarized here with emphasis on the difficult handling of liquefaction inside the levees and assessment of seismic performance. To mitigate the internal liquefaction, not only the field investigation technique but also numerical analysis for performance prediction has to be newly developed.

6.1 Introduction

The conventional philosophy of seismic design of river levees in Japan used to be based upon the concept of factor of safety greater than unity, being similar to the design of other structures. However, there have been two important issues to be considered in river levees. The first is the financial issue that the budget available for seismic retrofitting per unit length of levees is limited because the entire length of seismically vulnerable levee is substantial. The second is that the occurrence of the seismic factor of safety less than unity does not mean an immediate disaster or overtopping of water because the probability of the simultaneous occurrence of a strong earthquake and significant flooding has been considered low. Consequently, seismic factor of safety greater than unity has not been compulsory. Alternatively, it has been aimed to restore possible damage of river levees within a short period of time that is typically 14 days.

One of the earliest occasion that required deeper consideration of the situation above was the 1993 Kushiro-oki earthquake by which levees of Tokachi and Kushiro Rivers were substantially damaged by subsoil liquefaction (Sasaki et al. 1994).

I. Towhata (✉)
Department of Civil Engineering, University of Tokyo, Tokyo, Japan
e-mail: towhata@geot.t.u-tokyo.ac.jp

M. Maugeri and C. Soccodato (eds.), *Earthquake Geotechnical Engineering Design*, Geotechnical, Geological and Earthquake Engineering 28,
DOI 10.1007/978-3-319-03182-8_6,

Fig. 6.1 Distorted shape of Yodo River levee after 1995 Kobe earthquake

Because the total length of the damaged levee sections was enormous and the earthquake occurred in the middle of snowy winter, the arrangement and transportation of re-construction materials took a long time, and approximately 5 months were spent until completion of the restoration works. In spite of this unexpectedly long time of restoration, the situation did not change profoundly because the damaged levees were not immediately exposed to flooding or high river water.

Seismic design of a river levee was considered more seriously when the Yodo River levee in Osaka (Fig. 6.1) was considerably distorted by liquefaction during the 1995 Kobe earthquake (Matsuo 1996). Being situated near the river mouth, this part of the levee was subjected to high water level (high sea tide) as often as twice a day. Since then, seismic risk evaluation and seismic design, if necessary, of river levees have been considered very important. Because the practice of the risk evaluation is based on seismically-induced deformation of levees, it can be called seismic performance-based design. The short history of this practice, which is more or less 10 years only, means that the practice is not yet fully established and that lessons have to be learnt upon every earthquake event so that the design philosophy may be improved. In this regard, the 2011 East Japan gigantic earthquake was an important opportunity to learn new lessons. This paper is therefore intended to address new information and knowledge about seismic performance of river levees that were learnt during this earthquake.

6.2 Damage of River Levees Caused by the Earthquake on March 11, 2011

Because the author has been involved in different activities related with the seismic damage of river levees since the occurrence of the gigantic earthquake on March 11, 2011, this paper provides a good opportunity to summarize what happened and address their engineering significances.

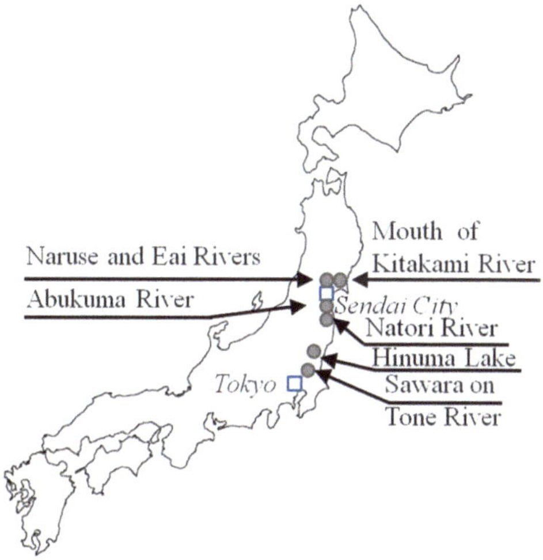

Fig. 6.2 Locations of levee damage of interest

Fig. 6.3 Significant distortion of levee of Tone River in Sawara

The earthquake caused levee damage at, in total, more than 1,900 sites and the induced damage extents have been classified into such three groups in accordance with the depth of the open cracks at the crest; "significant" when crack being deeper than the design high water level, "intermediate" when crack being shallower than the high water level, and "light" when only surface distortion occurred. Figure 6.2 indicates the locations of damage that are going to be discussed in this paper. Figure 6.3 illustrates an example of significant damage at Sawara. The subsidence at the crest was 1 m out of the 3.8-m height. This damage was caused by the subsoil liquefaction as evidenced by sand boils (Fig. 6.4). This situation is related with the geomorphological condition; Fig. 6.5 illustrates the original topography of the Sawara site where the present levee is situated on a former river channel. Thus, young reclaimed subsoil is vulnerable to liquefaction.

Fig. 6.4 Liquefaction and sand ejecta in front of the damaged levee in Sawara

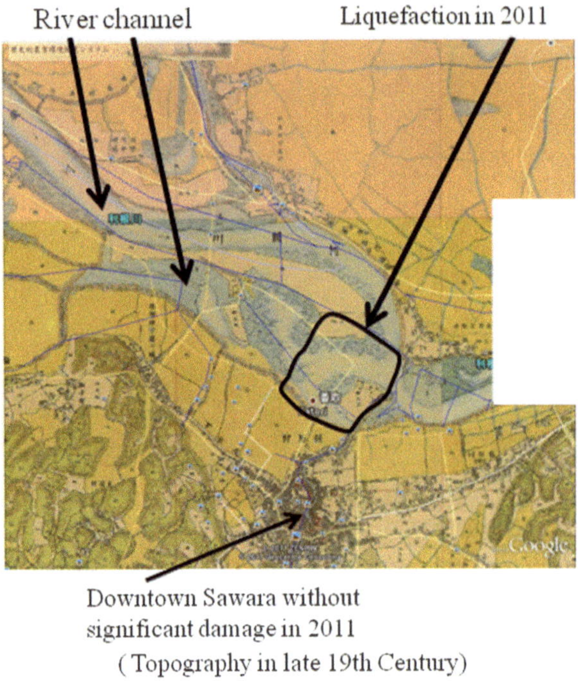

Fig. 6.5 Old river channels of the Sawara site (Zinsoku-Sokuzu 1/20000 by National Institute for Agro-Environmental Sciences)

A new technical problem was posed by such an event as shown in Fig. 6.6 where a levee of Hinuma Lake (Fig. 6.2) spread laterally, developed many longitudinal cracks, and subsided 1 m out of the original height of 2.6 m. Because the subsoil condition is clayey where liquefaction is unlikely (Fig. 6.7), it is supposed that liquefaction occurred within the body of the levee.

Figure 6.8 summarizes the geomorphological information from sites of levee damage in six river systems. As has been widely known, the majority of damage occurred in flood plains and former river channels where soft and probably liquefiable subsoil is abundant and also seismic amplification is significant. Another majority of damage occurred in natural levees where the subsoil condition is not very stiff but still better than the aforementioned two categories. Note, however,

Fig. 6.6 Significant distortion of the Hinuma levee at Shimo-Ishizaki

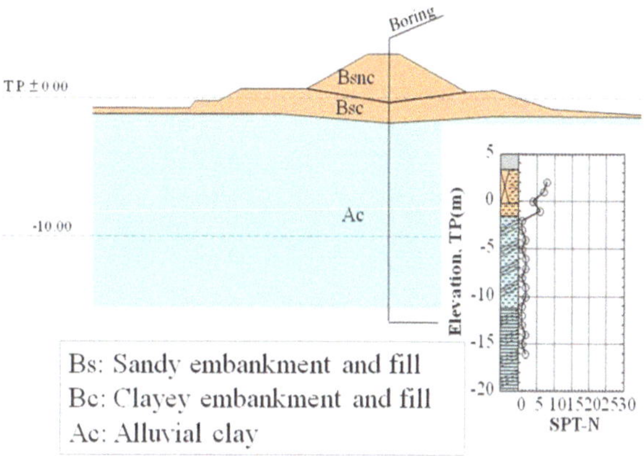

Fig. 6.7 Subsoil data at the Shimo-Ishizaki site of Hinuma Lake (After MLIT; Sasaki et al. 2012)

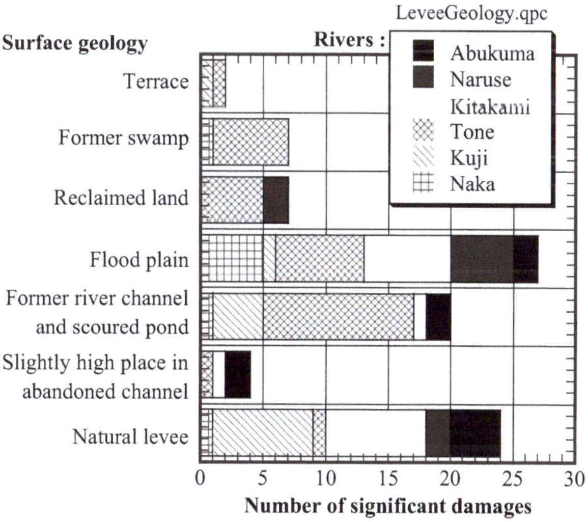

Fig. 6.8 Relationship between number of levee damage and geomorphology (Drawn after MLIT; Sasaki et al. 2012)

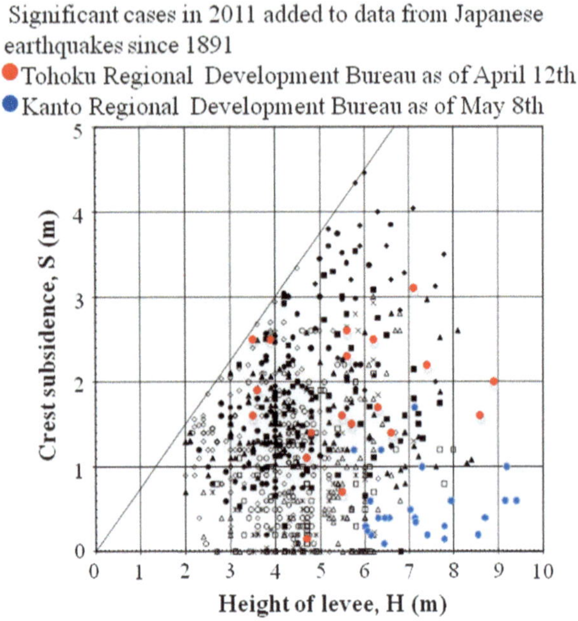

Fig. 6.9 Relationship between levee height and seismic subsidence at crest (new data in 2011 was added to the original diagram by MLIT (2007))

that the classified subsoil types are not so accurate as to precisely define the subsurface conditions at damaged sites.

The perspective of seismic performance-based design has been requiring the estimation of subsidence of levees caused by subsoil liquefaction. While the types of subsidence assessment vary from an empirical and simple one to more sophisticated numerical ones, practice has been conducting the simple type in the first stage and more sophisticated numerical analyses in the second stage. A typical empirical method is based on the past data that the subsidence never exceeds 75 % of the height of levees. Data from the earthquake in 2011 was added to the original data from many earthquakes from the end of the nineteenth century over one hundred years in Fig. 6.9 and it seems that the newly added data is consistent with previous data. Thus, it is reasonable to state that the subsidence of river levees is at the maximum 75 % of their original height. If the remaining 25 % is high enough to prevent flooding in the protected area, liquefaction countermeasure is not necessary.

6.3 On Tsunami and River Levee

Levees near a river mouth may be affected by tsunami that runs upstream along the channel. If this happens, the levee is expected to maintain its design height and prevent significant overtopping of water. Thus, the required seismic performance is remarkably more difficult.

Fig. 6.10 Kitakami River Levee at Kamaya under restoration after total destruction by the tsunami attack

Fig. 6.11 Appearance of the Yuriage Levee of Natori River after the tsunami attack

Figure 6.10 shows a levee of Kitakami River (Fig. 6.2) that was destroyed by run-up of tsunami. Figure 6.11 in contrast illustrates a Natori River levee (Fig. 6.2) that survived the overtopping of tsunami water. The different performance of the two levees may be interpreted as what follows and facilitates future tsunami mitigation;

1. The levee of the Kitakami River (Fig. 6.10) was aligned rather normal to the run-up of tsunami (Fig. 6.12) and was subject to significant dynamic tsunami pressure, while, in contrast, the Natori River levee in Fig. 6.11 is approximately parallel to the straight channel of the river and successfully avoided the dynamic tsunami pressure.
2. As reported by Sasaki et al. (2012), the area behind the Natori River levee liquefied prior to the tsunami attack and the ground surface had been covered by boiled water. This water absorbed the energy of the overtopping tsunami water and prevented erosion at the toe of the levee slope.

6.4 Discussion on Liquefaction Inside Levee Body

Liquefaction inside a levee is not rare. One possible cause is the infiltration of rain water that is not drained out and remains inside the levee. The saturated part of the levee got liquefied and produced minor sand boiling (Fig. 6.13). Being minor in size, this kind of liquefaction does not become fatal.

A more important mechanism of internal liquefaction is the one that was originally proposed by Sasaki et al. (1994) together with Kaneko et al. (1996) and is exhibited in Fig. 6.14. It is supposed herein that

Fig. 6.12 Aerial view of Kamaya site near Kitakami River mouth

Fig. 6.13 Liquefaction in levee caused by remaining rain water and sand boiling in the mid-height berm

Fig. 6.14 Schematic illustration of mechanism of liquefaction inside a levee

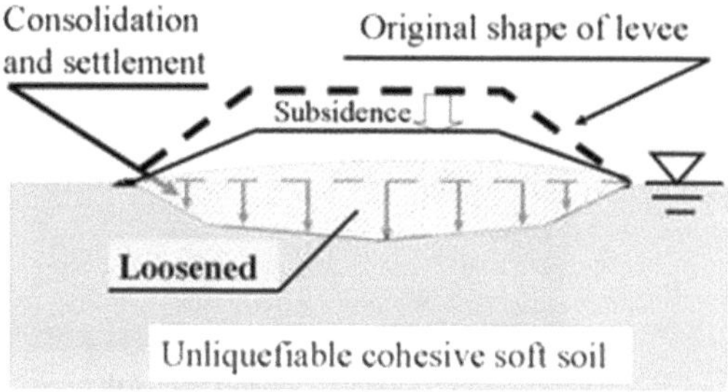

Table 6.1 Assessment of possibility of liquefaction in the sandy bottom part of levees

Rivers	Name of site	Surface acceleration k (G unit)	Depth (m)	SPT-N	F_L
Hinuma Lake	Shimo Ishizaki	0.546	4.3	6	0.588
Naruse River	Shimo Nakanome	0.458	2.3	3	0.162
			3.3	2	0.120
Abukuma River	Edano	0.411	3.3	3	0.264
			4.3	4	0.263
			4.8	2	0.207
			5.3	5	0.309
			5.8	6	0.321
Eai River	Fukunuma	0.571	3.3	3	0.182
			4.3	3	0.341
			5.3	4	0.341

1. a river levee sinks into soft clayey subsoil because of consolidation,
2. the sunken part has poor drainage of infiltrated rain water and seeping ground water because the elevation is lower than the ground surface and the surrounding soil is less pervious clay, and
3. the procedure of subsidence and consequent lateral expansion is possibly accompanied by loosening of soil and reduced liquefaction resistance.

The author has been working with engineers of the MLIT (Ministry of Land, Infrastructure, Transport and Tourism) on the liquefaction damage of levees since the onset of the gigantic earthquake. While more detailed report are going to be published in near future (Sasaki et al. 2012), the present paper attempts to introduce some part of it.

First, Table 6.1 examines the factor of safety against liquefaction, F_L, in sandy part of levees that distorted significantly in spite of clayey subsoil and are hence considered to have been subject to internal liquefaction. The values of F_L were calculated by using the JRA method of 2002 version and the surface ground acceleration at respective sites were assessed by using nearby K-NET records. It is seen here that the obtained F_L values are very low because of the low SPT-N values and that the internal liquefaction is likely to have occurred.

The government (MLIT) has been paying special attention to validate the idea of internal liquefaction by running more investigations in damaged levees. For

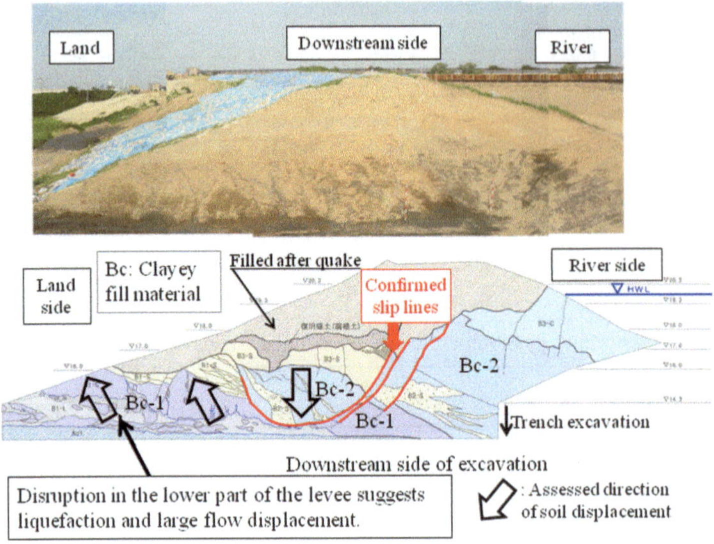

Fig. 6.15 Excavated cross section of the levee at the Shimo-Nakanome site on the Naruse River (Sasaki et al. 2012)

example, Sasaki et al. (2012) cited the excavation of the Naruse-river levee (Fig. 6.2) and the studied cross section is shown in Fig. 6.15. It is seen that the Bc-2 layer in the center of the levee sank down and the volume compatibility pushed the Bc-1 layer under the slope up towards the toe. This large distortion suggests the loss of bearing capacity in the lower part of the levee. Note, however, that the depth of excavation was limited by the ground water table and that the detailed study of the real liquefied soil was not possible.

6.5 Successful Mitigation of Liquefaction-Induced Damage of Levees

Because of the high concerns for seismic safety of levees in the past decade, mitigation measures have been executed. The earthquake in 2011 offered a good opportunity to validate their effects. Note that the mitigation measures were designed against Level-1 design earthquakes for which the recurrence period is about 50–70 years.

6.5.1 Gravel Drain

The first example is the case at Omigawa of Tone River (Fig. 6.2). As shown in Fig. 6.16, the land side of the levee had been reinforced by gravel drains and no

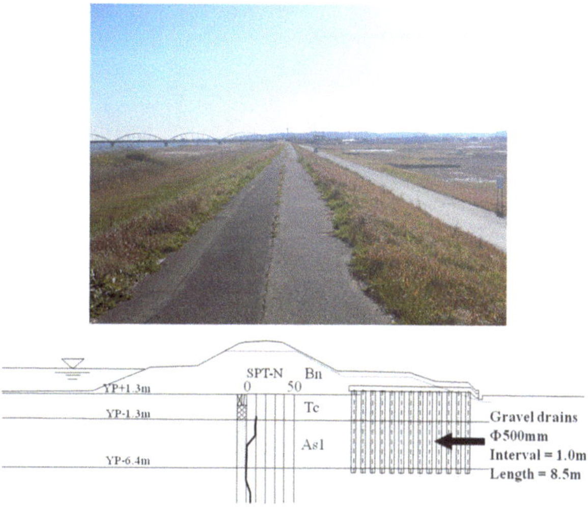

Fig. 6.16 Cross section of the Tone River levee at Omigawa (R 27.75 km) with soil improvement by gravel drains (Courtesy of MLIT)

damage occurred during the 2011 earthquake. After the 1995 Kobe earthquake, the use of gravel drain has been decreasing because design calculations against Level-2 design earthquakes suggest that gravel drains cannot keep the excess pore water pressure within a low level (typically less than 50 % of the initial effective vertical stress). Because of this successful case under a remarkable earthquake loading, it is reasonable that more attention should be paid to the use of gravel drains. The design of gravel drains take into account only pore pressure dissipation, while, in reality, the columns of gravel drains maintain rigidity during liquefaction of surrounding soils and may be able to keep the rigidity of the entire ground.

6.5.2 Sand Compaction Piles

The Naruse River levee (Fig. 6.2) at Naka-Shimo was damaged by the 1978 Off-Miyagi earthquake and it was partially improved by using sand compaction piles (Fig. 6.17) with the improvement ratio of 10 %. Because no damage occurred in 2011, there is not much information about its behavior except that the levee was safely used for evacuation from the tsunami attack. However, the experience during the 2003 North Miyagi earthquake ($M_{JMA} = 6.2$) provides some information. Figure 6.18 compares the records of excess pore water pressure in the sandy part during the 2003 event. Evidently, the excess pore water pressure in the densified part ("D" in Fig. 6.17) is lower than in the uncompacted part ("L"). It is reasonable to infer that the same situation occurred in 2011 as well, successfully preventing the onset of liquefaction and reducing the deformation.

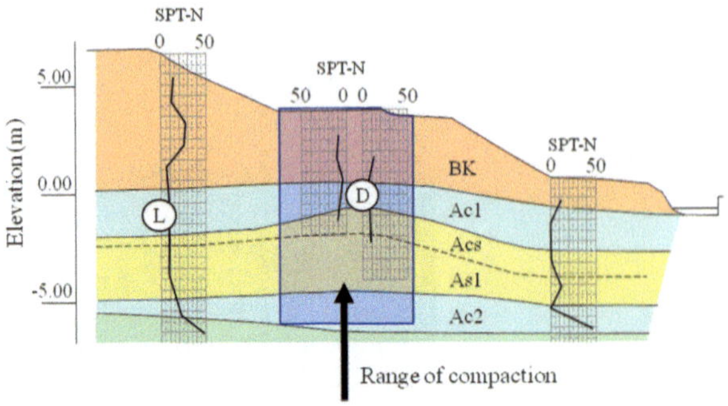

Fig. 6.17 Cross section of the Naka-Shimo levee with partial improvement by sand compaction pile (*L* and *D* stand for locations of pore pressure transducers in loose and dense part of the levee, respectively) (Drawn after documents owned by the Research Center for Disaster Risk Management belonging to the National Institute for Land and Infrastructure Management)

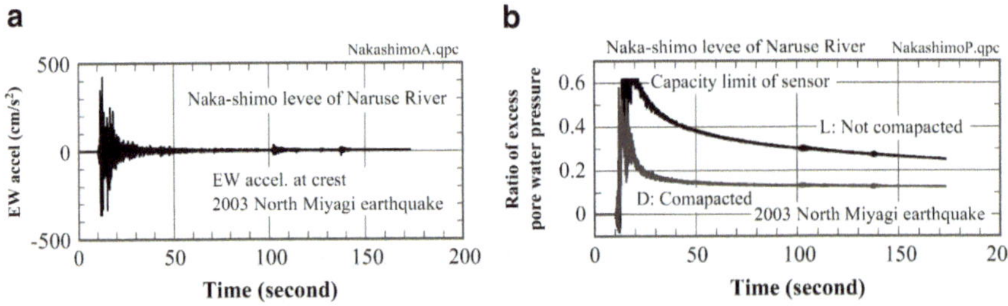

Fig. 6.18 Time histories of crest acceleration and excess pore water pressure ratio in Naka-Shimo levee during an earthquake in 2003 (Sasaki et al. 2012). (**a**) Acceleration. (**b**) Pore water pressure

6.5.3 Sheet Pile Wall

At Sakae of Ibaraki, the Tone River levee had been reinforced by underground sheet pile walls (for reducing seepage of water) accompanied by a wide berm on the river side. During the earthquake, no distortion occurred on the river side (Fig. 6.19). In contrast, the land side of the levee (Fig. 6.20) was distorted by liquefaction in the foundation soil in which the ground water level is within 50 cm from the surface (author's observation in April, 2012). The damaged section was restored by installing 8-m sheet piles along the slope toe.

6.6 On Restoration

The gigantic earthquake occurred on March 11th of 2011, while the rainy season starts normally at the beginning of June. Within the time shorter than 2 months, a complete restoration of damaged levees was impossible. For heavily damaged

Fig. 6.19 River side of Tone River levee at Sakae without damage (entire restoration of the levee going on in April, 2012)

Fig. 6.20 Reconstruction of land side of Tone River levee at Sakae by installing sheet pile walls with holes

levees, therefore, their height was recovered (see the top fill in Fig. 6.14) but the disruption inside the levees was not repaired. As an alternative way, it was decided to intensify the alert level of river water (alert being issued at lower water level than before) and a more number of personnel was allocated to watch the safety of levees. Because the affected levees in 2011 were not subject to significantly high water level in rivers, the above-mentioned strategy was successful. After the rainy season, the reconstruction procedure encountered another problem.

The recent practice employs two types of design earthquakes that are namely "Level 1" and "Level 2." Level 1 design earthquake is supposed to be what happens once during the life period of a designed structure and this period is typically 50–70 years. In contrast, Level 2 earthquake is a rare event whose recurrent period is more than 500 years according to the seismic activity in Japan. Certainly, Level

2 quake is more powerful than Level 1. Accordingly, the basic philosophy consists of two stages;

1. Existing levees and other river structures are seismically examined by using Level 2 design earthquake.
2. If the result is not satisfactory, seismic retrofitting is considered.

Because of the budget limitation, it takes time to start the required retrofitting.

After the 2011 gigantic earthquake, the first 2 months and half were spent on the emergency restoration in which, as mentioned before, the height of the heavily damaged levees was recovered, while the quality of soils inside the levee body could not be fully taken care of. Hence, during the rainy season from June to October, lots of efforts were made to conduct field watching and emergency action, if necessary. After the rainy season, the practice turned to design for a permanent restoration. The problem was that the Level-2 earthquake implied very thick liquefied layer at many damaged places and satisfactory mitigation such as soil improvement and deep mixing became too costly within the time (completion required by the next rainy season) together with budget limitation. Hence, it was decided to commence the restoration with design under Level-1 motion. This decision is partially supported by the fact that those levees designed with Level-1 were able to survive the gigantic earthquake as stated in the previous chapter. However, Level-2 retrofitting is considered to be necessary in near future.

6.7 Technical Demands

To date, heavy discussion has been conducted on the liquefaction inside levees; how to detect potentially liquefiable levee at a reasonable cost out of the long entire levees, and how to reduce the potential of internal liquefaction. This first question may be answered to some extent by considering the local surface geology: clayey subsoil causes consolidation and resultant subsidence of the overlying levee. Thus, the levee sections resting on a thick soft clay deposit should be focused on.

The first stage of screening as mentioned above is not sufficient, leaving long levee sections subject to further judgment. The author imagined that any record of subsidence and overlaying (filling earth at the crest) can be a good key to find vulnerable levee sections. However, there are hardly historical records of subsidence. Moreover, many levees used to be constructed until nineteenth century by using locally available clayey materials (Fig. 6.21). Because clayey material is not likely to liquefy, judgment of soil type is necessary.

At this moment, there are two possible solutions to the problem. The first candidate solution is the use of geophysical exploration by which the spatial distributions of moisture content and S-wave velocity are studied without drilling holes; moisture content by a survey of electric resistivity and S-wave velocity by a seismic survey. It is expected that the geophysical investigation is able to detect soft

Fig. 6.21 Excavated cross section of Tone River levee at Iijima

Fig. 6.22 Dynamic CPT with pore pressure recording (Sawada and Towhata 2011)

(low S-wave velocity) and water-saturated soils (low electric resistivity). A possible problem is the accuracy and resolution. The second solution may be the use of CPT or other simple soil exploration. Although time is needed to push the equipment into the levee body, useful information can be obtained; for example, the ratio of skin friction and cone tip resistance can suggest soil type (Begemann 1965; Robertson 1990), and the pore pressure generation during penetration implies the potential of liquefaction. Note, however, that the simple penetration test should be used as an interpolation of existing bore hole data and SPT profile because the current practice still trust these conventional ways of soil investigation. Figure 6.22 illustrates a recently-developed dynamic cone penetrometer that is quick in operation and is able to monitor pore pressure change during penetration (Sawada and Towhata 2011).

6.8 Numerical Analyses

The author proposed a viscous modeling of liquefied sand for performance (deformation) assessment of structures subject to strong shaking (Towhata et al. 1999, 2010) and developed a numerical program. Figure 6.23 is a levee model for an example analysis of which the performance (settlement at crest) is mitigated by embedded sheet pile walls, counter-weight berms, or both (Fig. 6.24). Results of a flow analysis as shown in Fig. 6.25 demonstrates that a combination of a sheet pile wall and a counter-weight berm can reduce the crest subsidence to less than 50 % after strong shaking and soil flow of 20 s.

An analysis was further continued on a levee model in Fig. 6.26 that is of an internal liquefaction. The analytical solution of flow that was developed by Towhata et al. (1999) and incorporated in the present analysis encountered a mathematical singular point at the ends of the internal liquefaction zone at which

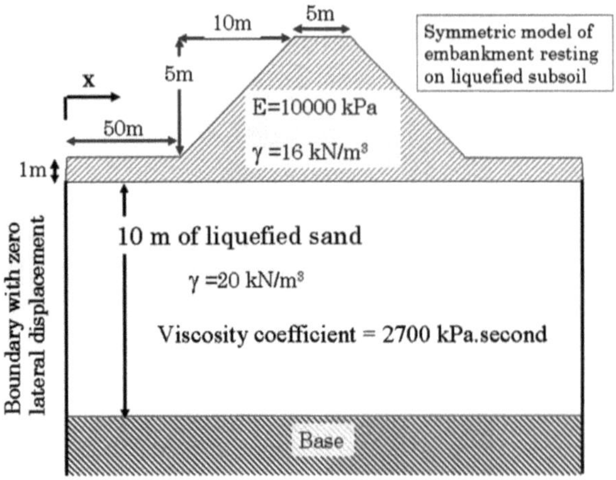

Fig. 6.23 Levee model subject to liquefaction in foundation for performance analysis

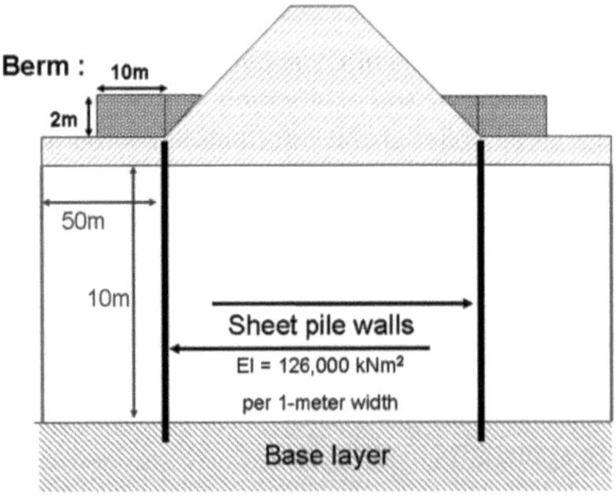

Fig. 6.24 Sheet-pile walls and counter-weight berm as mitigation measures

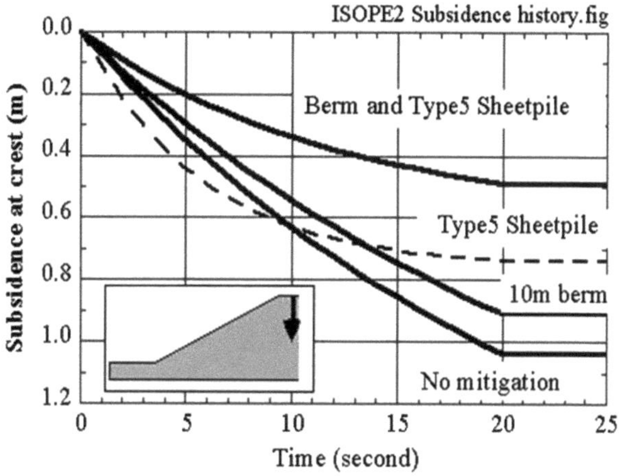

Fig. 6.25 Mitigation of subsidence of levee undergoing liquefaction in foundation

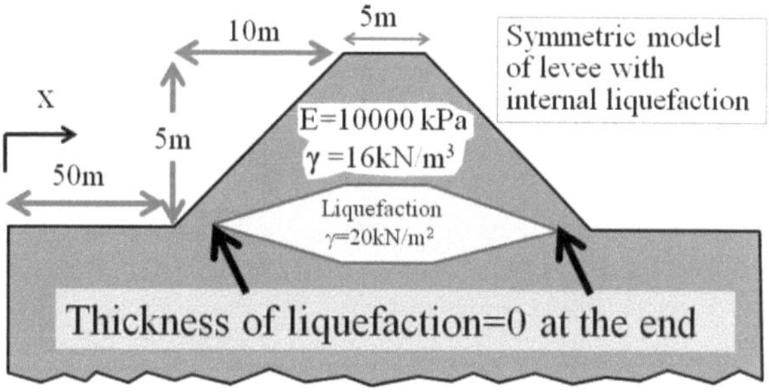

Fig. 6.26 Levee model subject to liquefaction inside its body

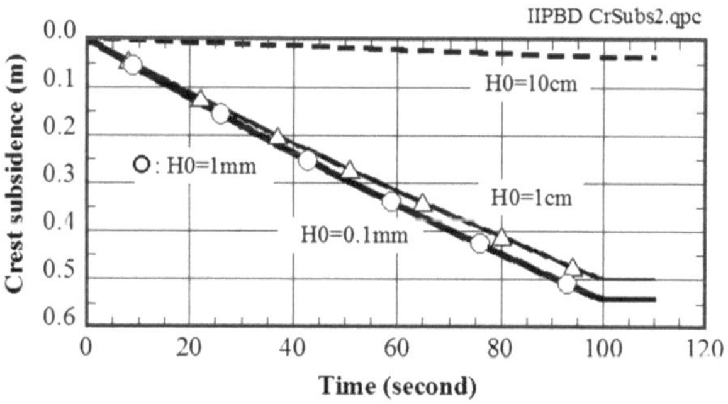

Fig. 6.27 Time history of subsidence at the crest for 100 s of flow

the liquefaction thickness = 0; see Fig. 6.26. To cope with this problem, the zero thickness was replaced by a very small thickness and Fig. 6.27 shows a good convergence at 0.1 mm thickness. The calculation was conducted with an open crack boundary condition at this end, which is expressed mathematically as

$$ET\frac{dF}{dx} + \frac{4\gamma H}{\pi^2}\left(H\frac{dF}{dx} + bF\right) = 0 \tag{6.1}$$

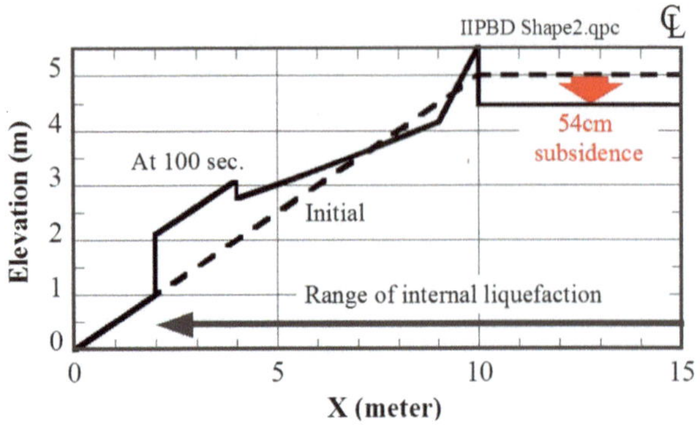

Fig. 6.28 Change of levee configuration after 100 s of flow

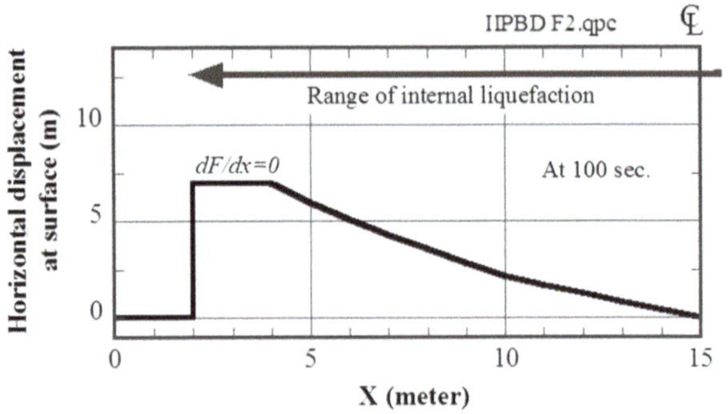

Fig. 6.29 Distribution of lateral displacement along the surface of modeled levee

in which x designates the horizontal coordinate, F the horizontal displacement at surface, T the thickness of surface unliquefied crust, H the thickness of liquefied layer, and b is defined as dH/dx. By substituting $H = 0$ in Eq. 6.1, we obtain

$$\frac{dF}{dx} = 0 \tag{6.2}$$

that is mathematically equivalent with what is called the infinite boundary. By using this boundary condition at the ends, while assuming a symmetric configuration of the embankment (lateral displacement = 0 at the axis of symmetry), the following results were obtained. This dynamic analysis was conducted by using the critical damping ratio of 40, that is most recommendable in the studies so far done.

Figures 6.27, 6.28 and 6.29 illustrate the deformation of the modeled levee after 100 s of flow. This long duration of flow corresponds to the long duration of strong shaking with M = 9. In particular the crest subsidence of 54 cm in Fig. 6.27 is realistic in this type of a levee. Moreover, Fig. 6.28 shows discontinuity in lateral displacement above the end of the internal liquefaction. This feature is consistent with the discontinuity in reality (Fig. 6.30). In case the assessed deformation is significant, mitigation is needed. However, drainage of pore water from the elevation below the local ground water table needs cost of pumping. Thus, there is no clear idea about mitigation yet.

Fig. 6.30 Surface distortion of Naruse River levee with internal liquefaction (at Shimo Niida site)

6.9 Conclusions

This paper concerns the seismic damage of levees during the 2011 gigantic earthquake in Japan. Lessons learned and future technical demands are described. Levees have to maintain some height even after the quake and this requirement seems particularly important in those prone to tsunami run-up. A special emphasis is placed on the new problem of liquefaction inside of a levee body, of which detection, mitigation and performance assessment is difficult.

Acknowledgement The collaboration works with MLIT of the government provided many in-situ information for this paper. The author is sincerely grateful to those officers who have been involved in the difficult works for damage restoration of river levees.

References

Begemann HKSP (1965) The friction jacket cone as an aid in determining the soil profile. In: Proceedings of the 6th ICSMFE, vol 1, pp 17–20

Kaneko M, Sasaki Y, Nishioka J, Nagase M, Mamiya K (1996) River dike failure in Japan by earthquake in 1993. In: Proceedings of the 3rd international conference on recent advances in geotechnical earthquake engineering and soil dynamics, pp 495–498

Matsuo O (1996) Damage to river dikes. Soils Found, Special issue 1 on geotechnical aspects of the January 17 1995 Hyogoken-Nambu earthquake, pp 235–240

Ministry of Land, Infrastructure, Transport and Tourism (2007) Guideline for inspection of seismic resistance of river structures, II, River levees (in Japanese)

Robertson PK (1990) Soil classification using the cone penetration test. Can Geotech J 27:151–158
Sasaki Y, Oshiki H, Nishikawa J (1994) Embankment failure caused by the Kushiro-Oki earthquake of January 15, 1993. Session publication during the 13th ICSMGE, Delhi, pp 61–68
Sasaki Y, Towhata I, Miyamoto K, Shirato M, Narita A, Sasaki T, Sako TS, Taguchi Y (2012) Prompt report on damage in and around river levees caused by the 2011 Great East Japan Earthquake (accepted by Soils and Foundations)
Sawada S, Towhata I (2011) Use of piezo drive cone for evaluation of subsoil settlement induced by seismic liquefaction. Bull ISSMGE 5(1):15–25
Towhata I, Orense RP, Toyota H (1999) Mathematical principles in prediction of lateral ground displacement induced by seismic liquefaction. Soils Found 39(2):1–19
Towhata I, Trinh Thi Lan Anh, Yamada S, Motamed R, Kobayashi Y (2010) Zero-gravity triaxial shear tests on mechanical properties of liquefied sand and performance assessment of mitigations against large ground deformation. In: Proceedings of the 5th international conference on recent advances in geotechnical earthquake engineering and soil dynamics, San Diego

Chapter 7
Earthquake Performance Design of Dams Using Destructiveness Potential Factor

G. Rodolfo Saragoni

Abstract The seismic performance design of dams is based in good estimates of sliding displacements and crest settlements. Theoretical result has shown a good correlation between sliding displacement of slope and destructiveness potential factor P_D. Chilean practice of seismic dam design considers performance design with three limit estates: operational, maximum credible earthquake and second maximum credible earthquake. Crest dam vertical settlement recorded from seismic performance in real Chile earthquakes for P_D between 50 and 60 $\times$ 10^{-4} [gs^3] confirm the good forecast of the theoretical values obtained in terms of P_D compare with PGA. In conclusion the use of P_D for performance design of dams is highly recommended.

7.1 Introduction

Chile large copper mining request many tall and large tailing dams as well as its agriculture important earth dam in a country with the largest seismicity of the world. Both conditions have forced the Chilean government to demand the seismic performance design of dams: SERNAGEOMIN (National Service of Geology and Mining), for tailing dams and Irrigation Department of Ministry of Public Works for agriculture earth dams.

In the Large Copper Mining production area, the existence of tailings dams has a capital importance since without their existence, the operation of copper mineral treatment plant would not be possible, due to the fact that environmental regulations and limitations do not allow mining companies to pour their debris into natural riverbeds.

G.R. Saragoni (✉)
Department of Civil Engineering, University of Chile, Santiago, Chile
e-mail: rsaragon@ing.uchile.cl

M. Maugeri and C. Soccodato (eds.), *Earthquake Geotechnical Engineering Design*, Geotechnical, Geological and Earthquake Engineering 28,
DOI 10.1007/978-3-319-03182-8_7,

The copper minerals that are currently being exploited in the country have an ore law of the order of 1 % and the country produces more than 5.5 million tons of fine copper every year. Therefore, the sterile material that tailing form, at present, gets to an annual volume that is higher than 550 million ton that have to be stored in secure places that guarantee they will not damage the environment. This is the role played by tailing reservoirs.

If we consider that the principal mining exploitations are located high in the mountains, this implies the need of constructing tailings deposits at a great altitude and dams of an important height, or transporting tailings to distant places with the purpose of finding sites that are topographically more adequate (Court 1996).

The seismic analysis of large dams of more than 40 m height is usually done considering performance design for two limit estates, corresponding to two types different of earthquakes: OBE "operational" (operational basic earthquake) and MCE "maximum credible earthquake", for which different levels of settlement of dams are prescribed as accepted seismic performance.

The estimation of these earthquakes is performed using probabilistic methods with 10 % probability of being exceeded in 50 years for the OBE and 2 % probability to be exceeded in 50 years for the MCE.

The probabilistic MCE is also compared with the maximum credible earthquakes obtained by deterministic methods for different seismic sources: subductive interplate thrust type earthquake, subductive intraplate intermediate depth earthquake, far subductive interplate thrust earthquake and crustal low depth earthquakes associated with surface active faults.

In the last decade a third limit estate has been introduced in Chilean practice especially for tailing dams, with abandon period by law of 200 years, considering the high seismicity of Central Chile which produces mega earthquakes $M = 8.5$ every 83 ± 5 years (Comte et al. 1986), two times in the dam abandon period, this corresponds to the "2nd MCE", "the second maximum credible earthquake".

The dam must stand for this "2nd MCE" with large deformation and settlement less than the dam freeboard.

The study of tension-deformation behavior and the dynamic response of dams are analyzed using finite difference or finite element methods with the corresponding artificial accelerograms for OBE, MCE and 2nd MCE limit estate condition, for 2D or 3D analysis depending of each case.

Since the seismic analysis of dams for large earthquakes considers plasticized and/or liquefied zones, the selection of the design artificial earthquakes must be done considering earthquake instrumental intensities that forecast the expected nonlinear behavior.

The Chilean dam design practice considers the destructiveness potential factor P_D defined by Araya and Saragoni (1984) to select the most damaging artificial accelerogram for each seismic scenario considered for each design estate limit (Saragoni 1996).

7.2 Destructiveness Potential Factor

An instrumental intensity based in the expected ductility of nonlinear simple elastoplastic one degree of freedom oscillators was defined by Araya and Saragoni (1984).

This method is based on non-linear accumulative deformations suffered by structures and earth works.

The destructives potential factor is defined considering a non-stationary process of acceleration and zero crossing intensity of this process:

$$P_D = \frac{\pi}{2g} \cdot \frac{\int_0^{t_0} a^2(t)dt}{\nu_0^2} \tag{7.1}$$

Where:

a(t) = ground acceleration.

t_0 = total duration of the accelerogram.

ν_0 = zero crossing intensity per second.

g = acceleration of gravity.

The capacity of the destructiveness potential factor P_D to order the recorded accelerograms according to the observed damage of large earthquakes was verified by Uang and Bertero (1989). They verified that P_D is the only index of the instrumental intensities: peak ground acceleration, Arias intensity, Housner intensity, root-mean square acceleration; that order the recorded accelerograms according to the real observed damage.

The correlation between P_D and modified Mercalli intensity MMI is given by Saragoni et al. (1989).

$$MMI = 4.86 + 1.35\log \cdot P_{DH} \tag{7.2}$$

With

$$P_{DH} = P_{DXX} + P_{DYY} \tag{7.3}$$

In which, PDXX and PDYY are the corresponding PD for two horizontal accelerograms measured in two orthogonal directions.

The correlation of Eq. (7.1) compared with observed Chile Maule 2010 and Valparaíso 1985 data is shown in Fig. 7.1.

Similar results were obtained by Decanini et al. (1993, 1995) considering the macroseismic scale MCS.

$$\log P_{DH} = 0.6IMCS - 0.005(IMCS)^2 - 3.00 \tag{7.4}$$

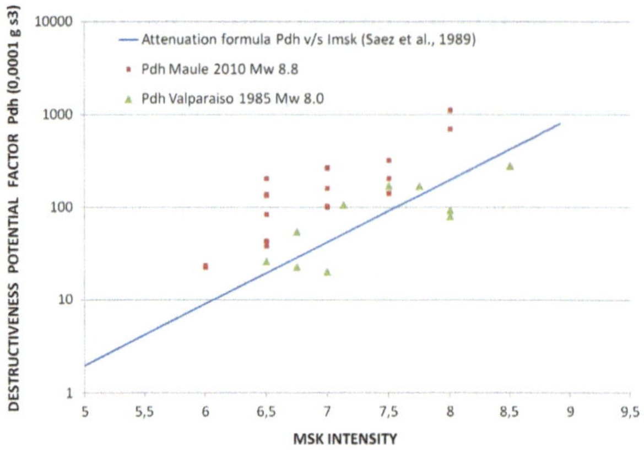

Fig. 7.1 Relation between destructiveness potential factor P_{DH} and MSK intensity scale including data of Chile Valparaiso 1985 and 2010 Maule earthquake (Saragoni and Ruiz 2012)

The destructiveness potential factor parameter can also be directly correlated with magnitude and hypocenter distance by the following formula (Schaad and Saragoni 1989):

$$P_D = \frac{3.02 \cdot 10^{-3} \cdot e^{2 \cdot M_s} e^{-0.001 \cdot R}}{(R+60)^{1 \cdot 39}} \left[10^{-4 \cdot} g s^3\right] \tag{7.5}$$

This formula is only valid for hard soils and it is not applicable to hard rock soils type. Since this formula was derived by Schaad and Saragoni (1989) only considering Central Chile 1985 subductive interplate thrust type earthquake, is not valid for subductive inslab intermediate depth type earthquake.

7.3 Destructiveness Potential Factor and Slope Displacements and Settlements

The most important requirement in geotechnical performance design is to evaluate the seismic displacement due to the different considered limit estates.

In the case of slopes the displacement can be estimated in terms of the PD (Crespellani et al. 1998), considering the displacement of a sliding Newmark block (Newmark 1965).

$$S_{50} = 0.011 P_D^{0 \cdot 977} K_c^{-1 \cdot 330} \tag{7.6}$$

Where
S_{50} = Expected displacement in cm.
P_D = Seismic destructiveness potential factor (10^{-4} gs^3).
K_c = Yielding acceleration factor.

Similar results has been obtained by Garini and Gazetas (2012) for sliding Newmark block for larger P_D values for near fault accelerograms with directivity and fling effects.

$$D\,[m] = 1.68 P_D\,[ms] + 0.33 \tag{7.7}$$

For Ac1 = 0.10 g and, where D is sliding displacement in m.

In both relations of Eqs. (7.6) and (7.7), the displacements are linear related with the destructiveness potential factor.

One important result of these researches is that displacements are not related with maximum ground acceleration PGA, commonly used for seismic design of dams.

Similar results were found by Vicents (1999) for soil seismic vertical settlements, in which the settlement is linear with P_D but in logarithmic scale.

Since all these reported results are theoretical, the scope of this paper is to show that this linearity with P_D is also observed for real dam subjected to large earthquakes.

7.4 Destructiveness Potential Factor Relation with αβγ Method for Nonstationary Characterization of Accelerograms

Considering the records of acceleration of an earthquake as samples of non-stationary stochastic process, it can be demonstrated that the expected quadratic function of accelerations of the process tends to a chisquare function of the type (Saragoni and Hart 1974; Saragoni 1977).

$$E\left\{a^2(t)\right\} = \beta \cdot e^{-\alpha \cdot t} \cdot t^{\gamma}, \tag{7.8}$$

Know as Saragoni and Hart curve, where:

$E\{\cdot\}$ = expected value.

$a(t)$ = ground acceleration.

t = time

β = intensity parameter.

Considering Eq. (7.8) the excepted destructiveness factor P_D can be expressed as

$$E\{P_D\} = \frac{\pi}{2g} \cdot \frac{\beta\Gamma(\gamma+1)}{\nu_0^2 \alpha^{\gamma+1}} \tag{7.9}$$

Where g is acceleration of gravity, ν_0 intensity of zero crossings and $\Gamma(\,)$ the gamma function.

Therefore when the parameters α, β, γ also ν_0 are estimated for each earthquake considered in the performance design, the E {P_D} can be estimated. In the case of deterministic earthquake scenarios, the most destructive earthquake for dam can be defined considering the accelerogram with the largest $P_{D.}$

7.5 Seismic Performance of Paloma Dam During Punitaqui Chile 1997 Earthquake

7.5.1 The Punitaqui Chile 1997 Earthquake

The Punitaqui earthquake was a subductive intraplate intermediate depth earthquake of $M_S = 6.7$ which struck the province of Coquimbo in land on October 14, 1997 with epicentre coordinates 30.93°S and 71.22°W and hipocenter depth 68 km. The earthquake epicentre was 32 km from Paloma dam. The earthquake had a maximum modified Mercalli intensity of VIII at Punitaqui causing 8 casualties.

The only accelerograms was obtained at Illapel with maximum horizontal acceleration 0.349 g and vertical 0.183 g (RENDIC 04/07 2004; Pardo et al. 2002).

Paloma Dam is located in the IV Region, about 20 km south of Ovalle city, on the Grande River, 410 m above the sea level; it was built with the purpose of irrigating and controlling flood on the Grande River by the Irrigation Directorate of Ministry of Publics Work of Chile.

The dam is of zoned earth type founded on old fluvial material and granite rock, with a maximum height of 96 m, a crest length of 1,000 m with storage capacity of 740 million m^3, with controlled spillway of 8 radial gates. The dam was completed in 1967. The dam was located in the modified Mercalli area of intensity VII and suffered seismic settlements which were investigated by Electricity du France (EDF) for the Ministry of Civil Works of Chile.

The analysis of the seismic performance of Paloma Dan is described in the next sections.

7.5.2 Earthquake Accelerograms

The only accelerogram obtained for this earthquake was at Illapel station which was characterized by using αβγ method. The obtained the values are indicated in Table 7.1.

For this accelerogram only a $P_D = 13.55 \cdot 10^{-4} [g \cdot s^3]$ was obtained.

Considering this result of the probabilistic characterization of the Illapel accelerogram it was estimated the probabilistic characteristic for the possible non recorded accelerogram at the foundation of Paloma dam, requested for the seismic performance of the dam.

This result is indicated in Table 7.2.

Table 7.1 Seismic αβγ parameters for Illapel 97 accelerogram

E $\{a_{max}(t)\}$ $[g]$	P_D $10^{-4} \cdot [g \cdot s^3]$	Δt_s $[s]$	t_1^* $[s]$	α $[1/s]$	β $10^{-2}[g^2 \cdot s^{-\gamma}]$	γ	ν_0	ν_m	P	Q
0.327	13.55	13.92	10.81	0.3668	0.00000497	6.5186	14.51	12.85	1.1672	0.0575

Table 7.2 Seismic αβγ parameters for possible accelerogram at Paloma dam for 1997 Punitaqui earthquake (no recorded)

E $\{a_{max}(t)\}$ $[g]$	P_D $10^{-4} \cdot [g \cdot s^3]$	Δt_s $[s]$	t_1^* $[s]$	α $[1/s]$	β $10^{-2}[g^2 \cdot s^{-\gamma}]$	γ	ν_0	ν_m	P	Q
0.661	43.59	14.00	10.68	0.3608	0.00002667	6.3792	16.22	14.04	1.3059	0.0542

Fig. 7.2 Vertical settlement of 68 cm of Paloma dam core for the Punitaqui, Chile 1997 inslab earthquake (Courtesy R. Verdugo)

The estimate accelerogram at the site of Paloma dam has PGA = 0.661 g and $P_D = 43.59\ 10^{-4}\ s^3$, for this accelerogram the dam suffered a vertical settlement of the core of 68 cm shown in Fig. 7.2.

In addition two maximum credible earthquakes were estimated for the subductive interplate thrust type earthquake and subductive intraplate intermediate

Table 7.3 Seismic parameters MCE and verification earthquakes

Design earthquake		Magnitude M_S	Focal depth $H[km]$	Epicenter distance $D[km]$	Hipocenter distance $R[km]$
A	MCE thrust	8.4	40	49	63
B	MCE interplate	7.8	70	0	70
C	Interplate Punitaqui	6.7	68	32	75

Table 7.4 Seismic αβγ parameters for MCE earthquake at Paloma Dam

MCE earthquake	$E\{a_{\max}(t)\}$ $[g]$	P_D $10^{-4} \cdot [g \cdot s^3]$	Δt_s $[g]$	t_1^* $[s]$	α $[1/s]$	β $10^{-2}[g^2 \cdot s^{-\gamma}]$	γ	ν_0	ν_m
Thrust	0.404	73.07	42.57	17	0.0845	0.00025336	3.2353	12.46	9.97
Intraplate	0.889	78.81	14.00	10.68	0.3608	0.00004822	6.3792	16.22	14.04

Table 7.5 Vertical crest Paloma dam settlements

Earthquake	Settlement (cm)	P_D (ms)
Punitaqui	68	5.0 10^{-2}
MCE thrust	83	7.6 10^{-2}
MCE interplate	109	8.9 10^{-2}

plate earthquake. The characteristics of these earthquakes are indicated in Table 7.3, including 1997 Punitaqui earthquake. In Table 7.4 the probabilistic seismic characterization of the accelerograms of both MCE are included.

In this case the maximum $P_D = 78.81 \cdot 10^{-4} gs^3$ is obtained for the MCE intraplate, with PGA = 0.889 g.

The $P_D = 73.07 \cdot 10^{-4} gs^3$ of the MCE interplate is similar to the MCE intraplate, but with the PGA = 0.404 g lower, almost half value.

7.5.3 Paloma Dam Settlement

The seismic analysis of the dam considering the artificial accelerograms for each design earthquakes produce the indicated settlement of the dam crest at the right profile indicated in Table 7.5.

The obtained settlements are almost proportional or logarithmic linear with the P_D as it is indicated in Table 7.4 and shown in Fig. 7.3.

Considering the relation between the maximum settlement of the core and the destructiveness potential factor it was possible to estimate the earthquake of rupture.

It was considered an operational limit estate with a settlement equal to one third of the freeboard equal 1.3 m for which $P_D = 100 \cdot 10^{-3} ms$ as it is shown with

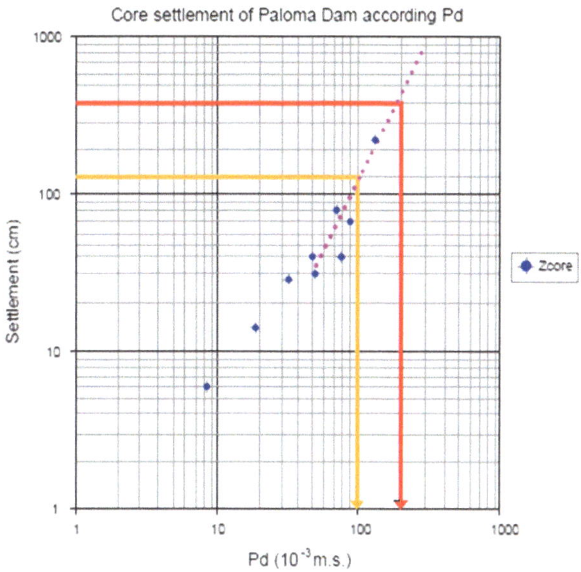

Fig. 7.3 Seismic vertical settlements of Paloma dam core for Punitaqui, Chile 1997 earthquake and operational (*yellow line*) and failure earthquake (*red line*) expressed in terms of a linear logarithmic relation of PD

yellow line in Fig. 7.3. For the rupture limit estate a vertical settlement equal to the total freeboard is adopted $P_D = 200 \cdot 10^{-3} ms$ which is double of recommended design P_D value for the MCE intraplate earthquake. This condition is show with red line in Fig. 7.3. Therefore Paloma dam is safe for seismic requirements.

The main conclusions of the analysis of the seismic performance of Paloma dam are:

1. The vertical settlement of the crest of the dam is logarithmic linear related with P_D values, similar to theoretical values of Vicents (1999).
2. The Paloma dam is seismic safe for an intraplate earthquake with PGA = 0.889 g showing that high PGA value in general does not control seismic performance of earth dam.

7.6 Seismic Performance of Convento Viejo Dam for Chile el Maule 2010 Earthquake

During the megaearthquake subductive interplate thrust type $M_W = 8.8$ El Maule Chile, 2010 earthquake the Convento Viejo dam was affected. Since this dam was instrumented with an accelerometer it makes possible to compare its seismic performance with the recorded destructiveness potential factor.

The Convento Viejo embankment dam is located above a stream of the same name. The earth dam is located 160 km south of Santiago nearby the Chimbarongo town. The height of this embankment is 32 m and corresponds to a zoned earth dam with an impervious clay core. The length of the crest of the dam is 500 m and its founded on fluvial material. The owner is the Irrigation Directorate of the Ministry of Public Works of Chile.

The accelerometer located at the foot of the dam, in alluvial soil recorded a maximum horizontal acceleration PGA = 0.38 g in the N-S direction and a maximum vertical acceleration of 0.27 g with a $P_{PH} = 63 \cdot 10^{-4} \cdot [g \cdot s^3]$.

The vertical settlement of the crest after the earthquake was lower than 35 cm which corresponds to 1 % of the total height, corresponding to a quite satisfactory behaviour of the earth dam. Superficial cracks were detected at the dam crest in the area where greater settlements were to be expected, according to the dynamic analysis carried out during the design stage (Campaña et al. 2010).

Observed vertical settlement for Convento Viejo dam for 2010 El Maule earthquake are in agreement with the one for Paloma dam for the 1997 Punitaqui earthquake for similar P_D values of order of $10 - 50 \cdot 10^{-4} \cdot [g \cdot s^3]$.

7.7 The Second Maximum Credible Earthquake Limit Estate

Chile is characterized by the largest seismicity in the world which produces strong earthquakes every 83 ± 9 years in the Central part of Chile, where it is located Santiago, the capital of Chile (Comte et al. 1986).

These earthquakes M = 8.5 have happened five times in the past in 1647, 1730, 1822, 1906 and 1985.

This short interval between large earthquakes magnitude 8.5 has conditioned the Chilean seismic design practice of building to achieve almost operational performance level, despite the fact that the Chilean Code declares a scope of life safe performance level.

Therefore Chilean practice assumes that for a given building at least one large magnitude earthquake will strike it in its life span.

This large seismicity of Chile leads to a deterministic strategy to assess seismic hazard for design of buildings despite the most used probabilistic approach considered in more low or diffuse seismicity regions of the world which considers return periods of 475 years (Lagos et al. 2012).

The probabilistic seismic hazard assessment technique (PSHA) assumes that seismic events are probabilistic independent; therefore this short interval of time between earthquakes is not well captured, considering only one possible maximum credible earthquake.

The second maximum credible earthquake that can happen during the abandon period of tailing dam or life period of earth dam for irrigation or power, must consider the dam already deformed by the first maximum credible earthquake but now for a rupture condition limit. This third performance design estate for the design of dams has been introduced in Chile in the last decade especially after 2010 El Maule earthquake.

7.8 Conclusions

The performance design of dams implies to estimate settlement and displacement for different seismic limit estates.

Seismic displacement and settlement seems to be better correlated with the destructiveness potential factor than peak ground acceleration.

Theoretical relations for displacements are linear correlated with the destructiveness potential factor and vertical settlements with logarithmic linear correlations with P_D.

Seismic performance of real dams for Chile earthquake confirm these results and shown that destructiveness potential factor looks to predict well the seismic performance of dams. These results must be confirmed with more information from performance of instrumented dams in future earthquakes.

Acknowledgements The author wish to thank the Irrigation Directorate of the Chilean of Ministry Public Works, for facilitating the reported information. This gratitude is also extended to Professor Jean-Jaques Fry of Ecole Centrale de Lyon, to EDF and Professor Ramon Verdugo of University of Chile.

References

Araya R, Saragoni GR (1984) Earthquake accelerogram destructiveness potential factor. In: Proceedings of the 8th world conference on earthquake engineering. San Francisco, USA

Campaña J, Valenzuela L, Bard E (2010) Seismic response of the Convento Viejo earth dam during the February 27th 2010 Chile earthquake. Obras y Proyectos ISSN 0718-2813 Universidad Católica Santisima Concepción. Departamento de Ingeniería Civil, Concepción, Chile, no 8, pp 37–46

Comte D, Eisenberg A, Lorca E, Pardo M, Ponce L, Saragoni GR, Singh S,K, Suárez G (1986) The 1985 central Chile earthquake: a repeat of previous great earthquakes in the región. Science 233:449–452

Court L (1996) Introduction of book Large Dams in Chile. Chilean National Committee of Large Dams (ed). Impresora Printer, S.A., Santiago, Chile

Crespellani T, Madiai C, Vannucchi G (1998) Earthquake destructiveness potential factor and slope stability. Geotechnique 48(3):411–419

Decanini LD, Gavarini C, Mollοioli e F (1993) Alcuna Considerazioni sol potenziale di danneggiamento del terremoto della Sicilia Orientale del 13 Dicembre 1990, 6° Convegno Nazionale: L'Ingegneria Sismica in Italia, Perugia, 3, 1091–1100

Decanini LD, Gavarini C, Molloioli e F (1995) Proposta di definizione della relazioni Tra intensite macrosismica e parametri del moto del suelo, 7° Convegno Nazionale. Ingenieria Sismica in Italia, Siena 1:63–72

Garini E, Gazetas G (2012) Destructiveness on earthquake ground motions: "Intensity Measures" versus sliding displacements. In: Proceedings of the 2nd international conference on performance-based design in earthquake geotechnical engineering, paper no 7.07, Taormina, Italia

Lagos R, Kupfer M, Lindenberg J, Bonelli P, Saragoni R, Guendelman T, Massone L, Boroschek R, Yañez F (2012) Seismic performance of high-rise concrete buildings in Chile. Int J High-Rise Build 1(3):181–194

Newmark NM (1965) Effects of earthquakes on dams and embankments. Geotechnique 15(2):139–159

Pardo MD, Comte T, Monfret R, Boroscheck M (2002) The October 15, 1997 Punitaqui earthquake (MW = 7.1): a destructive event within the subducting Nazca plate in Central Chile. Tectonophysics 335:199–210

RENDIC 04/07 (2004) Events recorded 1997 to 2000 (in Spanish) Red Nacional de Acelerógrafos Departamento de Ingeniería Civil. Universidad de Chile

Saragoni GR (1977) The αβγ Method for the characterization of earthquake accelerograms. In: Proceedings of the 6th world conference on earthquake engineering, New Delhi, India

Saragoni GR (1996) Seismic risk evaluation for the design of large dams in Chile. In: Chilean National Committee on Large Dams (ed) Large dams in Chile. Impresora Printer, S.A., Santiago, Chile

Saragoni GR, Hart GC (1974) Simulation of artificial earthquakes. Earthq Eng Struct Dyn J 2:249–268

Saragoni GR, Ruiz S (2012) Implications and new challenges for seismic design of 2010 earthquake accelerograms. In book "Mw = 8.8 Chile earthquake February 27, 2010", Civil Engineering Department, Faculty of Physic and Mathematical Sciences, University of Chile, Santiago, Chile

Saragoni GR, Holmberg A, Sáez A (1989) Destructiveness potential factor and destructiveness potential of the 1985 Chile Earthquake (in Spanish). In: Proceedings of the 5th Chilean congress of seismology and earthquake engineering, Viña del Mar Chile, vol I, pp 369–378

Schaad C, Saragoni GR (1989) Attenuation formulas considering the 1985 Chile earthquake (in Spanish). In: Proceedings of the 5th Chilean Congress on seismology and earthquake engineering, Viña del Mar, Chile, vol I, pp 379–388

Uang CM, Bertero VV (1989) Implications of recorded earthquake ground motions on seismic design of building structures. Report No UCB/EERC 88/13, Earthquake Engineering Research Centre, University of California, Berkeley, California, 100 pp

Vicents F (1999) Estimation of seismic soil settlements (in French). Ph.D. thesis Laboratoire de Tribologie et Dinamique des Systemes Ecole Centrale de Lyon, Lyon, France. No 99–10

Part IV
Foundations and Soil-Structure Interaction

Chapter 8
Seismic Response of Shallow Footings: A Promising Application for the Macro-element Approach

Claudio di Prisco and Michele Maugeri

Abstract In the last decades, the interest towards performance-based approaches in the field of seismic design and seismic adequacy assessment has rapidly grown, spreading an increasing awareness about the effects of the interaction between foundation and superstructure in particular under severe conditions. By the way, a lack of reliable methods for the seismic analysis of foundations is still apparent. To overcome this deficiency, the macro-element concept is in this paper suggested to be employed.

Although the macro-element approach is widely accepted to be very promising, it has not been supported so far by adequate experimental evidences, at least for seismic applications and few experimental results on the non-linear soil-foundation dynamic interaction are available in literature. Thus, in this paper, the macro-element theory in its different versions (elasto-perfectly plastic, elasto-strain-hardening plastic, bounding surface plastic and hypo-plastic) is first introduced and the mechanical response of shallow foundations under monotonic/cyclic loading, as it results from experimental tests, is outlined. All the critical issues concerning the employment of macro-element theory in Direct Displacement-Based approaches are then discussed and some application examples for solving practical problems are reported.

C. di Prisco
Dipartimento di Ingegneria Civile e Ambientale, Politecnico di Milano, Piazza Leonardo da Vinci 32, 20133 Milan, Italy

M. Maugeri (✉)
Dipartimento Ingegneria Civile e Ambientale, Università degli studi di Catania, Viale A. Doria 6, 95125 Catania, Italy
e-mail: mmaugeri@dica.unict.it

M. Maugeri and C. Soccodato (eds.), *Earthquake Geotechnical Engineering Design*, Geotechnical, Geological and Earthquake Engineering 28,
DOI 10.1007/978-3-319-03182-8_8,

8.1 Introduction

The interest towards performance-based approaches in the field of seismic design and seismic adequacy assessment, in the last decades, has rapidly grown, spreading an increasing awareness about the effects of the interaction between foundation and superstructure. However, while it is widely accepted that the role of foundations on the overall seismic capacity of structures (see e.g. ATC-40 1996; Martin and Lam 2000; Pecker 2006) cannot be neglected, on the other side a lack of reliable methods for the seismic analysis of foundations is still apparent. For this purpose, non-linear dynamic finite element (FE) simulations taking into account large numerical models, including the superstructure, the foundation and the surrounding soil, are likely not to be particularly suitable, because of their excessive computational costs, in particular when sophisticated constitutive laws are adopted for foundation soils. To overcome this shortcoming preserving a satisfactory description of the dynamic soil-structure interaction and accounting for the non-linearities of the soil mechanical behaviour, the macro-element concept is in this paper suggested to be employed. This approach is at present, at least in case of quasi-static conditions, a well-established theory (Nova and Montrasio 1991; Paolucci 1997; Cremer et al. 2001, 2002; Le Pape and Sieffert 2001) consisting in modelling the soil-foundation system as a unique non-linear macro-element characterised by a limited number of degrees of freedom (DOF). However, although the macro-element approach is widely accepted to be very promising, it has not been supported so far by adequate experimental evidences, at least for seismic applications and few experimental results on the non-linear soil-foundation dynamic interaction (Zeng and Steedman 1998; Maugeri et al. 2000; Negro et al. 2000; Faccioli et al. 2001; Gajan et al. 2005; PWRI 2005; Pitilakis et al. 2008; Abate et al. 2010; Massimino and Maugeri 2013) are unfortunately available in literature. Thus, in this paper, the macro-element theory in its different versions (elasto-perfectly plastic, elasto-strain-hardening plastic, bounding surface plastic and hypo-plastic) is first introduced and the mechanical response of shallow foundations under monotonic/cyclic loading, as it results from experimental tests, is outlined. All the critical issues concerning the employment of macro-element theory in Direct Displacement-Based approaches are then discussed and some application examples for solving practical problems are reported.

8.2 The Macro-element Theory

In general, the macro-element theory can be defined as an upscaling procedure conceived with the aim of solving soil-structure interaction problems by employing a sub-structuring strategy. As is well known, when a superstructure is analysed by assuming the constrains to be rigid, the previously mentioned sub-structuring

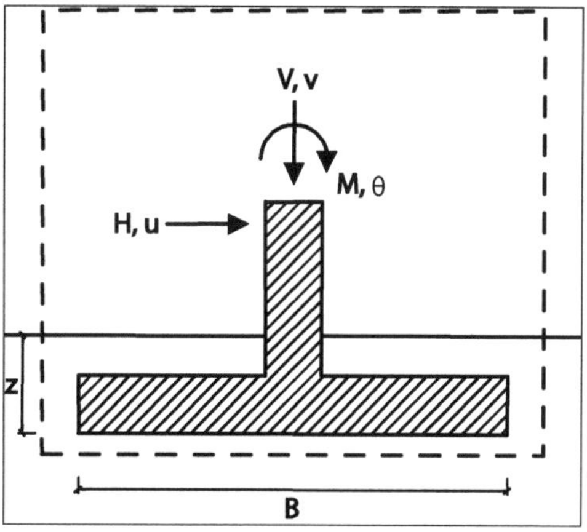

Fig. 8.1 Generalised stresses and strains for a shallow strip footing

procedure is so simplified to nullify the effect of deformability of one sub-structure (the foundation soil) on the other (the superstructure). In standard analyses, the influence of the foundation/soil system on the mechanical response of the superstructure is taken into account by introducing a certain number of linear springs, these summarising in a very simple manner the deformability of the foundation soil. In case the shallow foundation is further taken into account and inserted into the structural analysis, the problem becomes more complex, since the footing cannot be considered as a rigid block and the local interaction with the underneath soil has to be locally modelled.

In principle this theory can be employed even in case of extremely complex foundation systems (piles, grouted columns, etc.) without any severe restriction. Nevertheless, up to now, it has been fruitfully applied exclusively to very few cases like rigid shallow footings and/or pipelines.

In the practical applications developed up to now, the macro-element theory has been essentially employed to simulate the interaction of a rigid body with a deformable one, and, owing to the rigidity of the former, the interaction is described by introducing a suitable set of generalised stress and strain variables. Thanks to the infinite stiffness of the footing, the mechanical behaviour of the shallow foundation does not play any role and the complexity of the problem is abruptly reduced.

As is suggested in Fig. 8.1 with reference to shallow footings, the macro-element theory is thus based on the fundamental hypothesis, of considering as kinematics and statics unknowns displacements and actions, respectively, at the ground level, whereas footing and underneath soil are assembled and dealt with as a unique macro-system. The final objective is to define a suitable constitutive phenomenological relationship relating kinematics to statics generalized variables taking into consideration coupling and non-linearity.

The macro-element theory was initially conceived for rigid strip foundations on homogeneous dry sand strata under monotonic inclined/eccentric loads and subsequently it has been extended (i) to describe the response of rigid shallow footings under cyclic loading (di Prisco et al. 1998, 2003a, b; Cremer et al. 2001; Tamagnini and Salciarini 2009) and (ii) to deal with rectangular footings (Grange et al. 2008, 2009). At the macro- level, both local stresses σ'_{ij} and strains ε_{ii} can be disregarded, since the aim is to define a constitutive law where generalised loads Q and displacements **q** are directly related. As it is in all homogeneization theories, this relationship will be governed by the local mechanical response of the material as well as by the geometrical parameters (like, for instance, footing dimensions and relative foundation level depth) characterising the boundary value problem. The properties of the constitutive relationship, valid for the material at the micro/local level, are expected to be extended to the macro-level: as a consequence, the expected constitutive relationship will be non-linear and anolonomous, this implying (i) the history suffered by the system, as well as (ii) the sign of the mechanical perturbation to affect the system response. In the standard macro-element theory, for the sake of simplicity, the constitutive relationship between generalised stress $\dot{\mathbf{Q}}$ and generalised strain $\dot{\mathbf{q}}$ rate variables are thus assumed to be time-independent:

$$\dot{\mathbf{Q}} = D\left(\mathbf{Q}, \dot{\mathbf{Q}}, \boldsymbol{\psi}(\mathbf{q}^p)\right)\dot{\mathbf{q}} \tag{8.1}$$

where **D** stands for an incremental constitutive matrix depending on the current generalised stress vector **Q**, the generalised stress rate vector $\dot{\mathbf{Q}}$ and on a set of hidden variables $\boldsymbol{\psi}$, taking track of the previous history through the generalised plastic strain $\mathbf{q}^p$.

In case of shallow footings, under plane strain conditions, the mechanical interaction can be described in terms of three *generalised stresses* (the vertical load V, the horizontal load H and the overturning moment M), components of vector **Q**, and three *generalised strains* (the vertical displacement v, the horizontal displacement u and the rotation θ), components of vector **q** (Fig. 8.1).

If a homogeneous isotropic elastic soil is assumed, the stiffness matrix in Eq. (8.1) diagonalizes. In contrast, when irreversible strains progressively develop within the soil up to a local failure mechanism, the process concentrates and a sort of plastic hinge – as for metal beams– develops. Thus the stiffness matrix in Eq. (8.1) is intended to describe the response of the system during the entire process, from the very beginning up to the final collapse.

The upscaling procedure can be obviously developed by employing different approaches: (i) the experimental one (hereafter defined as phenomenological), (ii) the numerical (by solving boundary value problems by means of, for instance, either finite element or distinct element codes). In both cases the upscaling procedure, that is the macro-element theory, becomes useful for (a) interpreting either experimental/numerical results, (b) conceiving further tests/analyses (in this case it is used as a heuristic tool), (iii) designing/optimising the foundation itself.

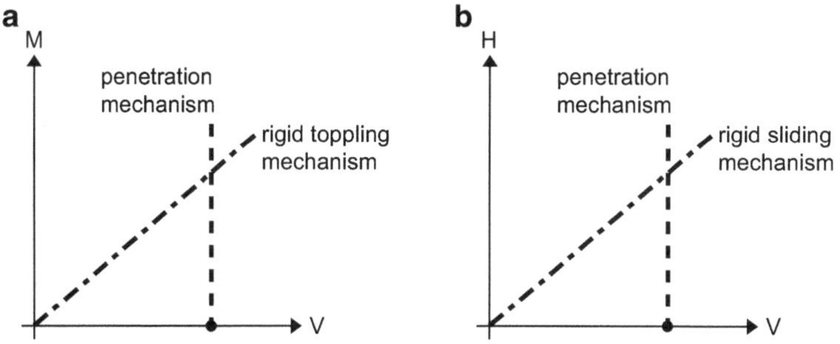

Fig. 8.2 Uncoupled interaction domains for a shallow rigid footing in the (**a**) $H-V$ and (**b**) $M-V$ planes

8.2.1 Phenomenological Evidence

In case of shallow footings, from quasi-static experimental test results the following hints can be inferred:

- under monotonic loading, the response of the footing is non-linear from the very beginning;
- the coupling between the different generalised stress/strain variables is evident from very low generalised stress levels and gets dominant at failure. For example, when a monotonic horizontal load is applied, vertical displacements develop even for a constant vertical load and a nil overturning moment;
- bearing capacity is severely affected both by the inclination and the eccentricity of the loads imposed; the so-call interaction domain describes this dependence. The interaction domain is a function of the nature of the foundation soil (the relative density severely influences its size and shape), as well as of the roughness, shape and embedment of the footing itself. As is suggested by Fig. 8.2, if the penetration mechanism was uncoupled with respect to either the sliding or the toppling mechanism, interaction domains in the two planes $H - V$ and $M - V$, respectively, would be uncoupled, too. Indeed, sliding would always concern uniquely the interface zone and toppling would consist solely in the result of the detachment between the footing and the underlying soil. Conversely, the compliance and the limited strength of the soil necessarily introduce the aforementioned coupling, so that interaction domains cannot be described by the four straight lines reported in Fig. 8.2;
- the geometry of failure mechanisms severely depends on the combination of generalised stresses: each point of the interaction domain corresponds with a unique failure mechanism;
- the experimental results for strip footings can be easily interpolated in a three-dimensional space by employing expressions, quite common in literature (Butterfield and Ticof 1979; Georgiadis and Butterfield 1988; Nova and

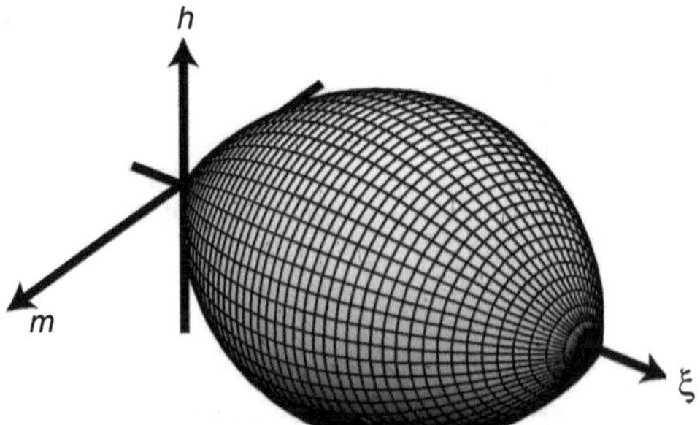

Fig. 8.3 Dimensionless interaction domain ($M = M/\psi B V_M$, $h = H/\mu V_M$, $\xi = V/V_M$)

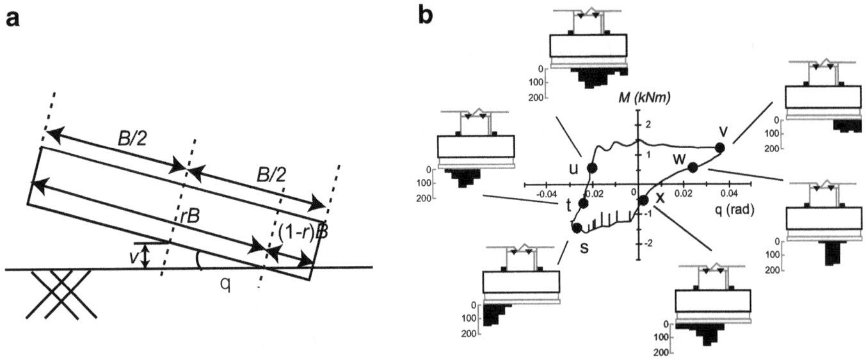

Fig. 8.4 Laboratory small-scale experimental results after Shirato et al. 2008: (**a**) uplift mechanisms, (**b**) vertical stress distribution during a cyclic test at constant vertical load and variable tilting angle

Montrasio 1991; Butterfield and Gottardi 1994; Montrasio and Nova 1997), such as that by Nova and Montrasio (1991):

$$F = \left(\frac{M}{\psi B}\right)^2 + \left(\frac{H}{\mu}\right)^2 - V^2\left(1 - \frac{V}{V_M}\right)^{2\beta} = 0 \tag{8.2}$$

- where ψ, μ, β and V_M are parameters describing shape and size of the failure locus. A geometrical representation in a suitable dimensionless fashion of expression (8.2) is given in Fig. 8.3;
- when overturning moments are applied (or better, under generalised strain controlled conditions, tilting angles are imposed), local measures testify a process of progressive concentration of vertical stresses under the footing. The uplift of the foundation dominates the response of the system and the footing progressively detaches from the underlying soil. In Fig. 8.4 the vertical stress distributions are reported as a function of the tilting angle θ imposed during a cyclic test performed on a dense sand stratum (PWRI 2005);

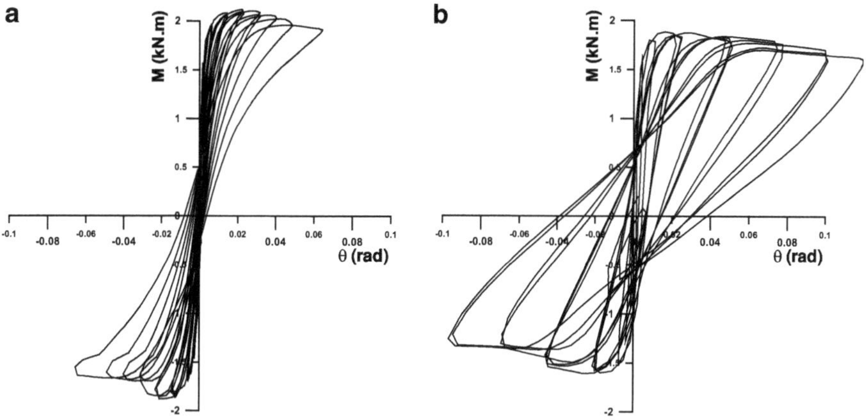

Fig. 8.5 Experimental data concerning a square shallow foundation cyclically tilted: (**a**) dense and (**b**) loose sand subgrade (After PWRI 2005)

- as is well known, standard general shear failures develop for stiff soils, while, as the soil stiffness reduces (loose sand strata), a punching mechanism is more likely to take place and the corresponding bearing capacity becomes hard to evaluate. A plateau can be envisaged solely if the foundation level is artificially maintained coincident with the ground level, as it can be artificially done in the laboratory;
- the shape of both failure mechanisms and the interaction domain is abruptly affected by the soil inhomogeneity. Moreover, the symmetry of the interaction domain with respect to the *V* axis is lost in the case of either inclined strata or anisotropic soils;
- under cyclic loading, the system dissipates energy and, in general, accumulates progressively irreversible generalized strains;
- if at each cycle the ultimate load is not attained, in case of both loose and dense sand strata, the mechanical response at each unloading phase is characterised by a monotonic decrease in the rotational stiffness;
- when the soil is sufficiently rigid (dense sands), during rotation-controlled tests with constant vertical load, the cycles in the $M - \theta$ plane assume for large θ the typical backbone shape in Fig. 8.5a and the uplift phenomenon is apparent. In the case of loose sands, the uplift is less pronounced and the reduction in the rotational stiffness during unloading seems to disappear (Fig. 8.5b). As is evident from Fig. 8.5, when the amplitude of cycles is significantly increased, the average stiffness progressively decreases, this suggesting a sort of progressive damage associated with a geometrical non linearity;
- if the experimental response of rigid footings to symmetric cycles is interpreted in terms of the well-known concepts of secant stiffness *K* and damping factor $\eta = D/4\pi\Delta W$ (where *D* is the energy dissipated in a cycle whereas ΔW the associated elastic stored energy), both a clear decay in *K* and an increase in η are evident at increasing values of either rocking or horizontal displacements. It is worth noting that even for relatively small foundation rocking, e.g. 1 mrad, the

reduction in foundation stiffness, depending on soil relative density, ranges from about 40 to 60 %. For rocking values up to 1 mrad, the equivalent damping ratio η ranges from 5 to 10 %, while it significantly increases for larger rocking angles, up to 20 % for dense sands and 30 % for medium dense sands. When the rocking angle is sufficiently large, the uplift dominates the response of the system and η stops evolving;
- the coupling between tilting/horizontal and vertical displacements is particularly severe for loose sands, but yet evident for dense sands as well;
- for a very large number of cycles, the accumulation rate of irreversible settlements progressively decreases: usually a sort of stabilization process takes place, at least in the absence of any damage phenomenon. When the generalized stress path is symmetric, only vertical displacements accumulate with the number of cycles, but when either the initial generalized stress state or the generalized stress path is asymmetric, the ratcheting phenomenon involves not only vertical settlements but also either horizontal displacements or rocking angles. The accumulation in generalized strains is observed to be essentially a function of (i) the cyclic stress path, (ii) the amplitude of the cycles and (iii) the image point in the generalized stress space which the loading path is imposed around and in particular of its position with respect to the failure locus.

8.2.2 Numerical Results

As was observed in the introduction here above, an alternative up scaling strategy consists in employing numerical codes and solving parametrically well-defined boundary value problems. For the sake of brevity, hereafter exclusively three exemplifying cases are mentioned: (i) the determination of the interaction domain by means of the limit analysis, the study of the interaction domain shape dependence on both (ii) the relative depth of the foundation plane by means of finite element numerical simulations and (iii) topographic conditions by means of discrete element numerical approaches.

Nevertheless, it is essential underling that when numerical analyses are performed to derive useful information at the macro-level, data have to be interpreted in the light of the hypotheses relative to the micro-level. For instance, if a finite element code is employed, hypotheses about the constitutive relationship describing the mechanical response of both interfaces and material as well as the choice of constitutive parameters can severely affect the response at the macro-scale. What is not sufficiently well appreciated is the role played in these cases by geometrical non-linearities on the system response, this often implying the necessity of performing particularly complex numerical analyses.

Since the late 1950s, the definition of both the $H - V$ and $M - V$ interaction domains has been a challenging issue for engineers and geomechanicians, so that a large number of solutions according to classical theories (e.g. Limit Equilibrium Method, Limit Analysis, Characteristic Line Method) has been proposed. Many of

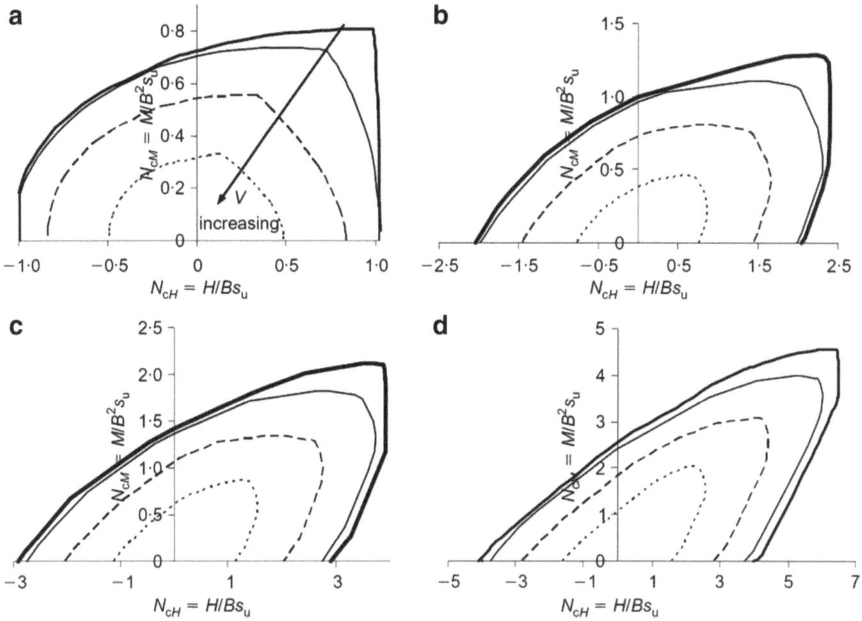

Fig. 8.6 Normalized failure $M–H$ envelopes for $\xi = 0$, 0.5, 0.75, 1: (**a**) $D/B = 0$ (surface foundations); (**b**) $D/B = 0.25$; (**c**) $D/B = 0.5$; (**d**) $D/B = 1$ (After Gourvenec 2008)

these earliest results are summarized in Vesic (1975), while it is worth citing the later contributions by Salençon and Pecker (1995a, b), who proposed improved Limit Analysis solutions for strip footings under inclined/eccentric loads on purely cohesive soils; further advances in the field of Limit Analysis approaches have been also recently developed by Randolph (2003). On the other side, in the last decade modern numerical techniques have been exploited to the same purposes. In particular, with special reference to Offshore Engineering applications, many authors (Bransby and Randolph 1998, 1999; Gourvenec and Randolph 2003; Bransby and Yun 2007; Gourvenec 2008) have been performing undrained FE numerical analyses to determine the interaction domain for shallow footings on clay strata under combined loading, by employing suitable both soil constitutive models and soil-footing contact laws.

Moreover, in the last decade, several authors have also been investigating how the $V - H - M$ failure locus can be affected by different mechanical/geometrical factors, such as the spatial inhomogeneity of soil properties (Gourvenec and Randolph 2003) or the embedment of the foundation (Bransby and Randolph 1999; Bransby and Yun 2007; Gourvenec 2008). For instance, an interesting – and less-studied – issue concerns the shape of the $M - H$ cross section, whose relevance, both for offshore and seismic applications, is self-evident. On this point, Gouvernec (2008) confirmed the intrinsically asymmetric shape of the $M - H$ envelope, highlighting its marked dependence both on the embedment ratio B/D and the normalized vertical load $\xi = V/V_M$ (Fig. 8.6).

Alternatively, the Discrete Element Method (Cundall and Strack 1979) has been also recognized as a suitable tool for studying soil-structure interaction problems (Calvetti et al. 2004). For instance, Gabrieli et al. (2009) numerically studied the influence of the local topography (sloping ground) on the vertical bearing capacity, focusing on the influence of the (a) footing-slope distance, (b) kinematical constraints (free versus fixed horizontal/rotational displacements) and (c) base roughness.

8.2.3 Elasto-plastic Macro-element Models

Here in the following, starting from the observations collected within the two previous sections, a series of elastoplastic constitutive models will be outlined and critically discussed. We will take into consideration first (i) perfect elastoplasticity, then (ii) isotropic and (iii) anisotropic strain hardening elastoplasticity and, eventually, (iv) generalized plasticity.

(i) Once the generalized stresses and strains are defined, the simplest way of describing the previous experimental evidence consists in setting up an homogenized constitutive relationship of elasto-perfectly plastic type. This requires (a) the evaluation of the stiffness coefficients (for this porpoise, the elastic formulas by Gazetas et al. (1985) and by Gazetas and Hatzikonstantinou (1988) can be employed), (b) the definition of a suitable failure locus $F(\mathbf{Q},\boldsymbol{\alpha}_F)$ and a plastic potential $G(\mathbf{Q},\boldsymbol{\alpha}_G)$, where vectors $\boldsymbol{\alpha}_F$ and $\boldsymbol{\alpha}_G$ are two sets of shape parameters. Within the failure/yield locus the mechanical behaviour is assumed to be elastic and uncoupled, while coupling exclusively is assumed to characterise the ultimate conditions. Either DEM or FEM codes can be employed to numerically evaluate the direction of the irreversible generalised strain vector at failure. In other words, this would imply a diagonal $\mathbf{D}$ matrix when either $F(\mathbf{Q},\boldsymbol{\alpha}_F) < 0$, or $F(\mathbf{Q},\boldsymbol{\alpha}_F) = 0$ and $\dot{F} < 0$, whereas matrix $\mathbf{D}$ becomes a full matrix exclusively if $F(\mathbf{Q},\boldsymbol{\alpha}_F) = 0$, $\dot{F} = 0$ and the plastic potential takes the coupling into account.

The plastic potential is also widely demonstrated not to coincide with the interaction domain: for instance, if an associated flow rule was accepted when sliding mechanisms are activated, that is when in the plane $M = 0$ the straight line $H/V = \mu$ is approached, a meaningless negative (i.e. upwards directed) unlimited vertical displacement would occur.

An example of elasto-perfectly plastic macro-element for rigid footings under seismic actions is given in Paolucci (1997); a practical application of Paolucci's model will be illustrated in a subsequent session of this paper.

(ii) A more sophisticated modelling approach consists in conceiving isotropic strain-hardening elasto-plastic macro-element models. This has to be done (a) by extending the well known Butterfield (1980) equation

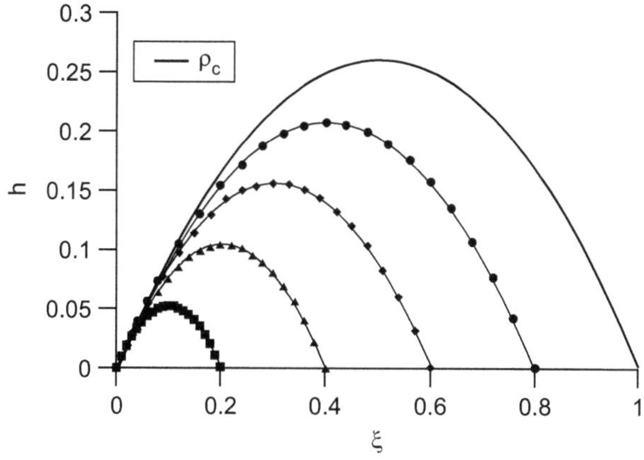

Fig. 8.7 Representation of the isotropic hardening in the dimensionless plane $h - \xi$

$$\frac{V}{V_M} = 1 - \exp\left(-\frac{R_0 V}{V_M}\right) \tag{8.3}$$

to multiple dimensional cases and (b) by conveniently defining both the loading function $f = f(Q, \ V_c)$ and the plastic potential g shapes to coincide with that of F and G (the failure locus is, therefore, the special yield locus for which a hidden variable V_c coincides with V_M (Nova and Montrasio 1991)). The development of plastic strains causes an increment in V_c according to a suitable hardening rule. For instance, Nova and Montrasio (1991) proposed the following for a strip shallow footing:

$$dV_c = \left(1 - \frac{V_c}{V_M}\right)\frac{R_0}{V_M}(|dv^p| + \alpha|du^p| + \gamma B|d\vartheta^p|) \tag{8.4}$$

where B stands for the footing width, R_0 is a constitutive parameter governing the stiffness of the system under pure vertical loads, whereas α and γ are constitutive parameters influencing the system response when either inclined or eccentric loads are applied. In Fig. 8.7, by defining $\rho_c = V_c/V_M$, such a hardening rule is illustrated in the $h - \xi$ plane.

As is testified by a very large number of data obtained by performing either FE numerical analyses and small scale experimental tests, this simple approach allows to satisfactorily reproduce the mechanical response of shallow footings under any monotonic loading path, at least in case the foundation soil is constituted of a homogeneous sand stratum.

Equation (8.4) implies an increase in size of the yield locus and, as a consequence, always a positive hardening. In contrast, some authors have also proposed more complex hardening rules to account for even a reduction in the yield function size and, consequently, to simulate a softening regime (Gottardi et al. 1999; Martin 1994; Cassidy et al. 2002; Nova et al. 2008). Furthermore, very recently Hodder and

Cassidy (2010) with regard to normally consolidated clays, whereas di Prisco and Vecchiotti (2010) for loose sands, modified the aforementioned hardening rules to reproduce the mechanical interaction between rigid bodies and largely deformable soils, that is to model large displacements as well.

In case the foundation level does not perfectly coincide with the ground level, the definition of the general failure locus $F(V,H,M) = 0$ is a non-trivial engineering matter. In di Prisco and Pisano (2011), for instance, the following modified expression for a non-negligible foundation embedment is suggested:

$$f = |h|^{\delta_1} + |m|^{\delta_2} - (\xi - \rho_c \xi_c)^{2\gamma} \left(1 - \frac{\xi}{\rho_c} + \xi_c\right)^{2\beta} \tag{8.5}$$

$$Q = [\xi \quad h \quad m]^T = \frac{1}{V_{M_1} - V_{M_2}} \left[V \quad \frac{H}{\mu} \quad \frac{M}{\psi B}\right]^T \xi_c = \frac{V_{M_1}}{V_{M_1} - V_{M_2}} \tag{8.6}$$

where $\mu, \psi, \delta_1, \delta_2, \beta, \gamma$ are six constitutive parameters, whereas, the bearing capacity under pure compression (V_{M1}), tension (V_{M2}), shear (H_0) and moment (M_0) are to be evaluated.

While expression (8.5) always guarantees that $F(\xi, h = m = 0) = 0 \Leftrightarrow V = V_{M_1}$ or $V = V_{M_2}; \mu, \psi$ and a unique relationship $\gamma(\beta,\xi_c)$ are set to ensure the $V - H$ and $V - M$ cross sections to attain the maximum at $\xi = 0$ (Bransby and Randolph 1999; Gourvenec 2008). Finally, the free parameters δ_1, δ_2 and β can be set to match the shape of $V - H$, $V - M$ and $M - H$ failure loci, as is derived through standard formulas for inclined/eccentric loads (Bowles 1996) and/or from FE analyses (swipe tests).

It is worth noting that the analytical expression (8.5) fails in reproducing the asymmetry of the $M - H$cross section discussed above, but a modified expression capable of accounting for such an asymmetry has been recently suggested by two of the authors (di Prisco et al. 2012).

(iii) An alternative approach is to employ anisotropic strain-hardening. This was the choice of Cremer et al. (2001), that, by following the theoretical approach conceived for cohesive strata by Salençon and Pecker (1995a, b), suggested two different expressions for the failure locus F and for the yield locus f. The complex analytical definitions introduced by the authors provide a convenient formulation for the anisotropic hardening as is schematized in Fig. 8.8, both in the $H' - V'$ and $M' - V'$ planes (where V', H' and M' are suitable non-dimensional variables similar to those introduced by Nova). An interesting application of this class of models has been recently published by Grange et al. (2007a, b, 2008, 2009). The most interesting peculiarity of such an approach consists in uncoupling (Fig. 8.8) the hardening induced by the first loading in one direction (for instance horizontal load rightward directed, $H > 0$) from that in the opposite direction ($H < 0$). This is really essential for capturing correctly the response of the system under the first cycle. Anisotropic hardening can be employed to capture the mechanical response of the system under cyclic loading, but in this case the effect of previous loading cannot be reproduced, since the size of the yield function has to be abruptly reduced.

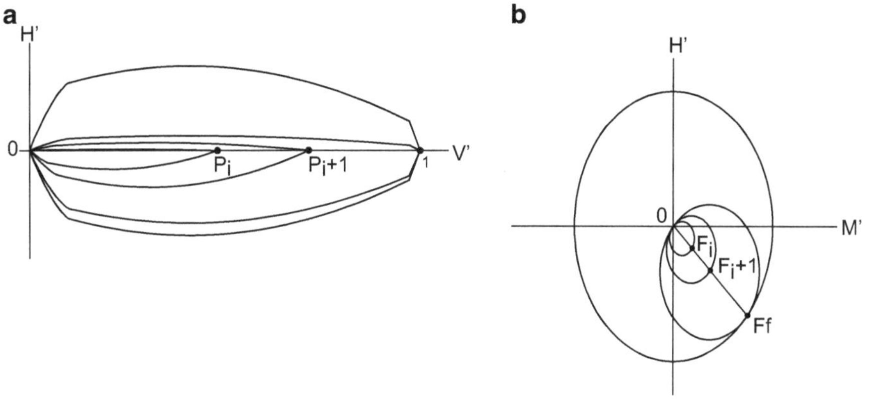

Fig. 8.8 Failure locus and evolving yield function according to the approach of Salençon and Pecker (1995a, b) (**a**) in the H'–V' and (**b**) in the H'–M' plane

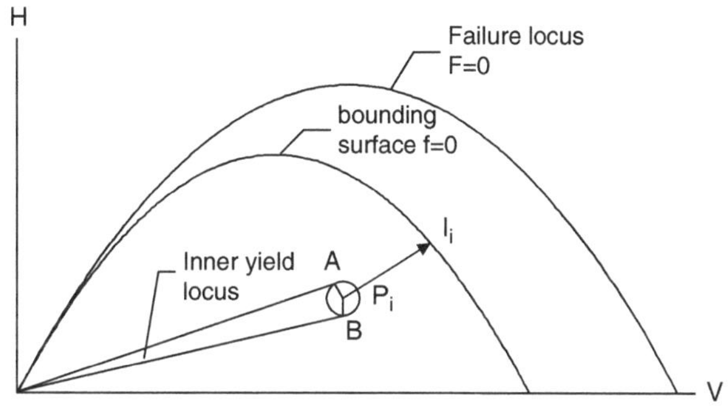

Fig. 8.9 Sub-loading function and bounding surface in the model by di Prisco et al. (2003b)

(iv) A way for overcoming the shortcomings of the previously defined class of models, as is done for soil REV constitutive relationships, is to conceive either generalized plastic or hypo-plastic constitutive models.

For instance, the above macro element model by Nova and Montrasio (1991) was modified by di Prisco et al. (1998), by introducing within the yield locus a sub-loading surface (Fig. 8.9), such that any intersection with the outer "bounding surface" is avoided. A convenient mapping rule allows to relate any point within the yield locus to an appropriate point on it. The plastic multiplier is evaluated for the bounding surface and suitably scaled according to the distance between the current point P_i and the corresponding image point belonging to the bounding surface I_i (Fig. 8.9). When such distance reduces to zero, the scaling function converges towards unity. The inner locus is therefore exclusively employed to define the elastic domain and to determine the image point I_i. In this way, both the occurrence of permanent generalized strains – even when the stress point is within what is usually considered a purely elastic region – and the accumulation of plastic distortions during cyclic or transient loading can therefore be simulated (di Prisco et al. 2003a, b). The model becomes inevitably more complex and a larger number of constitutive parameters become necessary.

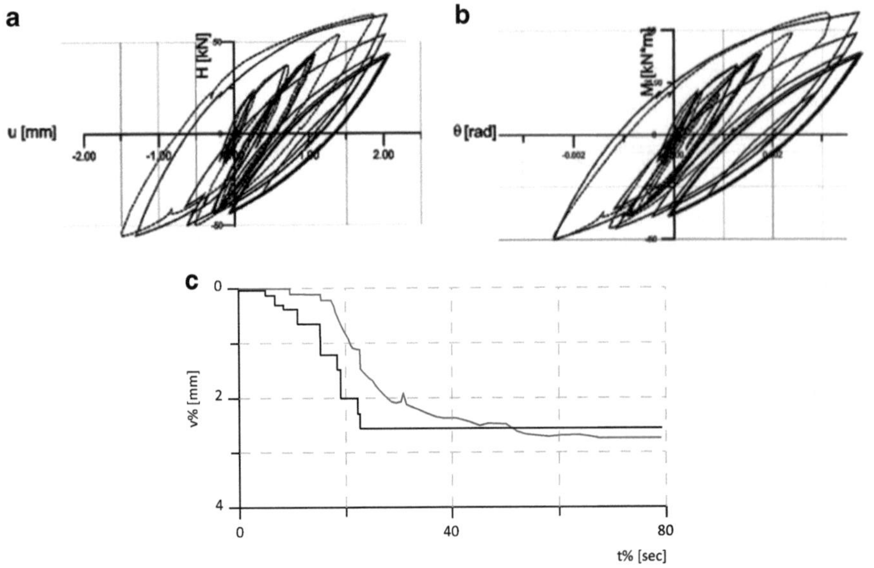

Fig. 8.10 Comparison of measured (*dotted lines*) and calculated (*solid lines*) displacements of a real scale foundation after Pedretti (1998); (**a**) horizontal load versus horizontal displacements; (**b**) overturning moment versus rotation; (**c**) accumulated vertical settlement versus time

A validation example of the model concerning the behaviour of a plinth, 1 m wide, founded on a dense sand stratum and subjected both to a cyclic horizontal load and overturning moment (experimental data after Pedretti 1998) is given in Fig. 8.10. Loads are applied at low frequencies, so that dynamic effects can be neglected; both the overturning moment and the horizontal force are varied at a constant ratio. It is evident that this constitutive approach is suitable for capturing in a satisfactory way the essential features of the experimental evidence and this conclusion regards both dense and loose sand strata (at least in the case of symmetric cycles applied on symmetric systems).

As was already mentioned, this class of models can be also employed to derive interesting heuristic information that obviously has successively to be experimentally confirmed. For instance, in case a loading path characterized by an ordered succession of different series of cycles, increasing in size, at $dV = 0$ and $d(M/H) = 0$ (Fig. 8.11) is considered, the aforementioned constitutive model (di Prisco et al. 1998) can be easily employed to justify the large scatter in the experimental data for instance reported by Paolucci et al. (2010). In Fig. 8.12, the different K/K_0 numerical curves concern two different relative densities and different vertical loads V. It is evident that at decreasing values of V/V_{MAX} the decay in K/K_0 becomes more rapid, while the increase in damping is less rapid. Unfortunately, bounding surface elastoplastic models fail in reproducing three very important aspects of the cyclic response of rigid shallow foundations: (i) the large settlements induced by the first unloading, (ii) the reduction in stiffness during the

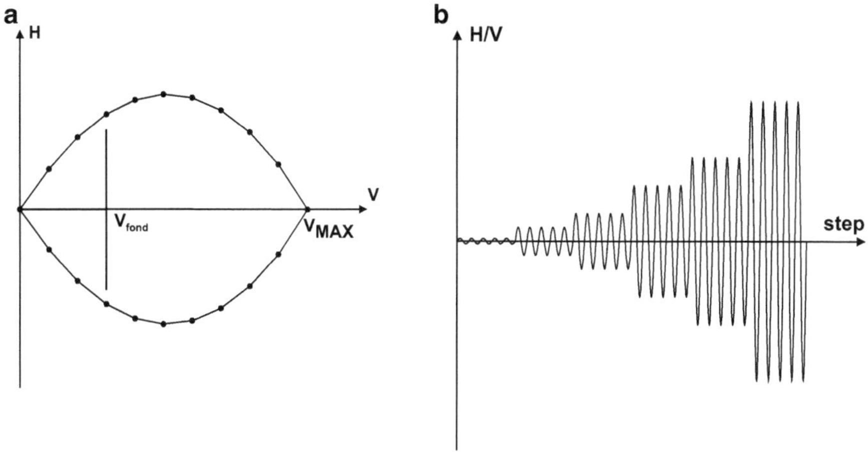

Fig. 8.11 (**a**) Constant load path and (**b**) loading history with symmetric cycles of increasing amplitude

first phase of the unloading when the footing is previously largely tilted, (iii) the ratcheting when asymmetric loading paths are imposed.

The first issue has been already commented above and is not quantitatively very important. In contrast the second aforementioned aspect, concerning the unloading rotational stiffness after getting large tilting angles, is essential and can be clarified by commenting the experimental results obtained on large scale models at Public Works Research Institute (PWRI 2005; Shirato et al. 2008) and already shown in Fig. 8.5. These experimental data have been obtained by imposing to a prototype rigid structure, placed on a caisson filled with sand, a horizontal cyclically varying displacement. During the cyclic phase, the footing is therefore loaded by a constant vertical load due to the steel frame weight and by cyclically varying overturning moment and horizontal force. As the test is performed under displacement-controlled conditions, even the reduction in generalized loads can be observed. As was already commented, when the soil density is sufficiently high, during the unloading a typical S-shaped trend is observed. This mechanical response during unloading in the case of dense sands is due to the uplift of the foundation: the reduction in the contact surface between the footing and the soil due to the detachment between the two materials generates a sort of damage of the system that could be described coherently for instance by introducing an elasto-plastic coupling. Recent efforts along this direction have been spent by Cremer et al. (2002), Shirato et al. (2008), Paolucci et al. (2008), and Grange et al. (2008, 2009). In particular, Paolucci et al. (2008) suggested a sort of pseudo-empirical damage rule, whereas Chatzigogos et al. (2011) a non-linear damaging elasticity accounting for the detachment.

Finally, as far as the third issue listed above is concerned, it is worth noting that both standard bounding-surface-plasticity and anisotropic strain hardening

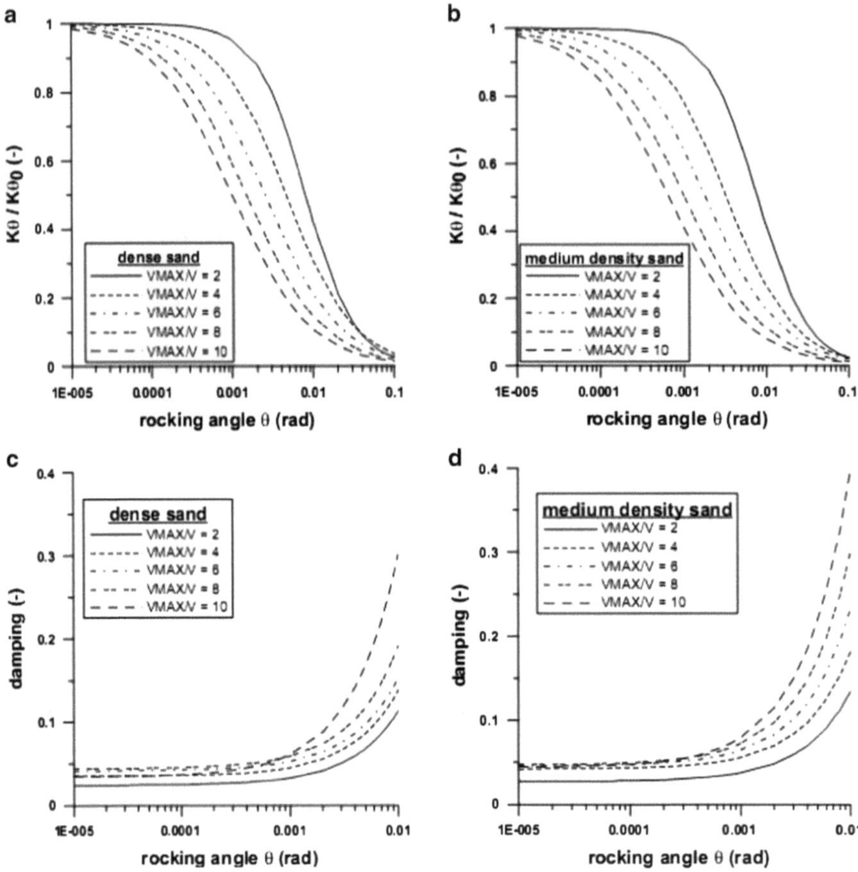

Fig. 8.12 Influence of the V/V_{MAX} ratio on the dependence of (**a**, **b**) the secant rotational stiffness K_θ and of (**c**, **d**) the damping factor for dense and medium dense sand on the rocking angle, respectively

constitutive models severely overestimate the phenomenon of ratcheting, when asymmetric loading paths are imposed. In the case of elastic isotropic strain-hardening, unloading/reloading cycles do not cause any accumulation in irreversible displacements, whereas the standard bounding surface approaches cannot generate any loop, so that ratcheting is too marked. To reduce the accumulation in irreversible generalized strains and to fit, even quantitatively, the experimental results, the previous di Prisco bounding surface model was further modified by the author by introducing a sort of artificial memory surface rotating in the $\xi - h - m$ space and "re-modulating" the $I_i - P_i$ distance.

8.3 Theoretical Issues Concerning the Use of the Macro-element Theory to Dynamic/Seismic Conditions

In the previous section, the use of the macro-element approach has been suggested with reference to a large variety of generalised stress paths, including virgin, complex and cyclic loading. What has been disregarded is the role played by inertial forces, inevitably developing in particular within the soil domain when perturbations are rapidly applied. In this case, just for the presence of inertial terms, the system mechanical response becomes severely dependent on the duration/frequency of the perturbation itself.

For the sake of clarity, here in the following two distinct conditions will be separately discussed: (i) the case of dynamic loading applied directly to the super-structure (like, for instance, wind actions) here below mentioned as dynamic actions; (ii) the case of seismic loading coming from the subsoil and coming back to the soil throughout the footing, after a transformation of the signal due to the mass of the super-structure, mentioned here below as seismic actions.

8.3.1 Dynamic Actions

Dynamic actions can be newly subdivided into impulsive loading and high frequency cyclic loading. As far as impulsive loading is concerned, in order to numerically simulate the impact of rigid boulders on granular materials (Fig. 8.13), an extension of the macro-element theory to visco-plasticity according to Perzyna's approach (Perzyna 1963) was suggested by di Prisco and Vecchiotti (2006). In this case, the dynamic interaction problem is approached by accounting for both the boulder mass and the inertial actions arising within the deformable sand stratum. This is obtained by means of: (i) a suitable definition for the viscous nucleus, (ii) the introduction, according to a radial mapping, of a sort of overstresses. As is suggested in the

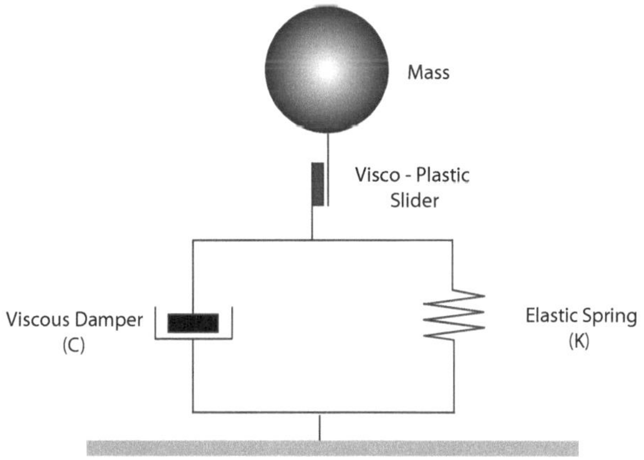

Fig. 8.13 Schematization of the elasto-viscoplastic rheological model introduced by di Prisco et al. (2006) for simulating the impact of rigid boulders on granular infinite strata

aforementioned figure, in the near field that is in the soil domain close to the footing, energy is prevalently dissipated for the developing of irreversibilities, largely inhibited by inertial effects, in contrast, in the far field, dissipation is prevalently due to the propagation of waves within the infinite domain.

According to the delayed plasticity theory, the generalized stress state can get external to both the yield function and the interaction domain. This implies that, during the evolution of time, the load applied by the boulder to the soil can be much larger than that can be statically reached. The comparison between numerical and experimental results has largely demonstrated the validity of the constitutive modeling hypothesis.

More complex is the interpretation of the mechanical response of shallow footings under high frequent cyclic loading by taking into consideration, as is suggested by the macro-element approach, the irreversibility and the coupling of the constitutive relationship. In fact in standard approaches, as is suggested by Gazetas et al. (1985) and Wolf (1985), either visco-plastic or elasto-histeretic constraints are inserted, but irreversibilities are disregarded. This is undoubtedly a challenging and unexplored domain. Nevertheless, with regard to this item what it follows can be observed:

- The parameter mainly affecting the response of the hypothetical macro-element mechanical response is the B/λ ratio, where B, as was already mentioned, in the case of strip footings is the footing width, whereas λ is the wave length characterizing the mechanical perturbation, defined as $v_{s/p}/f_r$ where f_r is the frequency of the applied perturbation, whereas $v_{s/p}$ are the shear and pressure velocities, respectively, of the propagating wave, mainly depending on the elastic deformability of the material.
- When the B/λ ratio is sufficiently small, that is when B is small, the soil is soft and the perturbation frequency is high, the soil under the footing will experience a synchronous motion. In these cases, as it was suggested in case of impulsive loading, inertial effects inhibit the development of irreversible strains and a delayed plastic constitutive strategy can be chosen again.
- In contrast when the B/λ ratio is large (that is when B is large, the soil is rigid and the perturbation frequency is low), the soil in the near field will experience an asynchronous motion. In this case, inertial effects can hardly affect the development of a global failure mechanism, although, however, the value of dissipated energy can be very large and local/not organized irreversible strains can develop in particular in the near field. This should suggest an increase in the amplitudes of cycles, that is in the value of damping defined previously.

8.3.2 Seismic Actions

In case of seismic conditions, the soil-structure interaction becomes more complex. In fact, the seismic input comes from the subsoil and during a first phase, the input

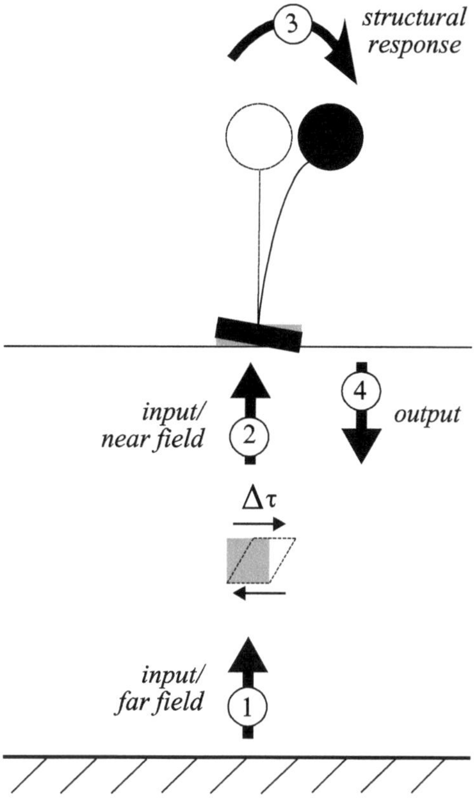

Fig. 8.14 Sketch of the seismic soil-structure interaction

motion (arrow 1 in Fig. 8.14), coming from the bedrock, propagates throughout the soil stratum. Its propagation is mainly influenced by the stratigraphic profile and the local topography. Factors of site amplifications are usually employed in national and European standards to account these effects for. Sometimes, when seismic waves cross saturated loose sand strata, the phenomenon of soil liquefaction can locally take place, with dramatic consequences for buildings and lifelines. In some other cases, seismic actions may cause catastrophic landslides, with even more disastrous effects on civil structures. In both cases, the triggering mechanisms of such events are assumed to be totally independent both of the structure and its foundation. In contrast, the inception both of soil liquefaction and slope failures is essentially driven by the local initial state of stress, that is in turn dominated by the presence of the superstructure. In this case, the soil-structure dynamic interaction markedly transforms the input motion (arrow 2 in Fig. 8.14).

If the input signal duration is sufficiently short and the natural period of the superstructure sufficiently large, the input (signal 2) and the output (signal 4) may be treated separately and any interference among them disregarded.

Under this hypothesis, by focusing on signal 2, we can state that system response is dominated again by the B/λ ratio, where obviously, in this case, the wave length

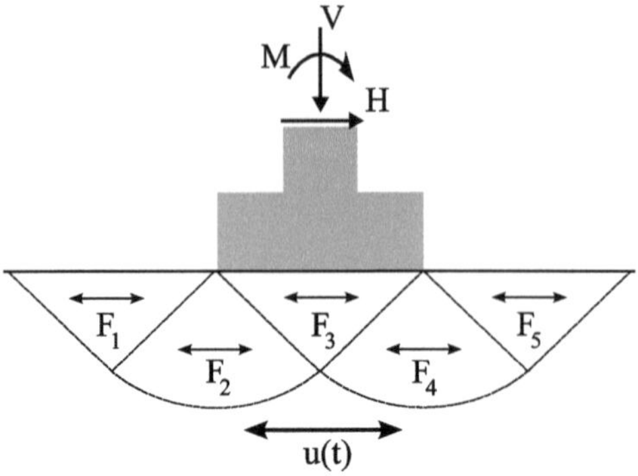

Fig. 8.15 Schematic representation of pseudo-static seismic forces

depends exclusively on the input motion. Here again, when B/λ is sufficiently small, the response of the soil in the near field is synchronous. In contrast with what inferred in case of dynamic loading coming from the structure, the type of wave involving the near field soil is practically (at least in the majority of the seismic cases) well defined: it can be simply schematized as a shear wave practically horizontally upward travelling. Under this simplifying hypothesis, we can easily imagine that, as was suggested by many authors in the past (Chen and Liu 1991; Paolucci 1997), the interaction domain is governed by inertial forces, as is schematically suggested in Fig. 8.15. In particular, both the shape and size (simply summarised this latter in the previous section by means of parameter V_M) of the interaction domain (Fig. 8.3) will change with time according to the activation of horizontal inertial forces F_i. In this case, the time dependence of the system mechanical response will thus be governed by the evolution with time of the interaction domain, while the stresses transmitted by the superstructure due to permanent/static loads are assumed to be constant.

Even in this case, these aspects have not yet accounted for by constitutive approaches based on the macro-element concept and uniquely when Ultimate Limit States are considered, this aspect is taken into consideration by introducing pseudo-static inertial forces: a further effort is then required even to consider this aspect.

More frequently, the incoming dynamic signal is assumed to get the foundation without any marked change neither in amplitude nor in frequency, so that the signal transformation is assumed to exclusively take place at the superstructure level; there, owing to the mass and the stiffness of the superstructure, the input motion is indeed modified in phase, amplitude and frequency. The superstructure will subsequently transmit to the footing, and this to the subsoil, a transformed dynamic signal (arrow 4 in Fig. 8.1), giving rise to the so-called inertial interaction. If the input signal has a "sufficient" duration and the natural period of the superstructure is not too large, the input (signal 2) and the output (signal 4) experience a constructive interference, this making the analysis and the interpretation of the interaction phenomenon further complex.

8.3.3 Examples of Application

8.3.3.1 Full-Coupled Approach

The most direct use of the macro-element theory consists in substituting either rigid constraints or standard elastic springs to the reference frame with a coupled/irreversible constitutive relationship. This can be done under seismic conditions by subdividing the whole spatial domain into three sub-domains: (i) the far field, (ii) the near field and (iii) the superstructure. The far field is the part of soil domain unaffected by the presence of the superstructure, i.e. where the displacement field can be assumed to be known. In this zone the seismic wave propagation is analysed and local/topographic factors have to be taken into account. Conversely, in the near field irreversible mechanisms due to the soil-structure interaction become dominant. From this perspective, the identification of the zone where significant plastic/irreversible strains develop is essential: for instance, in the case of shallow strip footings, the size of the domain is governed both by the foundation width and its embedment. The definition of the boundaries of this domain is somewhat ambiguous and its geometry may evolve with time.

The numerical solution can be thus judged to be acceptable, at least under the hypothesis of neglecting (i) the influence of soil masses in the near field and (ii) the interference between the input signal (arrow 2) and the output signal (arrow 4 of Fig. 8.14). For numerical analyses relative to the superstructure, the input signal will consist in the imposed displacement history assigned to the far field (and not affected by the structural dynamic response) and the output data will consist in the displacement time-history of both the superstructure and the foundation. The unique drawback of such an approach concerns the numerical solution of the system that, owing to the irreversibilities taking place at the macro-element level, will have to be found in the time domain and by employing non-linear integration techniques suitable for dealing with elasto-plastic constitutive relationships. The first simplified attempt conceived to solve the problem according to this strategy was done by Paolucci (1997), who considered an elasto-perfectly plastic constitutive relationship and applied the method to one degree of freedom superstructure. A very similar strategy, but developed by employing an anisotropic strain hardening constitutive relationship, was more recently chosen by Grange et al. (2008). An application for the reconstruction of Noto Cathedral (Italy), partially collapsed in 1996 as a consequence of a dangerous increase in stresses started during the 1990 Sicilian earthquake, was made by di Prisco et al. (2006) by employing the generalised plastic constitutive model previously described (Fig. 8.9).

An attempt of validating this approach by employing small scale experimental test results was recently done again by Paolucci et al. (2007, 2008). These authors compared the numerical simulations with the results of a series of tests performed at the PWRI (Japanese Public Work Research Institute of Tsukuba) by means of a laminar box positioned on a large scale shaking table equipment (Fig. 8.16). The shaking table was excited by means of four different seismic inputs.

Fig. 8.16 The experimental shaking table setup, including the footing prototype. Shaking table equipment at Public Works Research Institute, Tsukuba (Japan)

Numerical simulations seem to fit quite well the experimental data, when considering very large accelerations, while the accumulation in vertical settlements, prevalently accumulating during the entire seismic excitation, cannot be reproduced by the numerical model, since the constitutive model implemented is elasto-perfectly plastic.

The macro-element approach is also at present employed to reproduce the physical and numerical results of Massimino and Maugeri (2013). This application will lead to the validation of the macro element approach on the basis of extensively monitored shaking table test data and FEM numerical results.

8.3.3.2 Push-Over Method

An extremely simplified alternative to the complete approach, proposed here above, is the use of the push–over method. In this case, displacements/settlements are assessed by applying to the superstructure, under quasi-static conditions, loads equivalent to the inertial forces activated by the earthquake. This approach can be fruitfully employed, by taking the non-linear response of the footing into account, according to the macro-element theory as it follows:

- Evaluating the equivalent actions to be applied to the superstructure by disregarding the deformability of the macro-element;
- Calculating, by employing the macro-element approach, under known actions, rotations, settlements of the footing;

- Deriving, by employing again the macro-element approach, equivalent stiffness parameters for the footing and re-evaluating, point (1), the equivalent actions (remember that according to the push-over approach, the applied pseudo-static actions to be applied to the super-structure depend on the natural period of the superstructure, this again affected by the soil-structure interaction).

As is suggested in di Prisco et al. (2012) with reference to historical bell towers, when geometrical non-linearities are taken into account, the problem becomes even more complicated, since the change in the structural deformability can affect the inception of the superstructure instability.

8.3.3.3 DDBD Method

A further employment of the macro-element approach has been suggested by one of the authors in Paolucci et al. (2010). In this paper a visco-elastic adaptation of the macro-element approach has been proposed. In particular, curves in Fig. 8.12, describing the dependence of both equivalent secant stiffness modulus and damping parameter on the foundation rotation, have been first interpolated according to the following expressions:

$$\frac{K_f}{K_{f0}} = \frac{1}{1 + a\theta^m}; \quad \xi_f = \xi_{f,\min} + \left(\xi_{f,\max} - \xi_{f,\min}\right)[1 - \exp(-b\theta)] \tag{8.7}$$

where $\xi_{f,\min}$, $\xi_{f,\max}$, a, m and α are non-dimensional constitutive parameters dependent on the sand relative density and on the static safety factor V_{MAX}/V. To take the effect of uplift into account, the expression describing the dependence of ξ_f on θ was artificially modified with respect to that numerically provided by the constitutive model by introducing a saturation value for ξ_f. The saturation value for ξ_f, equal to $\xi_{f,\max}$, was experimentally evaluated from laboratory test results: equal to 0.25 for dense sands and to 0.37 for medium dense sands.

Once an analytical expression for K_f and ξ_f is provided, the displacement based method presented by Priestley et al. (2007), named as DDBD (Direct Displacement Based Design) has been used. This approach is particularly suitable for incorporating soil-structure interaction effects, since it is based on an equivalent linear visco-elastic description of the structure, which is characterized in terms of secant stiffness and equivalent damping, Considering the simplified 2 degree-of freedom (dof) oscillator of Fig. 8.17, interpreting the structure and the foundation as two springs in series, neglecting the foundation mass with respect to the superstructure mass and assuming the soil-structure interaction to be mainly affected by the rotational degree of freedom, Wolf (1985) demonstrated that the 2 dof system can be substituted by an equivalent 1 dof oscillator (see Fig. 8.17), characterized by equivalent properties (period, damping and base excitation). Using the equivalent values for stiffness and damper, the equivalent 1 dof oscillator response can be demonstrated to coincide with the response of the initial 2 dof system (Wolf 1985).

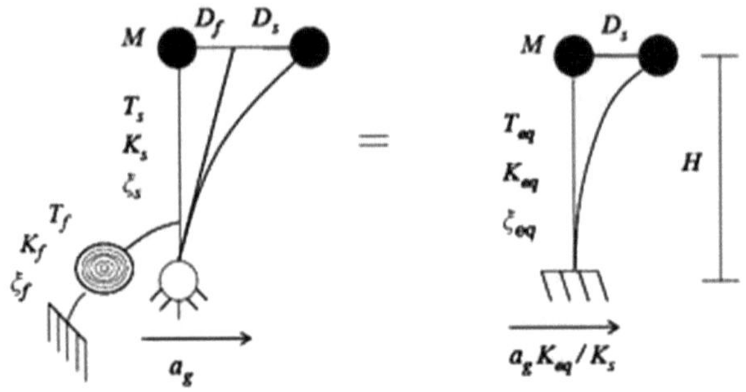

Fig. 8.17 *Left*: 2dof system representative of an harmonic oscillator subjected to base acceleration and its rotational compliance of the foundation. *Right*: equivalent 1 dof according to Wolf (1985) (where T_s, T_f, T_{eq} are vibration periods of the structure, of the foundation and of the equivalent 1 dof; K_s, K_f, K_{eq}: structural stiffness, soil-foundation impedance, stiffness of the equivalent 1 dof; ξ_s, ξ_f, ξ_{eq}: damping of the structure, of the foundation and of the equivalent 1 dof; H: effective height of the structural first vibration mode; a_g, $a_{g\ eq}$: ground acceleration and equivalent ground acceleration for the 1 dof system)

The procedure, which aims at introducing nonlinear soil-structure interaction effects into the displacement based design, is based on the idea that it is possible to apply the standard DDBD to the modified 1 dof oscillator. Analogously to point (b), in order to take non-linearities into account, iterations are needed, and at each iterative step soil stiffness and damping are updated as a function of the current foundation rotation, which is caused by the actions (base moment) calculated throughout DDBD.

8.4 Concluding Remarks

In this paper, the stimulating problem of the employment of the macro-element theory according to displacement based approaches, for both designing new and assessing the safety of already existing superstructures under seismic actions, has been tackled. To this purpose, the theory of the macro-element, by citing the large variety of constitutive models already available in the literature, has been outlined. To justify the effectiveness of the strategy, many experimental test results were cited and critically discussed in the light of the theory presented. A section was devoted to the critical analysis of the application of the macro-element theory to the interpretation of the soil-structure interaction under seismic actions. The authors have emphasized the role played by the soil mass/inertial contributions. In fact, at present inertial effects at the foundation soil level are totally disregarded and the constitutive relationships employed are, as a consequence, time independent. As far as this point is concerned, some theoretical suggestions to improve the currently available constitutive relationships have been provided. The main features of the practical strategies of employment of the macro-element approach for the evaluation of the displacements induced by seismic actions have been outlined in the last session: this task is not simple, since the approaches are numerous and the macro-element theory is applied differently according to the method taken into consideration. Three examples have been cited: the so called full-coupled approach – particularly suitable when the analyst has in mind the employment of a FEM dynamic numerical code -, the modified push-over, where the deformability of footings are taken into account; the modified DDBD approach, where more simply, the soil-structure interaction is interpreted by means of suitably adapted visco-elastic constraints. By summarising, although a lot of work is remaining to be done and new applications have to be found, we can state that the macro-element theory seems undoubtedly created just to be applied under seismic conditions: in fact, (i) under these conditions the importance of both horizontal loads and overturning moments cannot be contested, (ii) the dynamic superstructure response is in many cases severely influenced by the soil-foundation mechanical response and (iii) very often serviceability limit conditions are quite more crucial with respect to ultimate limit conditions and the evaluation of displacements/settlements cannot be avoided.

References

Abate G, Massimino MR, Maugeri M, Muir Wood D (2010) Numerical modeling of a shaking table test for soil-foundation-superstructure interaction by means of a soil constitutive model implemented in a FEM code. Geotech Geol Eng J 28:37–59

ATC-40 (1996) Seismic evaluation and retrofit of concrete buildings. Chapter 10: Foundations effects, Technical Rep. SSC 96-01, Seismic Safety Commission, State of California, 1996

Bowles JE (1996) Foundation analysis and design, 5th edn. McGraw Hill, New York

Bransby MF, Randolph MF (1998) Combined loading of skirted foundations. Geotechnique 48 (5):637–655

Bransby MF, Randolph MF (1999) The effect of embedment depth on the undrained response of skirted foundations to combined loading. Soil Found 39(4):19–33

Bransby MF, Yun G (2007) The horizontal-moment capacity of embedded foundations in undrained soil. Can Geotech J 44:409–424

Butterfield R (1980) A simple analysis of the load capacity of rigid footings on granular materials. Journée de Gotechnique 1:128–134

Butterfield R, Gottardi G (1994) A complete three dimensional failure envelope for shallow footings on sand. Géotechique 44(1):181 184

Butterfield R, Gottardi G (2003) Determination of yield curves for shallow foundations by swipe testing. In: Magnan JP, Droniuc N (eds) Fondations superficielles. Presses de l'ENPC/LCPC, Paris

Butterfield R, Ticof J (1979) Discussion: design parameters for granular soils. In: Proceedings of the 7th European conference on soil mechanics and foundation engineering, vol 4. Brighton, pp 259–262

Calvetti F, di Prisco C, Nova R (2004) Experimental and numerical analysis of soil-pipe interaction. ASCE J Geotech Geoenviron Eng 130(12):1292–1299

Cassidy M, Byrne B, Houlsby G (2002) Modelling the behaviour of circular footings under combined loading on loose carbonate sand. Geotechnique 52(10):705–712

Chatzigogos CT, Figini R, Salençon J, Pecker A (2011) A macroelement formulation for shallow foundations on cohesive and frictional soils. Int J Numer Anal Methods Geomech 35 (8):902–931
Chen WF, Liu XL (1991) Limit analysis in soil mechanics, Developments in geotechnical engineering. Elsevier Science, Amsterdam
Cremer C, Pecker A, Davenne L (2001) A model for landslide-pipe interaction analysis. J Earthq Eng 6:175–212
Cremer C, Pecker A, Davenne L (2002) Modelling of nonlinear dynamic behaviour of a shallow strip foundation with macro-element. Int J Numer Anal Methods Geomech 25:1257–1284
Cundall PA, Strack ODL (1979) A discrete numerical model for granular assemblies. Geotechnique 29:47–65
di Prisco C, Pisano F (2011) Seismic response of rigid shallow footings. Eur J Environ Civ Eng 15 (Suppl 1):185–221
di Prisco C, Vecchiotti M (2006) A rheological model for the description of boulder impacts on granular strata. Geotechnique 56(7):469–482
di Prisco C, Vecchiotti M (2010) Design charts for evaluating impact forces on dissipative granular soil cushions. ASCE J Geotech Geoenviron Eng 136(11):1529–1541
di Prisco C, Fornari B, Nova R, Pedretti S (1998) A constitutive model for cyclically loaded shallow foundations. In: Proceedings of the EUROMECH 385 – colloquium on inelastic analysis under variable loads: theory and engineering applications, Achen 8–11 Sept 1998, pp 107–111
di Prisco C, Nova R, Perotti F, Sibilia A (2003a) Analysis of soil-foundation interaction of tower structures under cyclic loading. In: Maugeri M, Nova R (eds) Geotechnical analysis of seismic vulnerability of historical monuments. Patron, Bologna
di Prisco C, Nova R, Sibilia A (2003b) Shallow footings under cyclic loading: experimental behaviour and constitutive modeling. In: Maugeri M, Nova R (eds) Geotechnical analysis of seismic vulnerability of historical monuments. Patron, Bologna
di Prisco C, Massimino MR, Maugeri M, Nicolosi M, Nova R (2006) Cyclic numerical analyses of Noto Cathedral: soil-structure interaction modeling. Ital Geotech J XL:49–64
di Prisco C, Pisano F, Lancellotta R (2012) Finite element analysis and macro element modelling of soil foundation interaction for laterally loaded historical towers (submitted for publication on Géotechnique)
Faccioli E, Paolucci R, Vivero G (2001) Investigation of seismic soil-footing interaction by large scale cyclic tests and analytical models. In: Proceedings of the 4th international conference on recent advances in geotechnical earthquake engineering and soil dynamics, Special presentation lecture, San Diego
Gabrieli F, Cola S, Calvetti F (2009) Use of an up-scaled DEM model for analysing the behaviour of a shallow foundation on a model slope. Geomech Geoeng Int J 4(2):109–122
Gajan S, Kutter B, Phalen J, Hutchinson TC, Martin GR (2005) Centrifuge modeling of load-deformation behaviour of rocking shallow foundations. Soil Dyn Earthq Eng 25:773–783
Gazetas G, Hatzikonstantinou E (1988) Elastic formulae for lateral displacement and rotation of arbitrarily-shaped embedded foundations. Geotechnique 38(3):439–444
Gazetas G, Tassoulas GL, Dobry R, O'Rourke MJ (1985) Elastic settlement of arbitrarily shaped foundations embedded in half-space. Geotechnique 35(3):339–346
Georgiadis M, Butterfield R (1988) Displacements of footings on sand under eccentric and inclined loads. Can Geotech J 25:192–212
Gottardi G, Houlsby G, Butterfield R (1999) Plastic response of circular footings under general planar loading. Geotechnique 49(4):453–469
Gourvenec S (2008) Effect of embedment on the undrained capacity of shallow foundations under general loading. Geotechnique 58(3):177–185
Gourvenec S, Randolph MF (2003) Effect of strength non-homogeneity on the shape of failure envelopes for combined loading of strip and circular foundations on clay. Geotechnique 53 (6):575–586

Grange S, Kotronis P, Mazars J (2007a) 3D macro-element for soil structure interaction. In: Proceedings of the 4th international conference on earthquake geotechnical engineering, Thessaloniki, June 25–28 2007, paper 1308

Grange S, Kotronis P, Mazars J (2007b) Extensive validation of the SSI macro-element using experimental and numerical results. European contract LESSLOSS, Proj. No. GOCE-CT-2003-505488, Risk mitigation for earthquakes and landslides, 6th Framework Prog., Deliverable (D8–2): http://hal.archives–ouvertes.fr/hal–00138210

Grange S, Kotronis P, Mazars J (2008) A macro-element for a circular foundation to simulate 3d soil-structure interaction. Int J Numer Anal Methods Geomech 32(10):1205–1227

Grange S, Kotronis P, Mazars J (2009) Numerical modelling of the seismic behaviour of a 7-story building. Mater Struct 42(10):1433–1442

Hodder MS, Cassidy MJ (2010) A plasticity model for predicting the vertical and lateral behaviour of pipelines in clay soils. Geotechnique 60(4):247–263

Le Pape Y, Sieffert JP (2001) Application of thermodynamics to the global modelling of shallow foundations on frictional material. Int J Numer Anal Methods Geomech 25:1377–1408

Martin CM (1994) Physical and numerical modelling of offshore foundations under combined loads. Ph.D. thesis, University of Oxford

Martin GR, Lam IP (2000) Earthquake resistant design of foundations – retrofit of existing foundations. In: Proceedings GeoEng 2000 conference. Melbourne, pp 19–24

Massimino MR, Maugeri M (2013) Physical modeling of shaking table tests on dynamic soil-foundation interaction and numerical and analytical simulation. Soil Dyn Earthq Eng J 49:1–18

Maugeri M, Musumeci G, Novità D, Taylor CA (2000) Shaking table test of failure of a shallow foundation subjected to an eccentric load. Soil Dyn Earthq Eng 20(5–8):435–444

Montrasio L, Nova R (1997) Settlement of shallow foundations on sand: geometrical effects. Geotechnique 47(1):46–60

Negro P, Paolucci R, Pedretti E, Faccioli R (2000) Large-scale soil-structure interaction experiments on sand under cyclic loading. In: Proceedings of the 12th world conference on earthquake engineering. Auckland, paper 1191

Nova R, Montrasio L (1991) Settlement of shallow foundations on sand. Geotechnique 41 (2):243–256

Nova R, Parma M, Castellanza R (2008) Settlements of foundations on soft rocks. Rivista Italiana di Geotecnica 42(2):10–22

Paolucci R (1997) Simplified evaluation of earthquake induced permanent displacements of shallow foundations. J Earthq Eng 1:563–579

Paolucci R, di Prisco C, Vecchiotti M, Shirato M, Yilmaz M (2007) Seismic behaviour of shallow foundations: large scale experiments vs numerical modelling and implications for performance-based design. In: Proceedings of the 1st US-Italy seismic bridge workshop, Eucentre, Pavia, Italy, 2007, pp 107–111

Paolucci R, Shirato M, Yilmaz MT (2008) Seismic behaviour of shallow foundations: shaking table experiments vs. numerical modelling. Earthq Eng Struct Dyn 37:577–595

Paolucci R, di Prisco C, Figini R, Petrini L, Vecchiotti M (2010) Interazione dinamica non lineare terreno-struttura nell'ambito della progettazione sismica agli spostamenti. Progettazione Sismica 1(2):83–103

Pecker A (2006) Enhanced seismic design of shallow foundations: example of the Rion Antirion bridge. In: Proceedings of the 4th Athenian lecture on geotechnical engineering, Athens

Pedretti S (1998) Non-linear seismic soil-foundation interaction: analysis and modelling methods. Ph.D. thesis, Politecnico di Milano

Perzyna P (1963) The constitutive equations for rate sensitive plastic materials. Q Appl Math 20:321–332

Pitilakis D, Dietz M, Muir Wood D, Clouteau D, Modaressi A (2008) Numerical simulation of dynamic soil-structure interaction in shaking table testing. Soil Dyn Earthq Eng 28:453–467

Priestley MJN, Calvi GM, Kowalski MJ (2007) Displacement based seismic design of structures. Iuss Press, Pavia

PWRI (2005) Experimental study on the residual displacements of shallow foundations. Public Work Research Institute, Technical note

Randolph MF (2003) 43rd Rankine lecture: science and empiricism in pile foundation design. Géotechnique 53(10):847–875

Salençon J, Pecker A (1995a) Ultimate bearing capacity of shallow foundations under inclined and eccentric loads, part I: purely cohesive soil. Eur J Mech A Solid 14(3):349–375

Salençon J, Pecker A (1995b) Ultimate bearing capacity of shallow foundations under inclined and eccentric loads, part II: purely cohesive soil without tensile strength. Eur J Mech A Solid 14 (3):377–396

Shirato M, Kuono T, Asai R, Nakatani S, Fukui J, Paolucci R (2008) Large-scale experiments on nonlinear behavior of shallow foundations subjected to strong earthquakes. Soil Found 48 (5):673–692

Tamagnini C, Salciarini D (2009) A hypoplastic macroelement model for shallow foundations under monotonic and cyclic loads. Acta Geotech 4(3):163–176

Vesic AS (1975) Bearing capacity of shallow foundations. In: Winterkorn HF, Fang HY (eds) Foundations engineering handbook, pp 121–147

Wolf JP (1985) Dynamic soil-structure interaction. Prentice Hall, Englewood Cliffs

Zeng X, Steedman RS (1998) Bearing capacity failure of shallow foundations in earthquakes. Geotechnique 48:235–256

Chapter 9
Seismic Response of a Large-Scale Highway Interchange System

Kyungtae Kim, Ahmed Elgamal, George Petropoulos, Aysegul Askan, Jacobo Bielak, and Gregory L. Fenves

Abstract A numerical simulation of a large-scale highway interchange system under seismic loading conditions is conducted. A three-dimensional (3D) Finite Element (FE) model of an existing bridge system at the Interstate 10/215 interchange (Riverside County, CA) is developed. This interchange is comprised of three connectors at different bridge superstructure elevations. Herein, focus is placed on one of these three connectors (the North-West connector), using the OpenSees FE framework. A strategy to incorporate ground response and soil-structure interaction (SSI) is implemented based on the Domain Reduction Method (DRM) for three-dimensional earthquake simulation. Modeling of this bridge-foundation-ground system is based on blue-prints that were provided by Caltrans (California Department of Transportation). Vibration properties and seismic response behavior for the connector and the soil domain are examined. Different scenarios of bridge response are considered and compared, including fixed-base

K. Kim (✉) • A. Elgamal
Department of Structural Engineering, University of California, San Diego, USA
e-mail: kyk002@ucsd.edu; aelgamal@ucsd.edu

G. Petropoulos
Department of Civil and Environmental Engineering, University of California, Berkeley, CA, USA

A. Askan
Department of Civil Engineering, Middle East Technical University, Ankara, Turkey
e-mail: aaskan@metu.edu.tr

J. Bielak
Department of Civil and Environmental Engineering, Carnegie Mellon University, Pittsburgh, USA
e-mail: jbielak@cmu.edu

G.L. Fenves
Cockrell School of Engineering, The University of Texas at Austin, Austin, USA
e-mail: fenves@mail.utexas.edu

M. Maugeri and C. Soccodato (eds.), *Earthquake Geotechnical Engineering Design*, Geotechnical, Geological and Earthquake Engineering 28,
DOI 10.1007/978-3-319-03182-8_9,

uniform excitation, and multiple-support excitation with and without the full ground/foundation soil domain.

9.1 Introduction

In the last decade, seismic analysis of full bridge-foundation-ground systems has been gradually gaining further attention (e.g., Jeremić et al. 2004, 2009; Kwon and Elnashai 2008; Zhang et al. 2008; Elgamal et al. 2008). Recent efforts have included more realistic nonlinear materials for the reinforced concrete and soil cyclic loading response, and analysis of liquefaction effects was undertaken (Zhang et al. 2008). Absorbing boundary conditions were addressed in earlier studies to limit fictitious reflections from incident seismic waves (e.g., Conte et al. 2002). Based on experimentation, further insights are being gained as well into the associated basic SSI mechanisms and the related numerical and analytical modeling frameworks (Pitilakis et al. 2008; Massimino and Maugeri 2013).

Efforts in 3D modeling and analysis of bridges were also reported. Soil vibration induced by high-speed train on a bridge was studied and isolation effects due to nearby building foundations and piles under the bridge were evaluated (Ju 2004). Separate models for an elevated highway bridge supported on deep foundations with and without soil-foundation-structure-interaction (SFSI) were considered to evaluate the SFSI beneficial and detrimental effects (Jeremić et al. 2004). Nonlinear-dynamic analysis for a bridge-foundation-ground system was conducted for estimation of settlement and longitudinal/transverse displacements of the abutments and pile foundations due to permanent ground deformation (Elgamal et al. 2008). Using recorded data, responses of a multiplatform simulation of a highway overcrossing bridge were compared for the integrated bridge-ground system as well as for substructures such as the pile foundations and the abutments (Kwon and Elnashai 2008). High-fidelity models for SFSI were developed to simulate seismic wave propagation (frequency up to 10 Hz) through nonlinear elastic-plastic, soil, piles, and bridge analyses (Jeremić et al. 2009). Difference in ground motions due to presence of the structure showed that treatment of free-field motions as input for the structural model can be inadequate.

Using the Domain Reduction Method (DRM, Bielak et al. 2003), a large localized ground response region of interest subjected to equivalent seismic forces from a causative fault was designed and developed via high-performance algorithms for simulation of SFSI. In addition, mixed time integration with explicit and implicit schemes was employed for efficiency of numerical simulation on parallel computers (Petropoulos 2008). More recently, effects of soil-structure interaction on the response of yielding single-story structures embedded in an elastic half-space were discussed (Jaremprasert et al. 2012).

This paper presents an evaluation of soil-structure interaction (SSI) on a common type of freeway bridge interchange subjected to incident seismic waves propagated through the soil medium. Emphasis is placed on modeling the bridges supported on a large-scale soil domain, with seismic responses induced by the spatially variable free-field motions. Efforts include simulation of the bridge-foundation-ground system, investigation of non-uniform support motions on the response of the bridge, and influence of foundation-ground flexibility on the response of the bridge compared with that obtained from analysis of the fixed-base structure.

9.2 Ground Model and Earthquake Motion

In the bridge-foundation-ground system (BFGS), the ground localized region of interest (Petropoulos 2008) was employed to include effects of soil-structure interaction on three connectors at a Highway interchange (Kim 2013). The region of interest was developed by the DRM (Bielak et al. 2003), shown earlier to be a robust approach for application of earthquake excitation over a large spatial domain. This approach was employed to create a 1,000 m by 500 m by 100 m localized ground region subjected to 3D incident seismic motion, using 20 m hexahedral elements. As a result of a Magnitude 7.1 (Moment magnitude) scenario earthquake due to rupture of the Puente Hills Fault, the DRM was applied to provide seismic input excitation generated by wave propagation resulting from the causative fault rupture. The broadband method developed by Graves and Pitarka (2010) was used to incorporate the effects of high-frequency excitation.

Lysmer-Kuhlemeyer (1969) absorbing boundary conditions were used at the side and bottom boundaries of the computational soil domain and a buffer zone is introduced in each direction to limit the effect of spurious wave reflections from the absorbing boundaries. This resulted in a total computational domain of 1.8 km by 1.3 km by 0.5 km in the X, Y, Z directions, respectively (Fig. 9.1). Based on the available geological data for the region (Olsen et al. 2003), mass proportional Rayleigh damping was used to represent energy dissipation in the soil.

9.2.1 Ground Model

Two types of soil profiles were used to evaluate the seismically-induced ground motions, namely stiff and soft soil profiles (with linear elastic properties). For the stiff soil profile, a near fault region in downtown Los Angeles was considered as the region of interest. In this region, the soil consists of two layers (Layers 1 and 2), with the properties indicated in Table 9.1 (Petropoulos 2008). The soil properties within the buffer zone are also shown in this Table. The finite element mesh was tailored on the wavelength of the shear waves so that there are at least eight nodes per wavelength throughout the computational domain.

In order to investigate the behavior under the conditions of a more compliant ground scenario, a second softer soil model was considered in which the shear wave

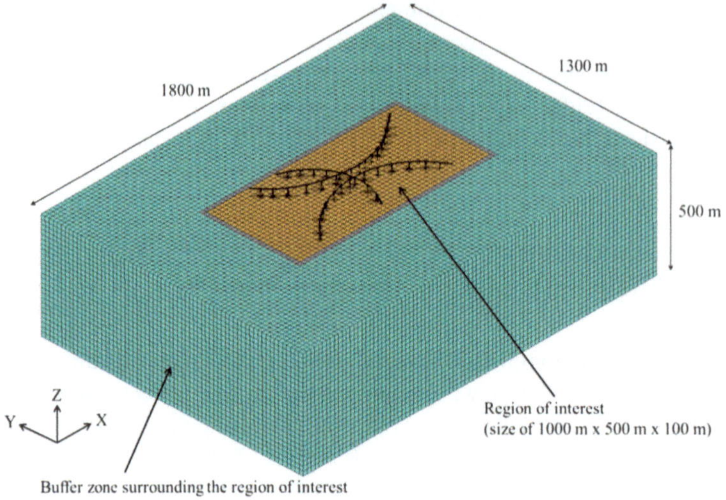

Fig. 9.1 3D view of the computational FE model of the bridge-foundation-ground system

Table 9.1 Material properties for the stiff soil profile

Layer	Density (t/m^3)	V_p (m/s)	V_s (m/s)	Thickness (m)	Region
1	1.5	1,337.3	482.2	40	Region of interest
2	1.714	1,622.4	584.3	60	Region of interest
3	1.714	1,622.4	584.3	340	Buffer zone
4	2.054	2,372.9	651.3	60	Buffer zone

velocity in the first layer is reduced to 100 m/s, and proportionately for the remaining layers and boundary dashpots parameters.

In addition, the soil mesh configuration near the ground surface was modified (below the location of each bridge column) in order to accommodate the structural model pile-group foundation geometry. As such, the mesh of the top 20 m thick layer in the region of interest was refined to crudely include the pile group geometry/stiffness effects below the bridge columns. Since a typical size of the original pile group foundation was 7.47 m by 7.47 m (in the horizontal plane), the uniform FE grid was modified in one single brick element or four brick elements depending on location of the particular bridge column (with respect to the soil mesh nodal coordinates). Except for the pile group locations, the rest of mesh remained tailored to the shear wave velocity of the soil, as mentioned already. The entire mesh is shown in Fig. 9.1. The free-field ground motion at the center node of the free surface is shown in the Appendix.

9.3 Bridge Interchange Structures and Numerical Model

For the BFGS simulation, the structural models were based on the existing highway overcrossing bridges at the interchange of interstate 10 and 215. The Interstate 10/215 interchange is located in the San Bernardino valley near the Santa Ana River basin (in southern California). This interchange mainly consists of three curved bridges. The North-West connector (Bridge No. 54-823G) is a 774 m long concrete box girder bridge with 16 spans supported by single column bents and diaphragm abutments (Kim 2013). The North-East connector (Bridge No. 54-824 F) is a 474 m long, curved, concrete box-girder bridge with 9 spans supported by single column bents and diaphragm abutments. The South-East connector is an 800 m long, curved, concrete box-girder bridge with 17 spans supported by single column bents and diaphragm abutments. An aerial photograph of the interchange is shown in Fig. 9.2.

The North-West connector, North-East connector, and South-West connector were strengthened under the Caltrans' Phase II seismic retrofit program considering that the interchange crossed a mapped fault (San Jacinto fault). The retrofit included field-welded steel jackets on most of the bent columns, strengthened pile caps and foundations for the bents with full jackets, new cable restrainer units for the hinges, and support blocks at the abutments. The techniques and the interpretation from experiments of the pertinent column retrofit can be found in Priestley et al. (1996) and more details concerning the implemented seismic retrofit are presented in Fenves and Desroches (1995).

Fig. 9.2 Aerial photograph of I-10/215 interchange (©2013 Google – http://maps.google.com)

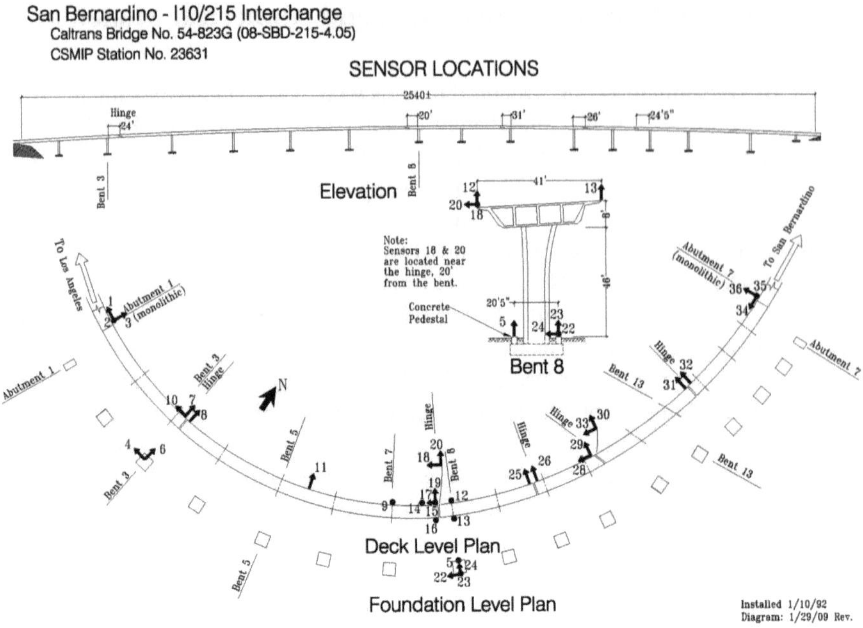

Fig. 9.3 Seismic instrumentation for North-West connector

As per guidelines for the instrumentation of highway bridges, the North-West connector was instrumented with a network of strong motion accelerometers by the California Strong Motion Instrumentation Program (CSMIP) in conjunction with the California Department of Transportation (Caltrans) in 1992. The CSMIP station for this connector is Station No. 23631 shown in Fig. 9.3. Since the instrumentation provides valuable information on the seismic response of the freeway bridge structure, substantial related research was already conducted (Fenves and Desroches 1995; Huang and Shakal 1995; Desroches and Fenves 1997; Mosquera et al. 2009). In the remaining sections of this paper, attention is directed towards this North-West connector only. Further discussion of the other connectors and the overall interchange response can be found in Kim (2013).

9.3.1 Structural Modeling

As mentioned earlier, the North-West connector is a 774 m long 16 span concrete box-girder bridge (2.4 m deep prestressed and conventionally reinforced box girders). The connector consists of six frames and each frame is connected at five intermediate hinges. The typical single column bent has an octagonal cross section with overall dimensions of 2.4 m by 1.7 m. The column is flared in both directions near the top. The column height from the top of the pile cap to the soffit of the box girder varies from 7.3 to 23.5 m. The original foundations for the bents are comprised of a pile cap and reinforced concrete piles. The largest foundations consist of a 7.3 m by 7.0 m pile cap (2.1 m in thickness) with 48 piles (piles are spaced 3–4 ft on center). The pile lengths vary from 6.4 to 15.2 m. At the diaphragm abutments, the box girder is integral with a 4.0 m high backwall and 5.5 m long tapered wingwalls. All hinges are straight, except for 23 and 13° skew for Hinges 11 and 13, respectively. Relative transverse displacement is prevented by a shear key. Seven cable restrainer units are installed in the longitudinal direction at the hinges (Fenves and Desroches 1995; Desroches and Fenves 1997).

9.3.1.1 Strategy of Structural Modeling

The FE model of the bridge structure consists of four components, which are the superstructure (including the deck and the columns), the piles, the intermediate hinges, and the abutments. The model properties were evaluated based on the blue prints provided by Caltrans (California Department of Transportation) and the earlier model of the North-West connector (Fenves and Desroches 1995; Desroches and Fenves 1997). Linear properties of the superstructure are considered. For the intermediate hinges, longitudinal properties are employed to simulate the presence of intermittent sharp spikes in earlier recorded accelerations (Huang and Shakal 1995). Rayleigh damping was included at a value of 5 % for this connector. The simplified abutment model (Aviram et al. 2008) was employed, with vertical translation assumed to be identical to translation of soil under the abutment. The piles are modeled as discussed earlier. Fixed base eigenvalue analysis of this model results in the first transverse motion mode shape at a period of about 1.5 s and a first longitudinal motion mode at a period of about 1.2 s.

9.3.1.2 Connection of the Structural Model to the Soil Domain

Since the topography under the connector varies significantly (Fig. 9.3), it was necessary to devise a procedure to couple the structural and soil finite elements in order to create a model of the complete soil–structure system. Figure 9.4 shows schematically how this coupling was performed between the bridge columns, the foundations (pile cap and piles), and the surrounding soil. The shaded volume, which is physically occupied by the pile group, was replaced by highly rigid brick elements embedded in the soil domain. Since the overall soil surface in the DRM soil model is at the same elevation, the lowest elevation of all pile caps was chosen as a datum level at which all pile caps were placed. Afterwards, beam–column elements representing the pile cap were connected to the solid (pile-group) elements. The connectivity was done only for the translational degrees-of-freedom (DOFs; three for each node), while the three rotational DOFs were left unconnected. For the connection between the column and the pile cap, translations

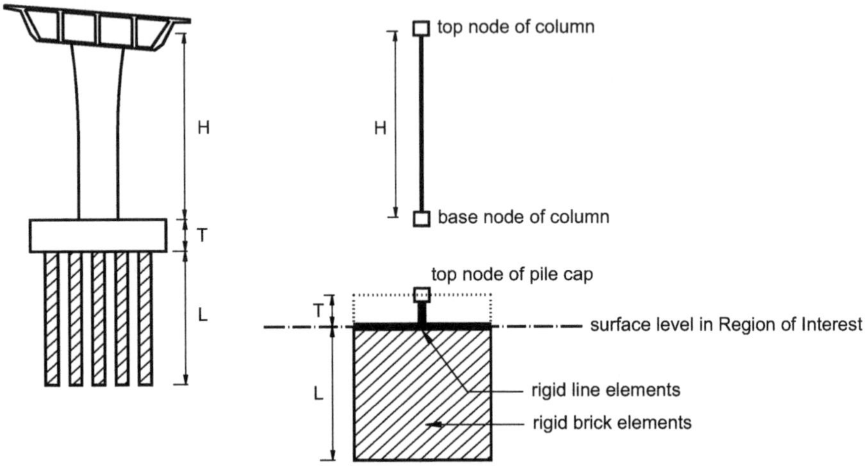

Fig. 9.4 Schematic view of connection between column and soil including pile groups

and rotations of the two nodes at the top face of the pile cap and at the base of the columns were constrained to be identical.

9.3.2 Numerical Model of the Bridge-Foundation-Ground System

Three connectors built at the interchange of interstates 10 and 215 were analyzed with an assumption that the connectors were placed on ground of either stiff or soft soil, as mentioned previously (Kim 2013). Due to the limited size of the region, the actual relative location of the connectors with respect to one another was slightly modified. The chord of the curved span of North-West connector and South-East connector were oriented parallel to the global X direction with the chord of the North-East connector taken in the global Y direction. Within the region of interest, the numerical model for the bridge-foundation-ground system is shown in Fig. 9.5.

9.4 Seismic Response of the North-West Connector Bridge

Results from the BFGS for the two types of soil profiles were obtained. Depending on the material properties of the soil and orientation of the incident seismic waves to the region of interest, the characteristic of the seismic waves propagated through the two soil profiles changed significantly. While horizontal peak ground accelerations in the soft soil profile were analogous to those in the stiff soil profile, the vertical component was significantly reduced in the soft soil profile. However,

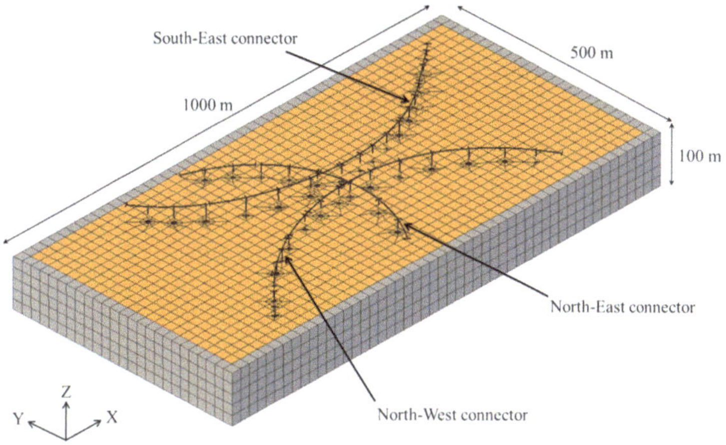

Fig. 9.5 Numerical model for the bridge-foundation-system

looking at the frequency content of the horizontal motion, a range of frequency changes occurred, with the main energy observed around 1 Hz. Moreover, more noticeable variation of ground motion over the soil surface was observed in the soft soil profile. These changes in the soft soil scenario ground motions caused in some differences in the bridge dynamic response.

9.4.1 DRM Results for the Bridge-Foundation-Ground System (North-West Connector)

Figure 9.6 shows the peak ground velocity at the bent pile cap locations with and without the connector in global X and Y directions for the soft soil profile. Deviation of ground motions in the presence of the structure from the free-field indicates the effects of soil-structure interaction. In the stiff soil profile, almost identical ground motions (about 0.5 % difference) were observed. For the softer ground, the effects were more pronounced and the maximum difference was about 6.4 % at Bent 12 in global X direction. Thus, the analysis of the bridge structure using the free-field motions without consideration of soil-structure interaction may lead to somewhat different estimates of seismic force demands, particularly in the softer soil case.

Due to the change of frequency content of the ground motion (soft versus stiff soil), the seismic base shears in the bridge columns were observed to change significantly. The base shears decreased 24 % for the X component and 5 % for the Y component in the North-West (NW) connector for the soft soil profile. Although horizontal peak ground accelerations were similar in the two different soil profiles, the seismic energy released at different frequencies plays a significant role in dictating the resulting seismically-induced column forces.

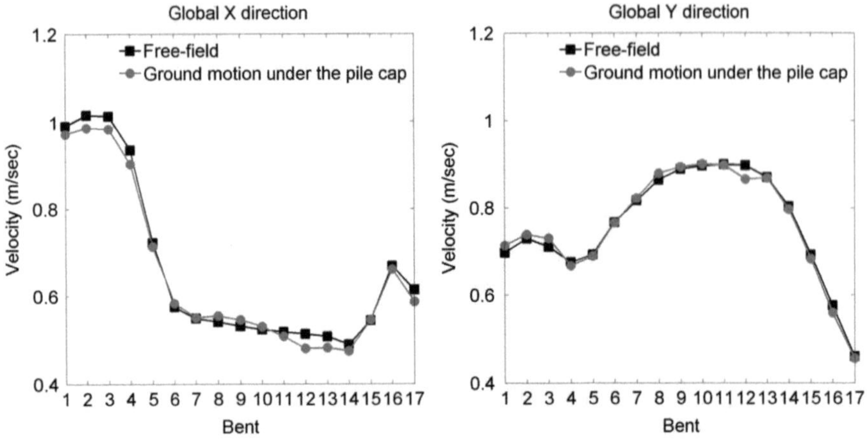

Fig. 9.6 Comparison of peak ground velocities with and without the connector (soft soil)

Drift ratios (column maximum lateral displacement divided by its height) were significantly dependent on the transmitted forces and flexibility of the soil domain. Figure 9.7 shows the deformed shape of the connector relative to the base of Bent 8 at the time instant of 17.1 s, when the relative displacement was 0.19 m at the top of Bent 7 in the global Y direction (soft soil case, without displaying the soil domain). For the North-West connector on soft soil, the maximum drift ratio decreased 32 % in the transverse direction at Bent 6 and increased 8 % in the longitudinal direction at Bent 4. On the other hand, the largest reductions of drift ratios were longitudinally 20 % at Bent 15 and transversally 46 % at Bent 15. As such, the beneficial or detrimental effects of soil-structure interaction may be examined. In the stiff soil profile, the effects were generally negligible.

9.4.2 Comparison of the Results Without Soil Domain

The DRM provides the advantage of capturing free-field ground motions at desired locations on the surface in the region of interest and enables one to analyze a fixed-base structure using the resulting multiple base excitations at the location of each column foundation. More crudely, a ground motion at a location near the structure may be as a representative base uniform excitation. In modeling such a fixed-base structure, potential relative motion between the pile groups are not accounted for.

On this basis, results for analysis of the DRM full bridge-foundation-ground system may be compared with those from uniform base excitation and multi-support excitation without the soil domain. Various seismic response quantities, such as support accelerations, base-shears, and column drift ratios/forces were considered to further examine the role of soil-structure interaction.

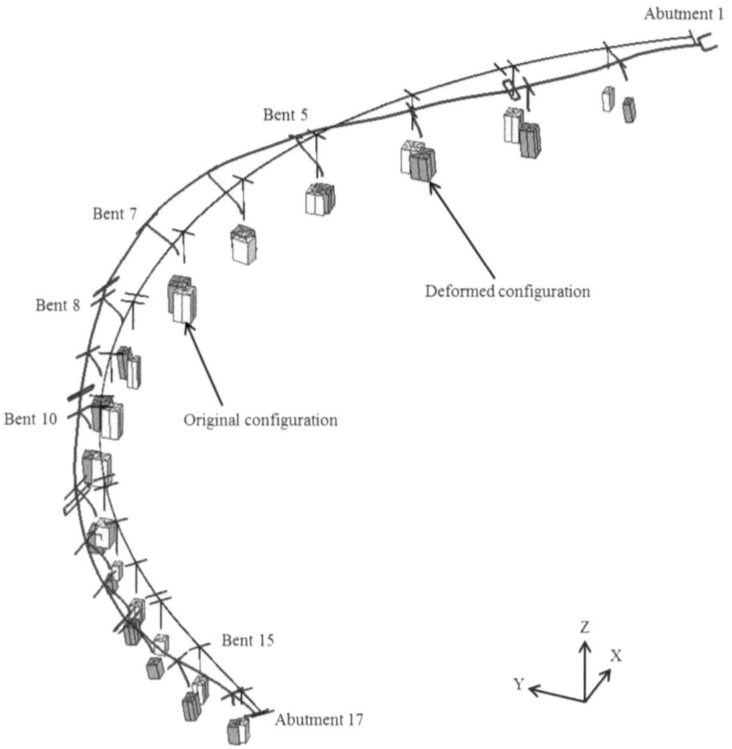

Fig. 9.7 Deformed shape of North-West connector with pile groups at the time instant of 17.1 s (soft soil case)

9.4.2.1 Multiple-Support Excitation

For the bridge super-structure model, spatially variable ground motion in terms of displacements were applied at the corresponding location of each column. Via the DRM approach, the ground motions were obtained for the two types of employed soil (stiff and soft) at the actual location of the bridge foundations in the absence of the structure (i.e., free-field motions). As such, base excitation was dictated by the free-field motions at the four corners of each pile cap, as a mechanism to include rotation of the pile cap into the conducted analysis.

9.4.2.2 Uniform Excitation

Computed from the DRM analysis of the region of interest in the absence of any structures, free-field motion at the center of the region of interest was taken as the representative structural uniform base excitation. This ground motion in terms of base acceleration, computed for the two soil profiles (stiff and soft) was used to

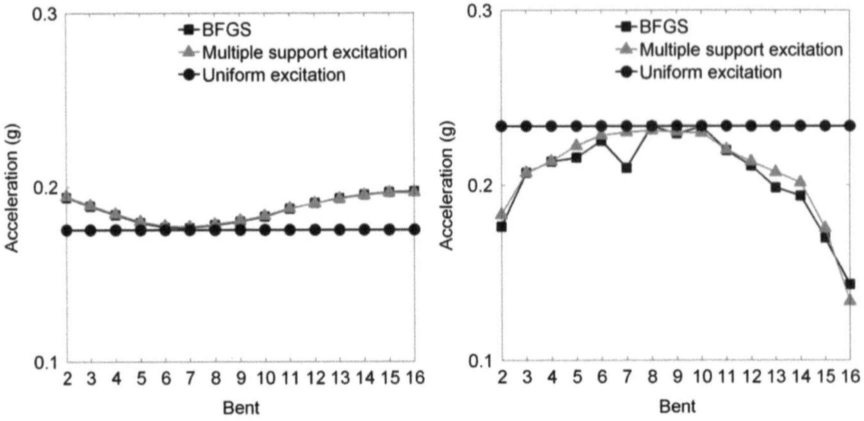

Fig. 9.8 Peak acceleration at the base of bents; stiff soil (*left*) and soft soil (*right*)

excite the structure uniformly in the horizontal and vertical directions. Ground acceleration in the three translational directions was applied at the base of each column and no rotations were induced at the supports unlike the above multiple-support excitation scenario.

9.4.3 Comparison of Responses

Figure 9.8 shows the maximum accelerations at the base for the three types of loading conditions (stiff and soft soil profiles) in the global Y direction, associated with overall transverse bridge response. The level of acceleration at the base in the stiff soil, induced by consideration of the soil domain (in the BFGS) and multiple support excitation, varies significantly along the longitudinal direction compared with the representative uniform excitation acceleration at the center surface near Bent 8. Comparison of the maximum accelerations at the base in the BFGS and under multiple support excitation shows that the values are very similar in the stiff soil profile. On the other hand, the values under multiple support excitation differ a bit more noticeably from those of the BFGS in the soft soil profile on account of soil-structure interaction induced by the presence of the structure. Variation of peak acceleration at the column bases in the soft soil domain scenario may be also noted to be quite significant.

9.4.3.1 Seismic Base Shears

The normalized total column base shear forces divided by the seismic weight of the bridge (about 160 MN) are compared with and without soil-structure interaction. Table 9.2 summarizes the total of peak base shears in the global directions in which

Table 9.2 Seismically-induced total base shear with and without the soil domain

		With SSI	Without SSI			
Soil profile	Base shear	BFGS	Multiple support excitation	Difference (%)	Uniform excitation	Difference (%)
Stiff soil	V_x/W	0.259	0.248	−4.25	0.263	1.54
	V_y/W	0.209	0.213	1.91	0.173	−17.22
Soft soil	V_x/W	0.196	0.197	0.51	0.314	60.20
	V_y/W	0.200	0.176	−12.00	0.243	21.50

the X direction is parallel to the chord of the curved configuration and the Y direction is perpendicular to the chord (Fig. 9.3).

Under the multiple support excitation conditions, the base shear is not greatly affected by the presence of the soil, except for the Y component in the linear-soft sol profile (12 % reduction). In the X direction, the base shear decreases 4 % for the stiff soil while it increases 0.5 % for the soft soil profile. In the Y direction, the base shear increases 2 % for the stiff soil and decrease 12 % for the soft soil profile. On the other hand, there is much variation of the base shear under uniform excitation. For the stiff and soft profile, the base shear in the Y direction decreases 17 % and increase 22 %, respectively, while the base shear in the X direction for the soft soil increases significantly, by as much as 60 %.

9.4.3.2 Drift Ratio

Drift ratio, calculated as the relative displacements at the top referenced to the base of each bent and divided by the height of the column, will be compared for the three cases. The relative displacement obtained from the BFGS is consistent with the deformation excluding a top displacement due to the rocking component of the foundation while the analysis of the fixed-base structure does not give rise to rocking of the foundation.

The maximum drift ratios are shown in Figs. 9.9 and 9.10. In general, the drift ratios are largest for the coupled soil-structure case and smaller for the fixed-base structure under uniform excitation in the transverse direction. However, this trend is not consistent in the longitudinal direction. Comparison between BFGS with soil domain and the multiple supports excitation case shows that the drift ratios increase due to soil-structure interaction, with the SSI effects being beneficial or detrimental depending on location of each particular column.

9.4.3.3 Column Forces

The computed column relative displacements induce significant shears and bending moments. Tables 9.3 (for the stiff soil profile) and 9.4 (for the soft soil profile)

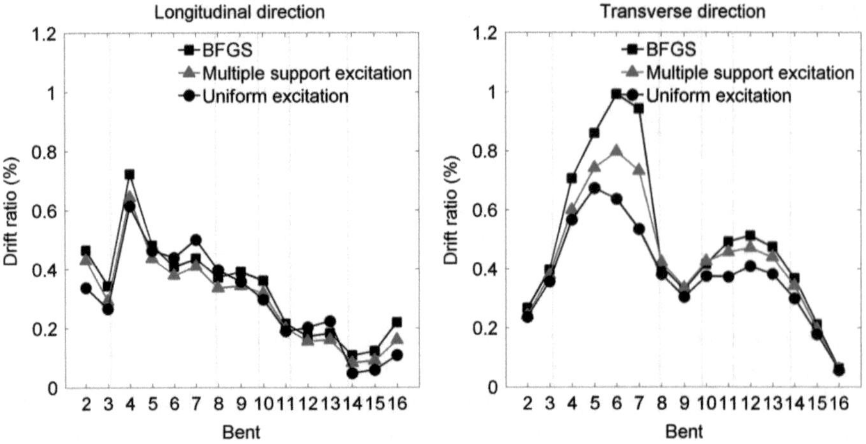

Fig. 9.9 Stiff soil profile; maximum drift ratios

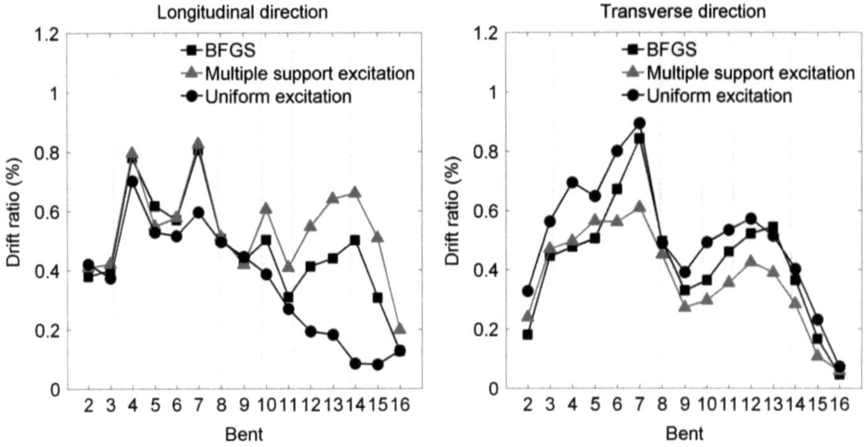

Fig. 9.10 Soft soil profile; maximum drift ratios

summarize the comparison of maximum column bending moment in the transverse direction at the representative locations of Bents 2, 5, 6, 7, and 15 (Fig. 9.3), computed under the BFGS, the multiple support excitation, and the uniform excitation scenarios.

In the stiff soil profile, with the effect of SSI, larger bending moments are computed. When the non-uniform ground motions were considered (multiple support excitation) without the effect of SSI, the computed moments were close to the ones with SSI, while the uniform excitation resulted in much deviation from the SSI scenario. The bending moments without the soil domain decreased up to

Table 9.3 Comparison of maximum column bending moment (MN-m) in the stiff soil profile

	Stiff soil profile				
Bent	BFGS	Multiple support excitation	Difference (%)	Uniform excitation	Difference (%)
2	38.1	34.5	−9.6	33.2	−12.9
5	62.1	53.8	−13.4	49.0	−21.0
6	75.9	61.2	−19.4	49.2	−35.1
7	92.9	72.1	−22.4	51.9	−44.2
15	25.1	23.1	−8.0	21.8	−13.0

Table 9.4 Comparison of maximum column bending moment (MN-m) in the soft soil profile

	Soft soil profile				
Bent	BFGS	Multiple support excitation	Difference (%)	Uniform excitation	Difference (%)
2	24.9	33.6	34.8	45.6	82.7
5	37.5	41.5	10.8	47.2	26.0
6	51.5	41.7	−19.0	60.5	17.5
7	84.5	60.9	−28.0	88.3	4.5
15	18.3	11.3	−38.7	27.9	51.9

22 % at Bent 7 under the multiple support excitation and 44 % under the uniform excitation scenarios, respectively.

In the soft soil profile, larger bending moments are generated under uniform excitation without SSI, providing an upper bound for the estimated column forces in this case. On the other hand, the moments analyzed by non-uniform ground motions (multiple support excitation) showed much variation depending on the deviation of the free-field motions from the ground motion in the presence of the structure. In particular, near Abutment 1, the moment increases by up to 35 %, while decreasing by as much 39 % at Bent 15. Thus, in analyzing the fixed-base structure, non-uniform free-field motion obtained from the soft soil may not provide an accurate estimation of the column forces. On this basis, it is necessary to consider the soil domain in order to obtain a more reliable prediction of structural response that takes into consideration the effects of SSI.

9.5 Summary and Conclusions

The influence of soil-structure interaction (SSI) on seismic response of a highway bridge with a large-scale soil domain has been investigated under excitation provided by three dimensional incident seismic waves propagated from a fault rupture scenario. Responses from the bridge-foundation-ground system were examined and compared with two types of soil profile. For comparison with the full SSI simulation, multiple support excitation and fixed-base uniform excitation were considered using free-field motions obtained from the soil domain in the absence of the structure. From the results of the present study, the following conclusions may be drawn:

- Beneficial or detrimental effects (in terms of peak column shear forces) of soil-structure interaction were discussed. For the stiff soil profile the differences were generally small, but peak shear forces were higher/lower depending on the particular column location. On the other hand, a beneficial effect was observed mostly in all connectors placed on the soft soil profile (for the cases studied).
- Base shear induced by the seismic excitation decreased under the multiple support excitation scenario, while the base shear significantly increased for the uniform excitation case primarily because the magnitude of the representative peak input acceleration (motion at the center of the soil domain) was highest (within the soil region of interest). For the soft case, analysis of the fixed-base bridge (and to some extent the multiple support scenario) resulted in outcomes that were noticeably different as compared to the full SSI simulation (e.g. drift ratios at the top of bents, bending moments in the columns). In this case, the free-field ground motions at the footprints of the supports in the absence of the bridge displayed some marked differences from the actual bridge SSI motions at these locations. Thus, in order to obtain satisfactory predictions of the response of bridge structures supported on soft soil, a detailed analysis that considers the full coupling between the structure and the surrounding soil may become necessary.

Acknowledgements This research was supported by the U.S. National Science Foundation (NSF) under grant OCI-0749227. Through a NSF allocation of advanced computing resources, the numerical simulations were performed on Ranger at the Texas Advanced Computing Center (TACC). Suggestions by Professor J. E. Luco (University of California, San Diego) concerning representation of the pile-group foundation stiffness/geometry are greatly appreciated.

Appendix

The ground responses of the center node at the surface are shown in Fig. 9.A1 for the stiff soil profile and in Fig. 9.A2 for the soft soil profile.

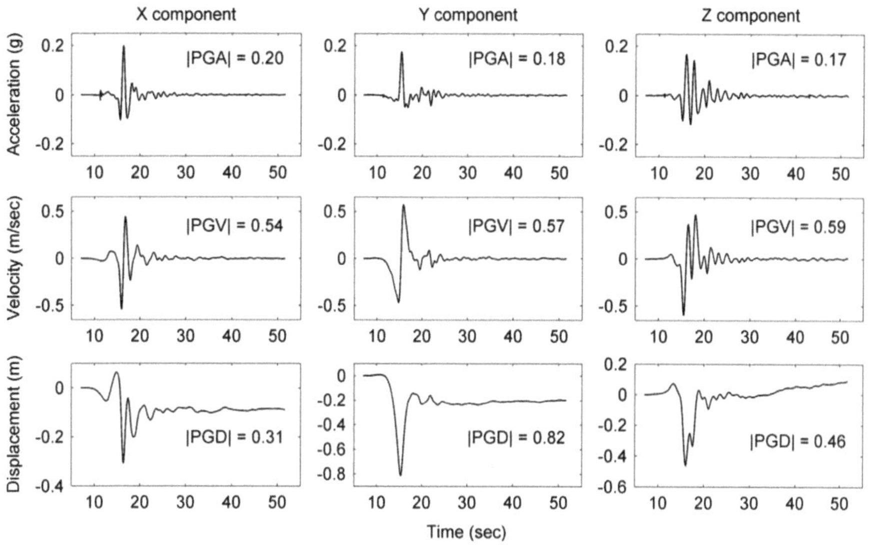

Fig. 9.A1 Ground responses at the surface center node in the stiff soil domain

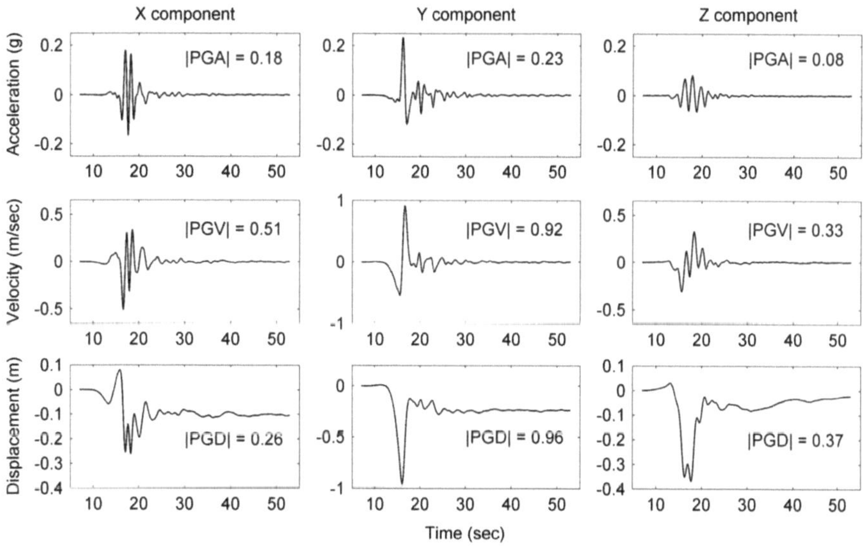

Fig. 9.A2 Ground responses at the surface center node in the soft soil domain

References

Aviram A, Mackie K, Stojadinovic B (2008) Effect of abutment modeling on the seismic response of bridge structures. Earthq Eng Eng Vib 7(4):395–402

Bielak J, Loukakis K, Hisada Y, Yoshimura C (2003) Domain reduction method for three-dimensional earthquake modeling in localized regions, part I: theory. Bull Seismol Soc Am 93(2):817–824

Conte J, Elgamal A, Yang Z, Zhang Y, Acero G, Seible F (2002) Nonlinear seismic analysis of a bridge ground system. In: Proceedings of the 15th engineering mechanics conference ASCE, Reston, VA

Desroches R, Fenves GL (1997) Evaluation of recorded earthquake response of a curved highway bridge. Earthquake Spectra 13(3):363–386

Elgamal A, Yan L, Yang Z, Conte JP (2008) Three-dimensional seismic response of bridge foundation-ground system. J Struct Eng ASCE 134(7):1165–1176

Fenves GL, Desroches R (1995) Evaluation of the response of I-10/215 interchange bridge near San Bernardino in the 1992 Landers and Big Bear earthquakes. Report no. CSMIP/95-02, California Strong Motion Instrumentation Program, Sacramento, CA

Graves RW, Pitarka A (2010) Broadband ground-motion simulation using a hybrid approach. Bull Seismol Soc Am 100(5A):2095–2123

Huang M, Shakal A (1995) CSMIP strong-motion instrumentation and records from the I10/215 interchange bridge near San Bernardino. Earthquake Spectra 11(2):193–215

Jaremprasert S, Bazan-Zurita E, Bielak J (2012) Seismic soil-structure interaction response of inelastic structures. Soil Dyn Earthq Eng. http://dx.doi.org/10.1016/j.soildyn.2012.08.008 (in press, to appear)

Jeremić B, Kunnath S, Xiong F (2004) Influence of soil–foundation–structure interaction on seismic response of the I-880 viaduct. Eng Struct 26(3):391–402

Jeremić B, Jie G, Preisig M, Tafazzoli N (2009) Time domain simulation of soil–foundation–structure interaction in non-uniform soils. Earthq Eng Struct Dyn 38(5):699–718

Ju SH (2004) Three-dimensional analyses of wave barriers for reduction of train-induced vibrations. J Geotech Geoenviron 130(7):740–748

Kim K (2013) Three dimensional seismic response of a large scale highway interchange system. Ph.D. dissertation, Department of Structural Engineering University of California, La Jolla, CA, (in preparation)

Kwon O, Elnashai A (2008) Seismic analysis of Meloland road overcrossing using multiplatform simulation software including SSI. J Struct Eng 134(4):651–660

Lysmer J, Kuhlemeyer R (1969) Finite dynamic model for infinite media. J Eng Mech Div ASCE 95(3):859–877

Massimino MR, Maugeri M (2013) Physical modelling of shaking table tests on dynamic soil-foundation interaction and numerical and analytical simulation. Soil Dyn Earthq Eng 49:1–18

Mosquera V, Smyth A, Betti R (2009) Utilization of strong-motion data for assessment of structural integrity in instrumented highway bridges. SMIP09: 65

Olsen K, Day S, Bradley C (2003) Estimation of Q for long-period (> 2 sec) waves in the Los Angeles basin. Bull Seismol Soc Am 93(2):627–638

Petropoulos G (2008) Soil-structure interaction analysis using high-performance parallel computation. Ph.D. dissertation, Department of Civil and Environmental Engineering, University of California, Berkeley, CA

Pitilakis D, Dietz M, Muir Wood D, Clouteau D, Modaressi A (2008) Numerical simulation of dynamic soil-structure interaction in shaking table testing. Soil Dyn Earthq Eng 8(6):453–467

Priestley MJN, Seible F, Calvi M (1996) Seismic design and retrofit of bridges. Wiley, New York

Zhang Y, Conte JP, Yang Z, Elgamal A, Bielak J, Acero G (2008) Two-dimensional nonlinear earthquake response analysis of a bridge-foundation-ground system. Earthquake Spectra 24(4):343–386

Chapter 10
Seismic Displacement Based Design of Structures: Relevance of Soil Structure Interaction

Gian Michele Calvi, Manuela Cecconi, and Roberto Paolucci

Abstract In this paper a brief summary of the seismic design method known as "*direct displacement – based design*" (DDBD) is presented, with some discussion of the appropriate seismic design input to be used, on the applicability to retaining structures and on the relevance of including nonlinear soil-structure interaction in the DDBD and the tools to account for it, with reference to shallow foundations.

10.1 Introduction

Viewed through the historical prism of the past 100 years, seismic structural design can be seen to have been in constant evolution – much more so than design for other load cases or actions such as gravity, wind, traffic etc. Initially, following structural damage in the seminal earthquakes of the early twentieth century (Kanto, Long Beach, Napier), seismic attack was perceived in terms of simple mass-proportional lateral forces, resisted by elastic structural action. In the 1940s and 1950s the influence of structural period in modifying the intensity of the inertia forces started to be incorporated into structural design, but structural analysis was still based on elastic structural response. Ductility considerations were introduced in the 1960s and 1970s as a consequence of the experimental and empirical evidence that well-detailed structures could survive levels of ground shaking capable of inducing inertia forces many times larger than those predicted by elastic analysis. Predicted

G.M. Calvi (✉)
Head of Graduate Studies, University Institute for Advanced Study (IUSS), Pavia, Italy
e-mail: gm.calvi@iusspavia.it

M. Cecconi
Department of Civil and Environmental Engineering, University of Perugia, Perugia, Italy
e-mail: ceccon@unipg.it

R. Paolucci
Department of Civil and Environmental Engineering, Politecnico di Milano, Milan, Italy
e-mail: roberto.paolucci@polimi.it

M. Maugeri and C. Soccodato (eds.), *Earthquake Geotechnical Engineering Design*, Geotechnical, Geological and Earthquake Engineering 28,
DOI 10.1007/978-3-319-03182-8_10, © Springer International Publishing Switzerland 2014

performance came to be assessed by ultimate strength considerations, using force levels reduced from the elastic values by somewhat arbitrary force-reduction factors, that differed markedly between the design codes of different seismically-active countries. Gradually this lead to a further realization, in the 1980s and 1990s that strength was important, but only in that it helped to reduce displacements or strains, which can be directly related to damage potential, and that the proper definition of structural vulnerability should hence be related to deformations, not strength.

This realization has led to the development of a large number of alternative seismic design philosophies based more on deformation capacity than strength. These are generally termed "performance-based" design philosophies. The scope of these can vary from comparatively narrow structural design approaches, intended to produce safe structures with uniform risk of damage under specified seismicity levels, to more ambitious approaches that seek to also combine financial data associated with loss-of-usage, repair, and a client-based approach (rather than a code-specified approach) to acceptable risk.

This paper does not attempt to provide such ambitious guidance as implied by the latter approach. In fact, it is our view that such a broad-based probability approach is more appropriate to assessment of designed structures than to the design of new structures. The approach taken in this paper is based on providing the designer with improved tools for selecting the best structural alternative to satisfy societal (as distinct from client-based) standards for performance, as defined in what we hope will be the next generation of seismic design codes.

The basis of this approach is the procedure termed "Direct Displacement Based Design" (DDBD), which was first introduced in 1993 (Priestley 1993), and has been subjected to considerable research attention, in Europe, New Zealand, and North America in the intervening years. The fundamental philosophy behind DDBD is that structures should be designed to achieve a specified performance level, defined by strain or drift limits, under a specified level of seismic intensity. As such, we might describe the designed structures as being "uniform-vulnerability" structures, which would be compatible with the concept of "uniform-hazard" spectra, to which we currently design, thus obtaining, in principle, a "uniform risk".

The research effort to develop a viable and simple design approach satisfying this goal has considered a wide range of structural types including, frame buildings, wall buildings, dual wall/frame buildings, bridges, seismically isolated structures and wharves, and a range of structural materials, including reinforced and pre-stressed concrete, structural steel, masonry and timber. The culmination of this research is a book published in May, 2007 (Priestley et al. 2007).

The research project behind the design approach, which can now be considered in a rather complete stage of formulation, has had to re-examine a number of long-held basic tenets of earthquake engineering. The consequences of this re-examination have been:

- A review of the problems associated with initial-stiffness characterization of structures expected to respond inelastically to the design level of seismicity.

- A review of seismological information to provide more appropriate input for displacement-based design.
- A re-examination of some of the fundamentals of inelastic time-history analysis, particularly aspects related to modelling of elastic damping.
- Development of equations relating equivalent viscous damping to ductility demand for different structural systems.
- Development of alternative methodologies for determining design moments in structural members from design lateral forces.
- A re-examination of capacity-design equations for different structural systems, and development of new, ductility-based equations and methodologies.

The focus throughout this lengthy research project has been on the development of practical and simple seismic design methodology, suitable for incorporation in codes in a format reasonably similar to that currently available for the "Equivalent Lateral Force" approach, but with much improved simulation of structural response. The book summarizing the research includes a chapter containing a draft building code, based entirely on DDBD procedures. It is hoped that this might become a platform for future development of seismic codes (Calvi and Sullivan 2009).

More recently, it has been attempted to extend this design approach to foundations, soil-structure interaction and retaining structures. In this paper a brief summary of the method is presented, with some discussion of the appropriate seismic design input to be used, on the applicability to retaining structures and on the relevance of including nonlinear soil-structure interaction in the DDBD and the tools to account for it, with reference to shallow foundations.

10.2 Problems with Seismic Design Using Initial Stiffness and Specified Ductility

Problems with initial-stiffness structural characterization in conventional force-based seismic design, and use of a code-specified ductility capacity have been identified in several previous publications (Priestley 1993, 2003) and will only be briefly listed in this paper:

(a) Initial stiffness is not known at the start of the design process, even if member sizes have been selected. This is because the stiffness depends on the strength. Increasing or decreasing reinforcement content to satisfy results of the force-based design proportionally changes the member stiffness. The same conclusion applies to steel members: changing the flange thickness to satisfy a strength requirement changes the member stiffness almost in direct proportion to the strength change.
(b) Distribution of lateral forces between different parallel structural elements (walls; frames, e.g.) based on elastic estimates of stiffness is illogical and

tends to concentrate strength in elements with greatest potential for brittle failure.

(c) Displacement-equivalence rules relating displacement demand based on initial elastic periods (which are likely, anyway, to be significantly in error) to expected inelastic response are invalid, in our view, being based on incorrect elastic damping estimates used in time-history analysis. This is discussed in more detail subsequently.

(d) Ductility capacity is a function of structural geometry, not just of structural type. Hence it is inappropriate to specify a displacement ductility factor for all structures of the same type (e.g. reinforced concrete frames).

(e) Seismic design of building structures will generally be governed by drift limits, when realistic estimates of building stiffness are used to determine displacements. Current design approaches require iteration to satisfy drift limits, and codes, such as the NZ concrete code, do not recognize the stiffening effect of added strength unless member sizes are also changed.

10.3 Direct Displacement-Based Seismic Design

The fundamentals of **DDBD** are very simple, and have also been presented in many earlier publications (e.g. Priestley 2000, 2003). Only a brief review is included here, with reference to Fig. 10.1 which considers a **SDOF** representation of a frame building (Fig. 10.1a), though the basic fundamentals apply to all structural types. The bilinear envelope of the lateral force-displacement response of the **SDOF** representation is shown in Fig. 10.1b.

While force-based seismic design characterizes a structure in terms of elastic, pre-yield, properties (initial stiffness $\boldsymbol{K_i}$, elastic damping), **DDBD** characterizes the structure by secant stiffness $\boldsymbol{K_e}$ at maximum displacement $\boldsymbol{\Delta_d}$ (Fig. 10.1b), and a level of equivalent viscous damping $\boldsymbol{\xi}$, representative of the combined elastic damping and the hysteretic energy absorbed during inelastic response. Thus, as shown in Fig. 10.1c, for a given level of ductility demand, a structural steel frame building with compact members will be assigned a higher level of equivalent viscous damping than a reinforced concrete bridge designed for the same level of ductility demand, as a consequence of "fatter" hysteresis loops.

With the design displacement at maximum response ($\boldsymbol{\Delta_d}$) determined as discussed subsequently, and the corresponding damping estimated from the expected ductility demand, the effective period $\boldsymbol{T_e}$ at maximum displacement response, measured at the effective height $\boldsymbol{H_e}$ (Fig. 10.1a can be read from a set of displacement spectra for different levels of damping, as shown in the example of Fig. 10.1d. The effective stiffness $\boldsymbol{K_e}$ of the equivalent **SDOF** system at maximum displacement can be found by inverting the normal equation for the period of a **SDOF** oscillator to provide

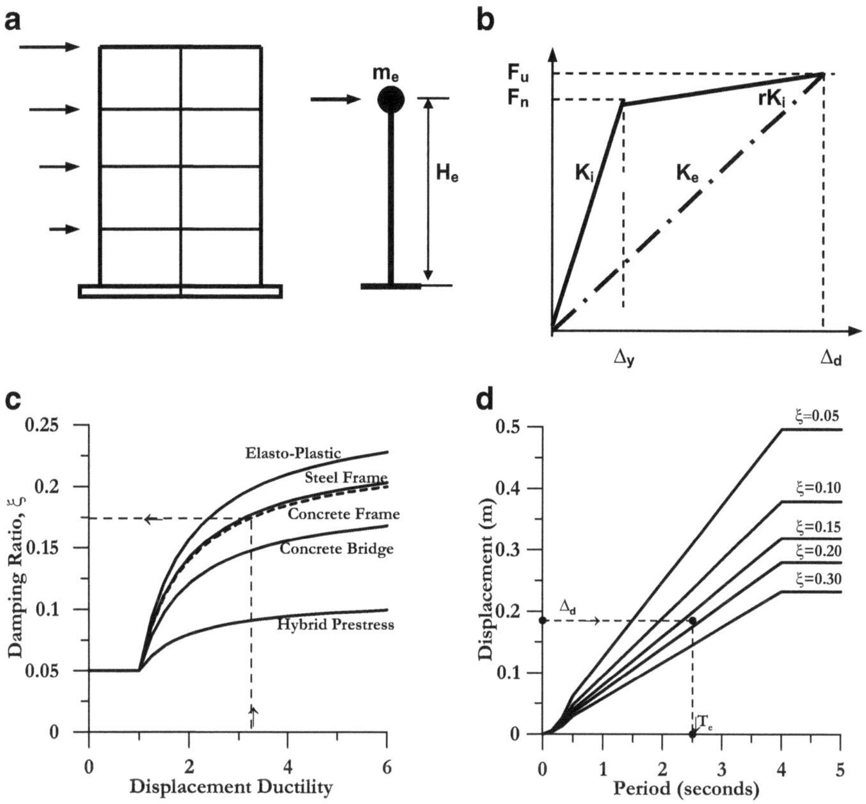

Fig. 10.1 Fundamentals of DDBD (**a**) SDOF simulation. (**b**) Effective stiffness K_e. (**c**) Equivalent damping vs. ductility. (**d**) Design displacement spectra

$$K_e = 4\pi^2 m_e / T_e^2 \tag{10.1}$$

where $\boldsymbol{m_e}$ is the effective mass of the structure participating in the fundamental mode of vibration. From Fig. 10.1b, the design lateral force, which is also the design base shear force is thus

$$F = V_{Base} = K_e \Delta_d \tag{10.2}$$

The design concept is thus very simple. Such complexity that exists relates to determination of the characteristics of the equivalent **SDOF** structure, the determination of the design displacement, and development of design displacement spectra.

10.3.1 Design Displacement

The characteristic design displacement of the substitute structure depends on the limit state displacement or drift of the most critical member of the real structure, and an assumed displacement shape for the structure. This displacement shape is that which corresponds to the inelastic first-mode at the design level of seismic excitation. Thus the changes to the elastic first-mode shape resulting from local changes to member stiffness caused by inelastic action in plastic hinges are taken into account at the beginning of the design. Representing the displacement by the inelastic rather than the elastic first-mode shape is consistent with characterizing the structure by its secant stiffness to maximum response. In fact, the inelastic and elastic first-mode shapes are often very similar.

The design displacement of the equivalent **SDOF** structure (the generalized displacement coordinate) is thus given by

$$\Delta_d = \sum_{i=1}^{n} \left(m_i \Delta_i^2\right) / \sum_{i=1}^{n} \left(m_i \Delta_i\right) \tag{10.3}$$

where $\boldsymbol{m_i}$ and $\boldsymbol{\Delta_i}$ are the masses and displacements of the $\boldsymbol{n}$ significant mass locations respectively. For multi-storey buildings, these will normally be at the $\boldsymbol{n}$ floors of the building. For bridges, the mass locations will normally be at the center of the mass of the superstructure above each column, but the superstructure mass may be discretized to more than one mass per span to improve validity of simulation. With tall columns, such as may occur in deep valley crossings, the column may also be discretized into multiple elements and masses.

Where strain limits govern, the design displacement of the critical member can be determined by integration of the curvatures corresponding to the limit strains. Similar conclusions apply when code drift limits apply. For example, the design displacement for frame buildings will normally be governed by drift limits in the lower storeys of the building. For a bridge, the design displacement will normally be governed by the plastic rotation capacity of the shortest column. Knowing the displacement of the critical element and the design displacement shape, the displacements of the individual masses are given by

$$\Delta_i = \delta_i \cdot \left(\frac{\Delta_c}{\delta_c}\right) \tag{10.4}$$

where $\boldsymbol{\delta_i}$ is the inelastic mode shape, $\boldsymbol{\Delta_c}$ is the design displacement at the critical mass, $\boldsymbol{c}$, and $\boldsymbol{\delta_c}$ is the value of the mode shape at mass $\boldsymbol{c}$. Specific details on structural mode shapes for **DDBD** of different structural types are given in the book (Priestley et al. 2007).

Note that the influence of higher modes on the displacement and drift envelopes is generally small, and is not considered at this stage in the design. However, for buildings higher than (say) ten storeys, dynamic amplification of drift may be important, and the design drift limit may need to be reduced to account for this. This factor is considered in detail in the relevant structural design chapters.

10.3.2 Effective Mass

From consideration of the mass participating in the first inelastic mode of vibration, the effective system mass for the substitute structure is

$$m_e = \sum_{i=1}^{n} (m_i \Delta_i)/\Delta_d \tag{10.5}$$

where $\boldsymbol{\Delta_d}$ is the design displacement given by Eq. (10.3). Typically, the effective mass will range from about 70 % of the total mass for multi-storey cantilever walls to more than 85 % for frame buildings of more than 20 storeys. For simple multi-span bridges the effective mass will often exceed 95 % of the total mass. The remainder of the mass participates in the higher modes of vibration. Although modal combination rules, such as the *square-root-sum-of-squares* (**SRSS**) or *complete quadratic combination* (**CQC**) rules may indicate a significant increase in the elastic base shear force over that from the first inelastic mode, there is much less influence on the design base overturning moment. The effects of higher modes are inadequately represented by elastic analyses and are better accommodated in the capacity design phase, rather than the preliminary phase of design.

10.3.3 Structure Ductility Demand

Determination of the appropriate level of equivalent viscous damping requires that the structural ductility be known. This is straightforward since the design displacement has already been determined, and the yield displacement depends only on geometry, not on strength. Relationships for yield curvature of structural elements, such as walls, columns, beams etc. have been established (Priestley 2003) in the general form:

$$\phi_y = C_1 \varepsilon_y / h \tag{10.6}$$

where $\boldsymbol{C_1}$ is a constant dependent on the type of element considered, $\boldsymbol{\varepsilon_y}$ is the yield strain of the flexural reinforcement and $\boldsymbol{h}$ is the section depth. Yield drifts for concrete and steel frames can be expressed as

$$\theta_y = C_2 \varepsilon_y \frac{L_b}{h_b} \tag{10.7}$$

where $\boldsymbol{L_b}$ and $\boldsymbol{h_b}$ are the beam span and depth, respectively, and $\boldsymbol{C_2} = 0.5$ and 0.65 for concrete and structural steel respectively. For building structures, Eqs. (10.6) and (10.7) can readily be integrated to obtain the displacement at the effective height, $\boldsymbol{H_e}$, given by

$$H_e = \sum_{i=1}^{n} (m_i \Delta_i H_i) / \sum_{i=1}^{n} (m_i \Delta_i) \tag{10.8}$$

where $\boldsymbol{H_i}$ are the heights of the $\boldsymbol{n}$ storeys.

The displacement ductility demand for the structures is thus known at the start of the design, by Eq. (10.9), even though the strength is not yet established:

$$\mu = \Delta_d / \Delta_y \tag{10.9}$$

From relationships between structural type, ductility demand, and equivalent viscous damping, discussed in the following section, the appropriate level of elastic damping to use in Fig. 10.1d can be directly obtained, and hence the base shear force calculated (Eq.10.2). This base shear force is then distributed to the structural masses in accordance with Eq. (10.10), and the structure analyzed, as also discussed subsequently.

$$F_i = V_{Base}(m_i \Delta_i) / \sum_{i=1}^{n} (m_i \Delta_i) \tag{10.10}$$

10.4 Equivalent Viscous Damping

A key element of **DDBD** is that hysteretic damping is modeled by equivalent viscous damping (**EVD**), using relationships such as those presented in Fig. 10.1c. The total equivalent damping is the sum of elastic, $\boldsymbol{\xi_{el}}$ and hysteretic, $\boldsymbol{\xi_{hyst}}$ damping:

$$\xi_{eq} = \xi_{el} + \xi_{hyst} \tag{10.11}$$

Both components need some examination.

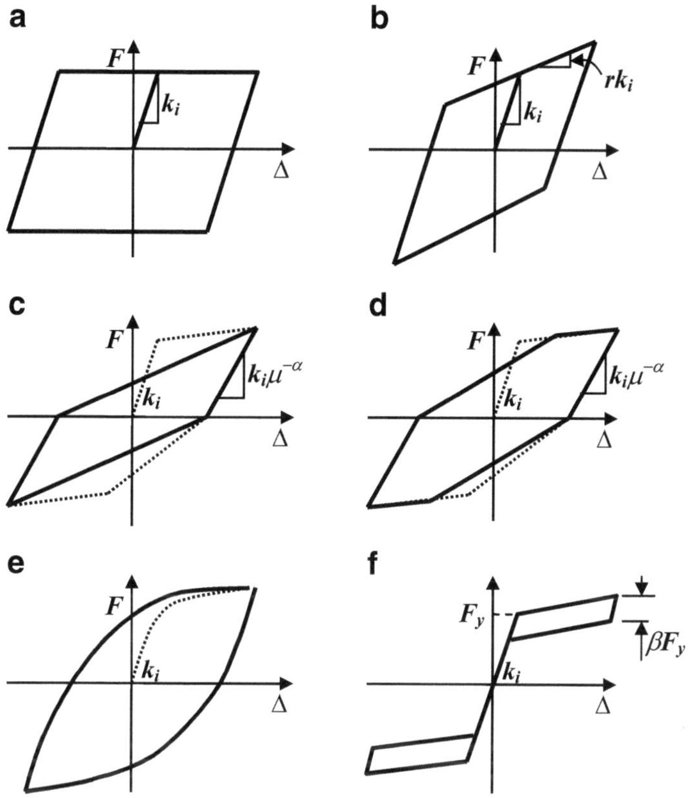

Fig. 10.2 Hysteresis rules considered in time-history analyses (**a**) Elasto-plastic (*EPP*). (**b**) Bi linear, r = 0.2. (**c**) Takeda "Thin" (*TT*). (**d**) Takeda "Fat" (*TF*). (**e**) Ramberg-Osgood (*RO*). (**f**) Flag Shaped (*FS*)

10.4.1 Hysteretic Component

The approach adopted has been to use values of **EVD** that have been calibrated for different hysteresis rules (see Fig. 10.2, for examples) to give the same peak displacements as the hysteretic response, using inelastic time-history analysis. Two independent studies, based on different methodologies were used to derive the levels of equivalent viscous damping. The first involved the use of a large number of real earthquake accelerograms (Dwairi et al. 2007), where the equivalent viscous damping was calculated for each record, ductility level, effective period and hysteresis rule separately, and then averaged over the records to provide a relationship for a given rule, ductility, and period. The second study (Grant et al. 2005), using a wider range of hysteresis rules was based on a smaller number of spectrum-compatible artificial accelerograms where the results of the elastic and inelastic analyses were separately averaged, and compared. In each case the equivalent viscous damping was varied until the elastic results of the equivalent substitute structure matched that of the real hysteretic model.

The damping coefficient, and hence the damping force depends on what value of stiffness is adopted. In most inelastic analyses, this has been taken as the initial stiffness. This, however, results in large and spurious damping forces when the response is inelastic, which, it has been argued in a previous NZSEE conference (Priestley et al. 2007) is inappropriate, and that tangent stiffness should be used as the basis for elastic damping calculations. With tangent stiffness, the damping coefficient is proportionately changed every time the stiffness changes, associated with yield, unloading or reloading, etc. This results in a reduction in damping force as the structural stiffness softens following yield, and a reduction in the energy absorbed by the elastic damping.

The two studies initially were carried out without additional elastic damping, for reasons that will become apparent in the following section. It was found that the approaches resulted in remarkably similar relationships for equivalent viscous damping for all hysteresis rules except elastic-perfectly plastic (**EPP**), where the discrepancy was about 20 %. It is felt that the difference for the **EPP** rule is a consequence of the use of real records, with comparatively short durations of strong ground motion in the first study, and artificial records, with longer strong ground motion durations in the second study. **EPP** hysteresis is known to be sensitive to duration effects.

10.4.2 Elastic Component

Elastic damping is used in inelastic time-history analysis to represent damping not captured by the hysteretic model adopted for the analysis. This may be from the combination of a number of factors, of which the most important is the typical simplifying assumption in the hysteretic model of perfectly linear response in the elastic range (which therefore does not model damping associated with the actual elastic non-linearity and hysteresis). Additional damping also results from foundation compliance, foundation non-linearity and radiation damping, and additional damping from interaction between structural and non-structural elements.

Since the hysteretic rules are invariably calibrated to model the full structural energy dissipation subsequent to onset of yielding, this approach to characterization of the elastic damping is clearly more appropriate than is initial-stiffness elastic damping. The significance to structural response of using tangent-stiffness rather than initial-stiffness damping has been discussed in detail in Priestley et al. (2007).

However, in **DDBD**, the initial elastic damping ratio adopted in Eq. (10.11) is related to the secant stiffness to maximum displacement, whereas it is normal in inelastic time-history analysis to relate the elastic damping to the initial (elastic) stiffness, or more correctly, as noted above, to a stiffness that varies as the structural stiffness degrades with inelastic action (tangent stiffness). Since the response velocities of the "real" and "substitute" structures are expected to be similar under seismic response, the damping forces of the "real" and "substitute" structures, which are proportional to the product of the stiffness and the velocity will differ significantly, since the effective stiffness $\boldsymbol{k_e}$ of the substitute structure is approximately equal to $\boldsymbol{k_{eff}} = \boldsymbol{k_i}/\boldsymbol{\mu}$ (for low post-yield stiffness). Priestley and Grant (2005) have determined the adjustment that would be needed to the value of the elastic damping assumed in DDBD (based on either initial-stiffness or tangent-stiffness proportional damping) to ensure compatibility between the "real" and "substitute" structures. Without such an adjustment, the verification of DDBD by inelastic time-history analysis would be based on incompatible assumptions of elastic damping.

The adjustments depend on whether initial-stiffness damping (conventional practice), or tangent-stiffness damping (correct procedure, we believe) is adopted for time-history analysis. If initial-stiffness damping is chosen, the elastic damping coefficient used in **DDBD** must be larger than the specified initial-stiffness damping coefficient; if tangent-stiffness is chosen, it must be less than the specified tangent-stiffness coefficient. It is possible to generate analytical relationships between the substitute-structure and "real" structure elastic damping coefficients that are correct for steady-state harmonic response. However, as with the hysteretic component, these are not appropriate for transient response to earthquake accelerograms, though the trends from time-history response follow the form of the theoretical predictions. Hence, to obtain the appropriate correction factors, it is again necessary to rely on the results of inelastic time-history analyses. Priestley and Grant (2005) compared results of elastic substitute-structure analyses with inelastic time history results to determine the correction factor to be applied to the elastic damping coefficient for the assumptions of either initial-stiffness or tangent-stiffness elastic damping. The form of Eq. (10.11) is thus slightly changed to:

$$\xi_{eq} = \kappa \xi_{el} + \xi_{hyst} \tag{10.12}$$

The correction factor $\boldsymbol{\kappa}$ is plotted for different hysteresis rules, and for different initial damping assumptions (initial-stiffness or tangent-stiffness) in Fig. 10.3. It is possible to include the ductility dependency of the elastic damping inside the basic form of the equivalent viscous damping equations. With the usual assumption of 5 % elastic damping, the damping –ductility relationships can be expressed in the general form:

$$\xi_{eq} = 0.05 + C_3\left(\frac{\mu - 1}{\mu\pi}\right) \tag{10.13}$$

where the coefficient C_3 varies between 0.1 and 0.7 (for the assumption of tangent-stiffness elastic damping) depending on the hysteresis rule appropriate for the structure under design (Priestley et al. (2007)).

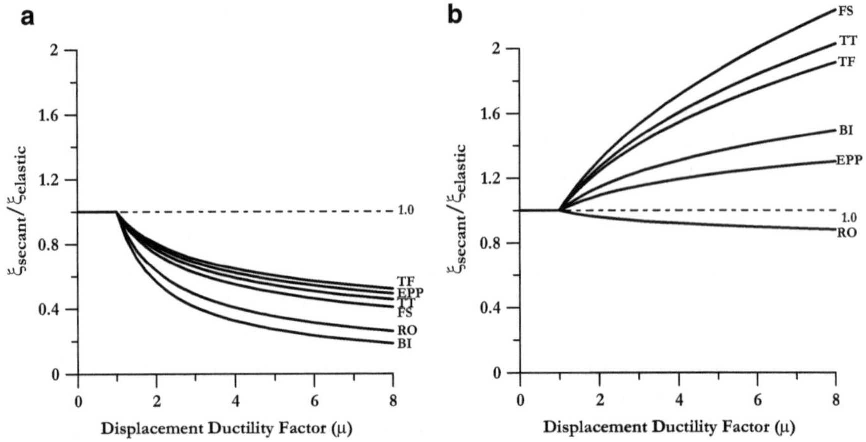

Fig. 10.3 Secant stiffness equivalent elastic viscous damping related to initial elastic stiffness and elastic damping model. (**a**) Tangent stiffness damping. (**b**) Initial stiffness damping

10.4.3 Damping in Mixed Systems

In conventional force-based design there is some difficulty in assessing the appropriate level of ductility to use in determining required base-shear strength for mixed systems, such as wall/frames, or even wall structures with walls of different length in a given direction. In **DDBD** the procedure is straightforward, with the **EVD** of the different structural lateral-force-resisting elements being separately calculated and then combined, weighted by the proportion of base-shear force (or overturning moment) carried. Details are presented in the Priestley et al. (2007).

10.5 Displacement Response Spectra

10.5.1 Displacement Spectra for Design

Displacement-based seismic design uses displacement-defined limit states rather than acceleration or force-based considerations to determine the required strength. This is because damage is strain related (structural damage) or drift related (non-structural damage). It is straightforward to determine the design displacement corresponding to given limit state, such as the serviceability limit state (no damage requiring interruption of normal operations) or damage-control (economically repairable damage). There is, however, no valid relationship between strength and damage potential.

A fundamental aspect of the displacement-based design approach is to use displacement response spectra, defined for different levels of equivalent viscous

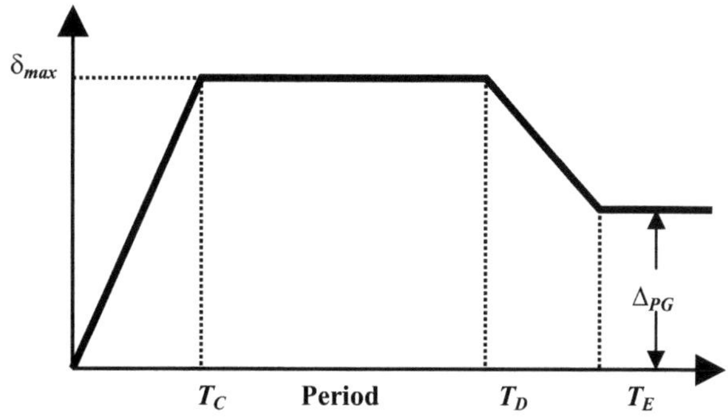

Fig. 10.4 General characteristics of elastic displacement response spectra

damping, rather than acceleration response spectra as the means for defining design seismic intensity. Although many codes do not define displacement spectra, they are becoming more common. Ideally these should be developed separately, though using the same data, from acceleration spectra. However, most code-based design displacement spectra are generated from the acceleration spectra assuming that the peak response is governed by the equations of steady-state sinusoidal response. Thus the relationship between displacement and acceleration can be expressed as

$$\Delta_{(T)} = \frac{T^2}{4\pi^2} S_{A(T)} g \tag{10.14}$$

where $\boldsymbol{g}$ is the acceleration due to gravity, and $\boldsymbol{S_{A(T)}}$ is expressed as a multiply of $\boldsymbol{g}$. The general, but idealized, characteristics of displacement spectra are illustrated in Fig. 10.4. Apart from an initial period range up to about 0.5 s, the displacement spectra can be considered to increase linearly up to a period $\boldsymbol{T_C}$, termed the corner period, for obvious reasons, with a subsequent plateau of displacement up to a period $\boldsymbol{T_D}$, followed by a decrease in displacement up to a period $\boldsymbol{T_E}$, at which stage the response displacement has decreased to the peak-ground displacement (***PGD***). Information on the period at which the response displacement starts to decrease is less reliable than data on other key periods in Fig. 10.4, and is of little interest to the designer of other than extremely long-period structures such as suspension bridges and large-diameter fluid storage tanks, where convective modes of vibration may be very long.

10.5.2 Hazard Maps for Italy in Terms of Spectral Displacements

Recently, considerable research attention has been focused by seismologists on improving the accuracy of prediction of displacement spectra, traditionally limited at long periods by insufficient signal-to-noise ratio of seismic instruments. Such

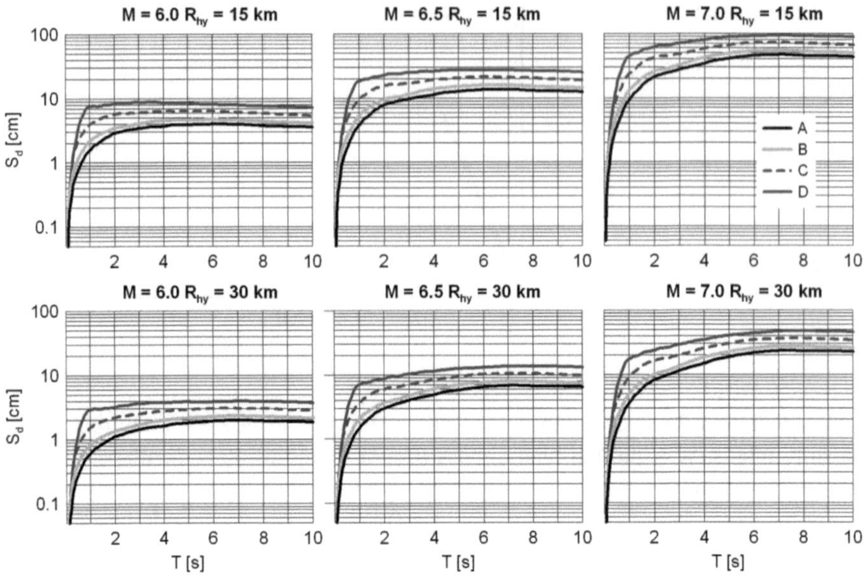

Fig. 10.5 Influence of magnitude, distance and site conditions, defined in terms of the EC8 site classes, on 5 % damped displacement spectra, based on the Cauzzi and Faccioli (2008) broadband attenuation relationship

research was supported by an ever increasing number of high quality digital records, especially coming from the Japanese K-Net (www.k-net.bosai.go.jp) and Kik-net (www.kik.bosai.go.jp) strong motion networks and from other regions worldwide as well. After the study by Faccioli et al. (2004), who proposed simplified theoretical models to explain the observed features of displacement spectra, Cauzzi and Faccioli (2008) calibrated a broadband, up to 20 s, empirical prediction equation of displacement spectral ordinates, based on a worldwide digital strong motion dataset, consisting of about 1,200 three-components records. This research was supported by several studies (e.g., Paolucci et al. 2008a, b; Akkar and Boore 2009), who proved that response spectral ordinates from digital records are reliable up to periods much longer than usually thought, beyond at least 10 s in most cases. Median values of 5 % spectral displacements according to the Cauzzi and Faccioli (2008) ground motion prediction equation for different values of magnitude, distance and site conditions (EC8 site classes) are illustrated in Fig. 10.5.

Taking advantage of the previous results, and as the main product of the Project S5 (Faccioli and Rovelli 2007) funded by the Italian Department of Civil Protection (DPC), hazard maps in terms of spectral displacements were produced for the Italian territory, with the objective to provide uniform hazard spectra at long periods suitable for applications of DDBD. The hazard maps for Italy in terms of PGA and of the 10 s spectral displacement (D_{10}), referring to the same probability of exceedance of 10 % in 50 years, are compared in Fig. 10.6. It is immediate to note the relevant differences of the two maps, where the areas with frequent

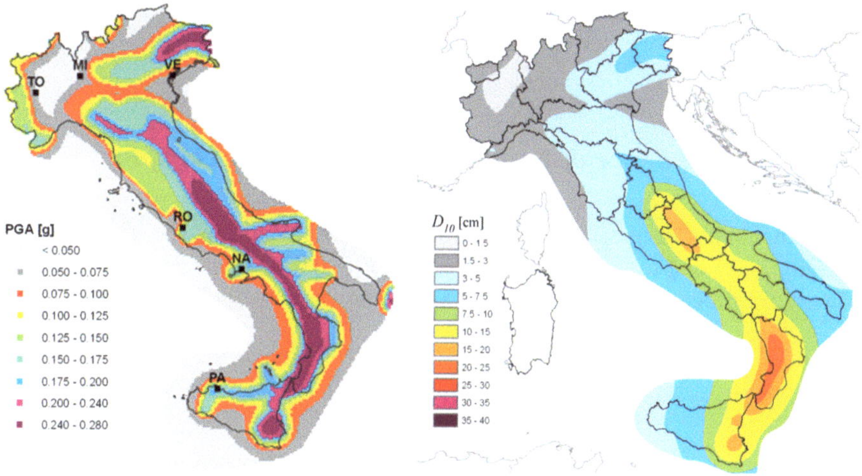

Fig. 10.6 Seismic hazard maps of Italy in terms of acceleration (*left*) and displacement (*right*). PGA and D_{10} are the peak ground acceleration and the spectral displacement at a period of 10 s, expected with a probability of exceedance of 10 % in 50 years (Faccioli and Villani 2009)

seismicity but low values of the maximum expected magnitude (like Piemonte, Liguria and Northern Appennines) tend to provide relatively large values of PGA but low D_{10}. The interested reader is referred to Faccioli and Villani (2009) who provided a comprehensive introduction of the hazard maps for Italy in terms of spectral displacements.

10.5.3 Ground Motion Selection for Displacement-Spectrum Compatibility

A further step towards the definition of seismic input for DDBD is presently under way in the framework of the RELUIS (Italian Network of University Labs for Earthquake Engineering) research activities funded by the DPC (2010–2013).

The REXEL-DISP software (Iervolino et al. 2012) has been produced for computer-aided real record selection based on displacement spectral compatibility and made available in the beta version in the Internet (http://www.reluis.it). Among the main features of REXEL-DISP there are: (i) definition of a target displacement spectrum based on the results of Project S5, in addition to user-defined spectra; (ii) implementation of a strong motion database made of records with high signal-to-noise ratio at long periods, with magnitude ranging from 5 to 7.3 and epicentral distances up to 30 km. The application of REXEL-DISP to real cases in Italy (Smerzini et al. 2013) has made clear that combining in the search procedure well calibrated target displacement spectra together with a suitable dataset provides a rational approach to get suites of displacement-spectrum compatible accelerograms with no need of amplitude scaling and broadband spectral compatibility, suitable both for short and long periods analyses.

10.5.4 Dependence of Displacement-Spectra on Damping

A commonly used expression for relating the displacement response spectrum for a damping ratio of ξ to the elastic spectrum for $\xi = 0.05$ was presented in the 1998 edition of Eurocode EC8 (CEB 1998), and is shown below in Eq. (10.15):

$$R_\xi = \left(\frac{0.07}{0.02 + \xi}\right)^{0.5} \tag{10.15}$$

Although a more recent edition of EC8 (CEB 2003) revised the coefficients of numerator and denominator of Eq. (10.15) to 0.10 and 0.05 respectively, our analyses of real and artificial records convince us that the form given in Eq. (10.15) gives a better representation of the damped spectra.

Equation (10.15) applies to sites where forward directivity effects are not apparent. It would also be desirable to have an equivalent expression for sites where forward directivity velocity pulse characteristics might be expected. It has been suggested (Priestley 2003) based on limited data, that a modification to the 1998 EC8 expression given by:

$$R_\xi = \left(\frac{0.07}{0.02 + \xi}\right)^{0.25} \tag{10.16}$$

might be appropriate. Some qualified support for Eq. (10.16) is available in Bommer and Mendis (2005) which provides additional discussion of this topic. The influence of Eqs. (10.15) and (10.16) on the elastic displacement spectra is illustrated in Fig. 10.7. It will be noted that the influence of damping in reducing the elastic displacement spectrum is much less apparent for the velocity pulse condition.

10.6 Some Remarks on DDBD for Retaining Structures

At present, the application of DDBD to geotechnical systems is limited to shallow foundations (Paolucci et al. 2009, 2013) and retaining structures (Pane et al. 2007; Cecconi et al. 2009). As a matter of fact, the main difficulties concerning the possible use of DDBD for retaining structures lie in:

- simplifying and reducing the seismic behaviour of the complex soil/structure system to a simple equivalent single-degree-of-freedom (SDOF) system;

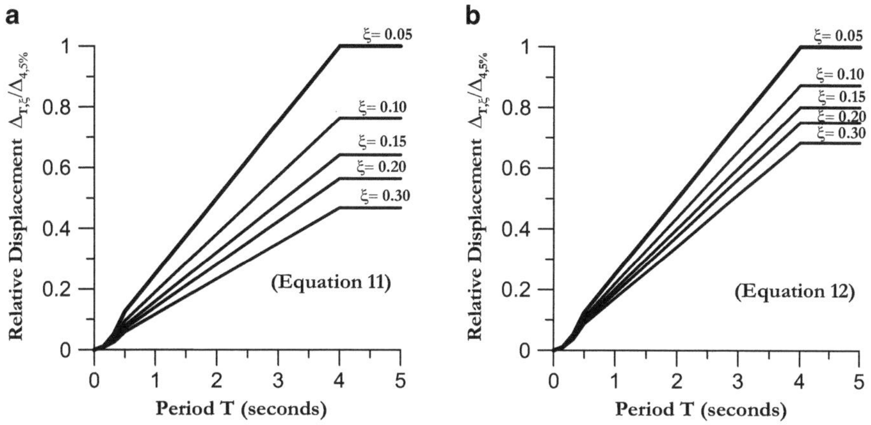

Fig. 10.7 CEB (1998) damping reduction factor, and tentative factor (Eq.10.16) for forward directivity effects. (**a**) "Normal" conditions. (**b**) Velocity pulse conditions

- the definition of the soil masses interacting with the structure during the seismic motion;
- the evaluation of equivalent damping (ξ_e) of the soil/wall system.

On the other hand, the major benefit of DBD method for retaining structures can be actually recognized in providing a rational seismic design approach for retaining walls, based on a convenient reduction of seismic thrusts associated to the capacity of the system of dissipating energy (ductility capacity).

With regard to embedded cantilever walls, the first two issues are discussed in a paper submitted to this Conference that focuses on emphasizing the effects of a proper choice of participating masses (active and passive wedges), of different soil shear strength (ϕ_k) and seismic demand on DDBD results (Cattoni et al. 2012). In that paper a simplification of the design process is proposed, by providing non-dimensional charts for the evaluation of "reduction factors" of seismic thrust, as a function of seismic design displacement.

The application of DDBD procedure to anchored retaining structures (Calvi et al. 2008; Cecconi et al. 2008) is synthetically presented in the following in order to exemplify the development of the procedure and discuss some other pending and debatable questions which still need further study.

10.6.1 Application of DDBD to Anchored Diaphragm Walls

Figure 10.8 represents the position of the problem for DBD analysis. An anchored diaphragm wall retains a vertical excavation (h) in coarse-grained soils (Fig. 10.8a). For simplicity, it is assumed that the anchor/strut is located at the top of the wall and

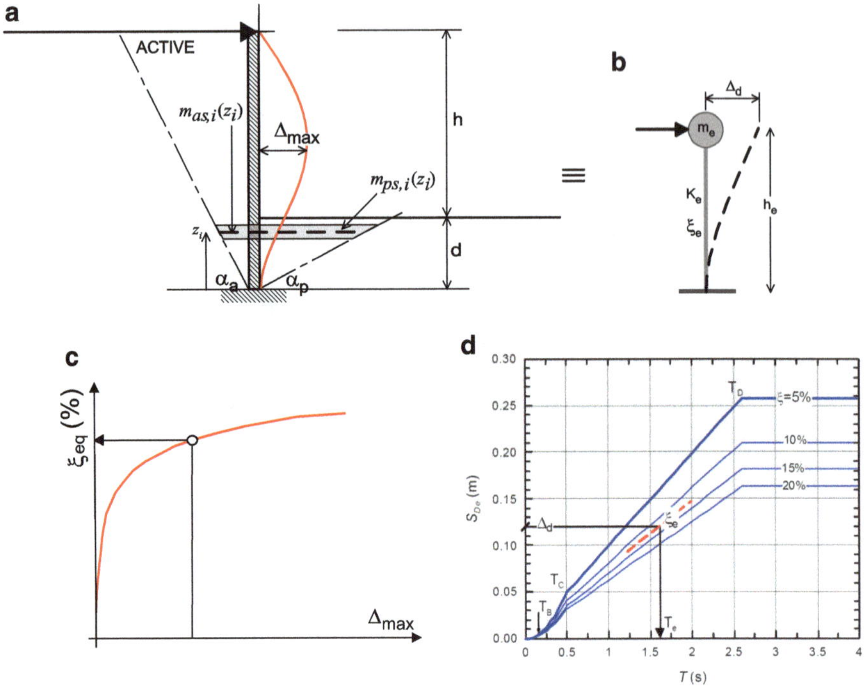

Fig. 10.8 Fundamentals of DDBD and application to anchored retaining walls. (**a**) Section of the retaining wall. (**b**) One DOF model. (**c**) Equivalent damping vs. maximum displacement demand. (**d**) Displacement spectra as a function of equivalent damping

that ground water is absent. An appropriate embedment depth, *d*, satisfies the fundamental equation $E_\mathrm{d} \leq R_\mathrm{d}$ in seismic conditions.

As a fundamental of DBD, the starting point of the procedure consists in modeling the structure as an equivalent SDOF system defined by equivalent values of mass, m_e and damping, ξ_e (Fig. 10.8b). The design process is then developed by assuming that the seismic performance of such system is characterized by its secant stiffness (K_e) at maximum displacement, rather than its initial tangent elastic properties.

For the case at hand, an assumed displaced shape has been chosen according to the *fixed earth support* scheme, with no rotation or horizontal/vertical displacements being permitted at the toe. Thus, the shape of the displacement pattern of the system has been assumed to match with the elastic 1st mode of vibration of a supported-cantilever beam. The maximum seismic displacement (Δ_{max}) – which approximately occurs at a location of 0.4 of the total wall length from the top of the wall – can be set by the designer to ensure an acceptable level of displacement. Obviously, the choice of an acceptable displacement value results from due consideration of the damage/collapse of both structural (wall, anchors, struts, etc.) and non-structural items, and it is often governed by the surrounding environment.

For the given displaced shape of the structure, the profile is scaled until the most critical element in the structure reaches its maximum seismic displacement Δ_{max}. The design displacement Δ_d induced by the earthquake on the SDOF equivalent system of given period, T_e, and damping, ξ_e is then given by Eq. (10.3).

The participating masses are now conventionally represented by the active soil wedge behind the wall and the passive wedge in front of it (see Fig. 10.8a); the soil wedges are bounded by planar surfaces with inclinations α_a and α_p which – generally – are both a function of soil friction angle (ϕ), friction angle (δ) at the soil/wall contact and horizontal seismic coefficient (k_h). Also, the weight of the structure is considered, even if its contribution is generally small with respect to the soil masses. The equivalent mass of the system can be evaluated through Eq. (10.5).

The equivalent damping ratio of the soil/wall system, ξ_e, is estimated from the displacement demand assuming that appropriate damping can be calculated for different structural and non-structural elements of the system, *i.e.* the soil and the wall. For retaining structures, the evaluation of equivalent damping – corresponding to an expected level of ductility demand – is a crucial point of the design process, considering that energy dissipation is mostly provided by cyclic deformation of soil in both passive and active wedges, as further discussed in what follows. The equivalent damping for the soil/wall system can be calculated through Eq. (10.17), by scaling the soil damping in the active and passive wedges (ξ_{as}, ξ_{ps}) and the wall damping (ξ_w) over their relative masses,

$$\xi_{eq} = \frac{\xi_{as}\sum_{i=1}^{n} m_{as,i} + \xi_{ps}\sum_{i=1}^{n} m_{ps,i} + \xi_w\sum_{i=1}^{n} m_{w,i}}{\sum_{i=1}^{n}\left(m_{as,i} + m_{ps,i} + m_{w,i}\right)} \tag{10.17}$$

or by using Eq. (10.18), formerly developed for combined structural systems with different and independent (n) sources of energy dissipation:

$$\xi_{eq,sys} = \frac{\xi_1 V_1 \Delta_1 + \xi_2 V_2 \Delta_2 + \ldots. + \xi_n V_n \Delta_n}{V_{Base}\Delta_d} \tag{10.18}$$

where the shear components $V_1, \ldots V_n$ are referred to the resistance offered to the equivalent SDOF system (Model Code, DBD09). For the case at hand, Eq. (10.18) can be further arranged, thus resulting into:

$$\xi_e = \xi_{as} \cdot \frac{\sum_{i=1}^{n} m_{as,i} \cdot \Delta_i^2}{\sum_{i=1}^{n} m_i \cdot \Delta_i^2} + \xi_{ps} \cdot \frac{\sum_{i=1}^{n} m_{ps,i} \cdot \Delta_i^2}{\sum_{i=1}^{n} m_i \cdot \Delta_i^2} + \xi_w \cdot \frac{\sum_{i=1}^{n} m_{w,i} \cdot \Delta_i^2}{\sum_{i=1}^{n} m_i \cdot \Delta_i^2} \tag{10.19}$$

From a numerical point of view, Eqs. (10.17) and (10.19) lead to comparable

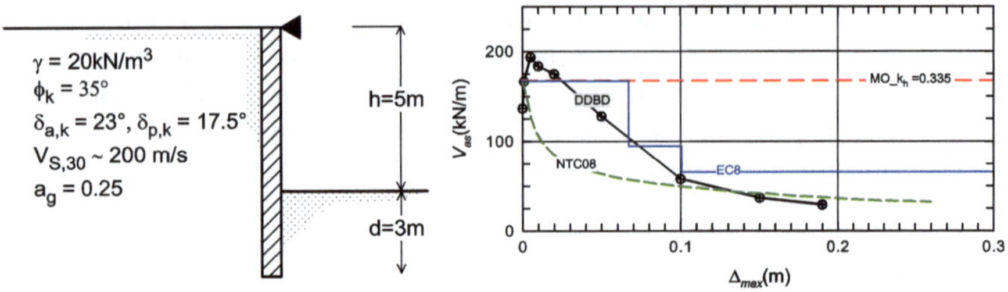

Fig. 10.9 Anchored retaining wall: geometry, material properties and DBD results (Adapted from Cecconi et al. 2009)

results. The evaluation of soil damping (ξ_{as}, ξ_{ps}) is further dealt with in a following section.

To evaluate the wall damping, ξ_w, Eq. (10.13) is used, with the displacement ductility demand given by Eq. (10.9), considering the yield displacement Δ_y as a function of the yield curvature of the wall at yield, ϕ_y (Priestley 2003).

Once the equivalent damping of the system is estimated from the expected ductility demand (Fig. 10.8c), the effective period T_e at maximum displacement response measured at the effective height, h_e (Fig. 10.8b and Eq. 10.8) can be directly read from a set of displacement spectra for different levels of damping. The effective stiffness of the equivalent single-degree of freedom system is then estimated using Eq. (10.1):

Finally, the procedure comes to an end with the evaluation of the design active seismic increment, V_{as}, and passive seismic reduction, V_{ps}, respectively given by:

$$V_{as} = \sum_{i=1}^{n} V_{as,i} \qquad V_{as,i} = K_e \Delta_d \cdot \frac{m_{as,i}\Delta_i}{\sum_{i=1}^{n} \left(m_{as,i} + m_{ps,i} + m_{w,i}\right)\Delta_i} \tag{10.20a}$$

$$V_{ps} = \sum_{i=1}^{n} V_{ps,i} \qquad V_{ps,i} = K_e \Delta_d \cdot \frac{m_{ps,i}\Delta_i}{\sum_{i=1}^{n} \left(m_{as,i} + m_{ps,i} + m_{w,i}\right)\Delta_i} \tag{10.20b}$$

Equations (10.20a and 10.20b) also provide the distribution of the active/passive seismic forces along the wall or, alternatively, the position of their application point.

An example of the results obtained from the DDBD procedure is shown in Fig. 10.9 for an anchored embedded wall retaining a vertical excavation of height, $h = 5$ m, in a cohesionless coarse-grained soil classified as Ground Type C (Italian Codes, D.M. 2008); the peak ground acceleration is assumed as $a_g = 0.25$. For such a wall, the embedment depth resulting from for stability calculation, according to EC8 (Combination C2 of Design Approach 1) is $d = 3$ m.

In Fig. 10.9, DDBD results are compared with those obtained from Mononobe-Okabe (MO) pseudo-static analyses (horizontal dashed line, represented for horizontal seismic coefficient, $k_h = 0.335$). For rather small displacements ($\Delta_{\max} < 0.05$ m), a satisfactory agreement is found between the seismic increment V_{as} of active thrust, and the corresponding pseudo-static MO increment.

For larger displacements, the DDBD procedure, by taking into account the ductility capacity of the soil-structure system, leads to lower seismic forces.

The reduction is rather significant; in fact for $\Delta_{\max} = 0.1$ m, the active seismic increment is approximately 40 % of the one calculated with the MO equation. On the other hand, the seismic thrust reduction obtained from DDBD analysis appears in a good agreement with that calculated applying the reduction factors, r, proposed by EC8 for rigid cantilever and gravity walls, and – to some extent – more conservative than that evaluated following the recent Italian code (D.M. 2008). Similar results have been obtained for the case study of a cantilever embedded retaining wall (Cecconi et al. 2009).

From this analysis, it can be concluded that the DDBD method for retaining structures leads to a clear relationship between the seismic (active and passive) thrusts and the wall displacement, providing a significant reduction of the seismic demand with respect to common pseudo-static approaches.

10.6.2 Simplified Procedure for Seismic Thrust Reduction

The procedure described above for anchored retaining walls – which can be generalized for other wall types and geometries – requires the definition of the participating soil masses and of the damping ratio of the soil/wall system. A great advantage, in terms of applicability of the method for earth retaining structures and related feasibility, is obtained by introducing normalization criteria and non-dimensional charts for the evaluation of "reduction factors" of seismic thrust, as a function of allowable seismic displacement Δ_{max}. This is briefly described in this section with regard to a cantilever embedded wall (Fig. 10.10a), whereas it is fully described in (Cattoni et al. 2012).

In order to find out the appropriate non-dimensional groups, the inclination of the active and passive Coulomb soil wedges (representing the participating masses) has been varied in a wide range of values, by introducing a non-dimensional mass ratio "a" given by:

$$a_{as} = \frac{m_{as}}{m_{as\_case\ study}} \qquad a_{ps} = \frac{m_{ps}}{m_{ps\_case\ study}} \tag{10.21}$$

for the active and passive wedge, respectively (see Fig. 10.10b). Despite of the large difference exhibited by the resulting V_{as} curves, when the non-dimensional ratio $V_{as}/V_{as,max}$ is plotted against the maximum seismic displacement, Δ_{max}, the results fall approximately on a single curve (see Fig. 10.10c). The normalization appears

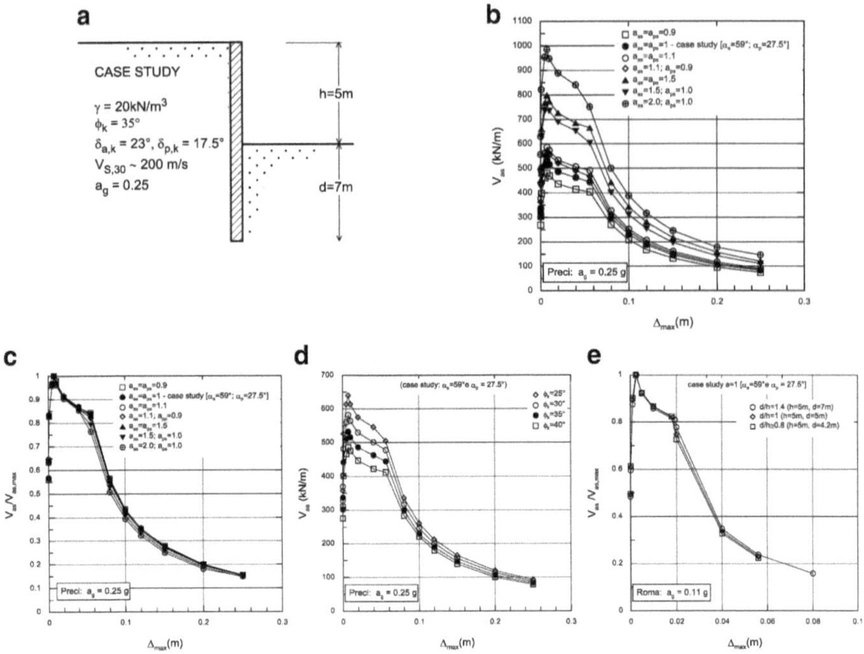

Fig. 10.10 Example of normalization of active seismic thrusts (1st step) (From Cattoni et al. 2012). (**a**) Section of the retaining wall. (**b**) Active shear force. (**c**) Active shear force normalized to the peak shear force demand. (**d**) Active shear force as a function of the soil friction angle. (**e**) Active shear force normalized to the peak shear force demand as a function of the soil friction angle

suitable also when considering – for a given set of seismic input parameters and soil-wedge geometry – the effects of different soil friction angles (ϕ_K) and embedment ratios (d/h). This is shown in Fig. 10.10d, e, where the $V_{as}/V_{as,max} - \Delta_{max}$ curves, obtained for widely different values of soil friction angles (Fig. 10.10d) and embedment ratios (Fig. 10.10e), are approximately described by a single reduction curve.

Moreover, in order to account for the effects of seismic input in the evaluation of the seismic thrust (see Fig. 10.11), a non-dimensional displacement ratio ($\Delta_{max}/\Delta_{max\_TB}$) is further introduced. Displacement Δ_{max_TB} represents a *reference displacement* given by the seismic displacement for $T = T_B$ (see Fig. 10.8d). Figure 10.11 shows the effect of normalization when two different locations – with two widely different sets of seismic input parameters – are considered.

The DDBD procedure can thus be conveniently applied to evaluate the *reduction* of the seismic thrust (active and passive) with the displacement capacity, rather than the seismic thrust itself. In other words, whatever is the method adopted for the evaluation of the seismic thrust (*e.g.*: the active seismic increment, $V_{as,max}$, provided by pseudo-static Mononobe-Okabe solution), a reduction factor of the thrust itself ($V_{as}/V_{as,max}$) can be directly obtained from a normalized plot like the one represented in Fig. 10.12, in which the displacement demand of the system is

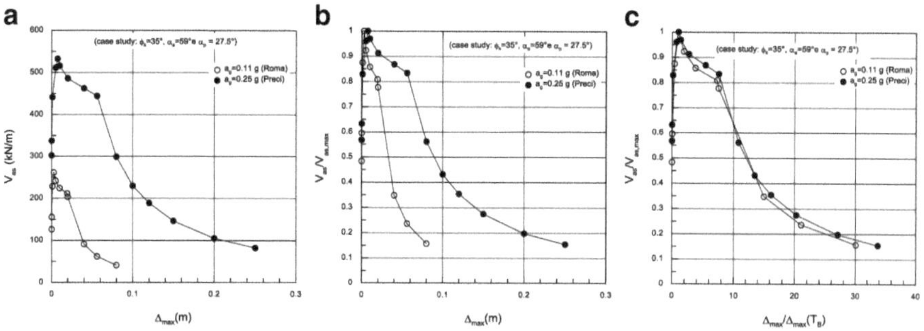

Fig. 10.11 Example of normalization of active seismic thrusts (2nd step) (From Cattoni et al. (2012)). (**a**) Active shear force vs. maximum displacement. (**b**) Active shear force normalized to maximum shear force vs. maximum displacement. (**c**) Active shear force normalized to maximum shear force vs. maximum displacement normalized to displacement at period TB

Fig. 10.12 Example of active seismic thrust reduction curves

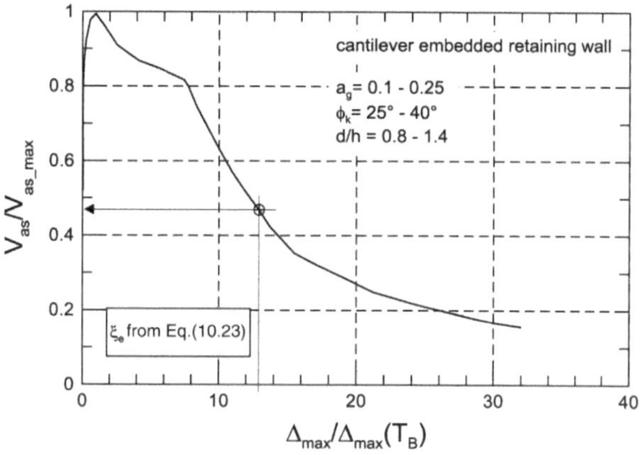

introduced through the normalized displacement ratio ($\Delta_{max}/\Delta_{max\_TB}$). A practical example of application of this procedure is given in Cattoni et al. (2012).

However, the calculation of the normalized displacement ratio requires – as a first step – the evaluation of the equivalent damping of the system. As mentioned above, such equivalent damping corresponds to an expected level of ductility demand and, in particular, increases with the maximum expected displacement Δ_{max}.

10.6.3 *Evaluation of Soil Damping for DBD Procedure*

In the previous paragraphs, it has been stated that an important point of the DBD design process lies in the evaluation of the equivalent damping ratio, ξ_e, as a function of the deformation levels of the soil/wall (see Fig. 10.8c). For earth retaining structures, in addition to energy dissipation related to structural plastic

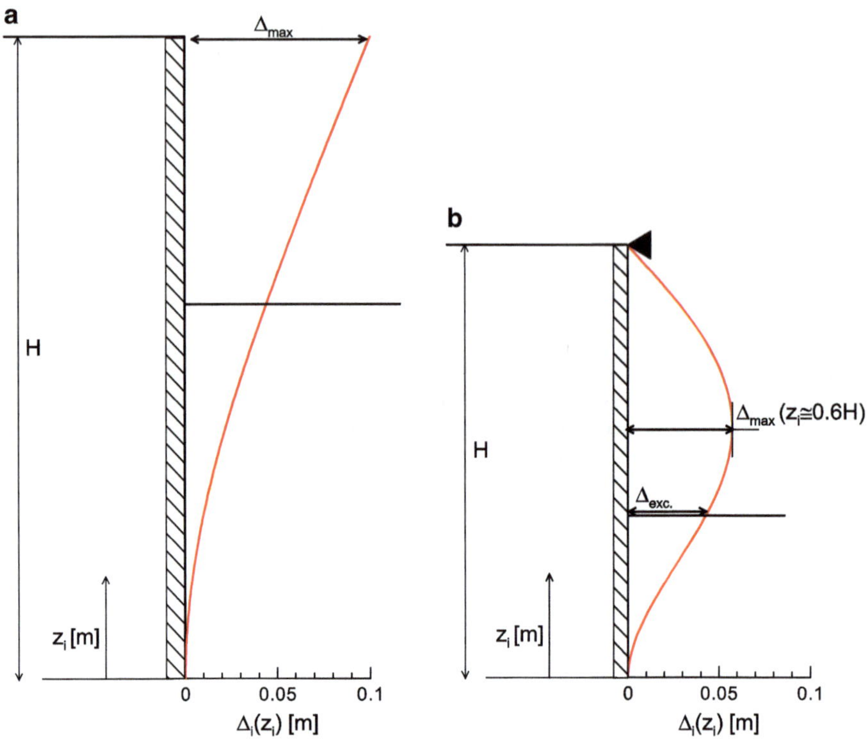

Fig. 10.13 Displacement profile for (**a**) cantilever embedded wall; (**b**) anchored retaining wall (Adapted from Cecconi et al. (2009))

hinges (structural wall damping, ξ_w), cyclic deformation of the soil in both active and passive wedges and associated shear strain (γ_s) presumably provide the more relevant source of energy dissipation for the equivalent system.

Assuming the displacement profiles sketched as an example in Fig. 10.13 for the two wall types considered, the average shear strain (γ_s) within the two soil wedges can be simply estimated as:

$$\text{cantilever embedded wall} \quad \gamma_{as} \cong \frac{\Delta_{\max}}{H} \quad (\textit{active side}) \quad \gamma_{ps} \cong \frac{\Delta_{exc}}{d} \quad (\textit{passive side}) \tag{10.22a}$$

$$\text{singly-anchored wall} \quad \gamma_{as} \cong \frac{\Delta_{\max}}{0.5H} \quad (\textit{active side}) \quad \gamma_{ps} \cong \frac{\Delta_{exc}}{d} \quad (\textit{passive side}) \tag{10.22b}$$

where $\Delta_{max.}$ and $\Delta_{exc.}$ represent the horizontal displacements at the top of the wall (Δ_{max}) and at the excavation level respectively (see Fig. 10.13b).

Fig. 10.14 Variation of damping ratio with cyclic shear strain amplitude and plasticity index (Adapted from Vucétic and Dobry (1991))

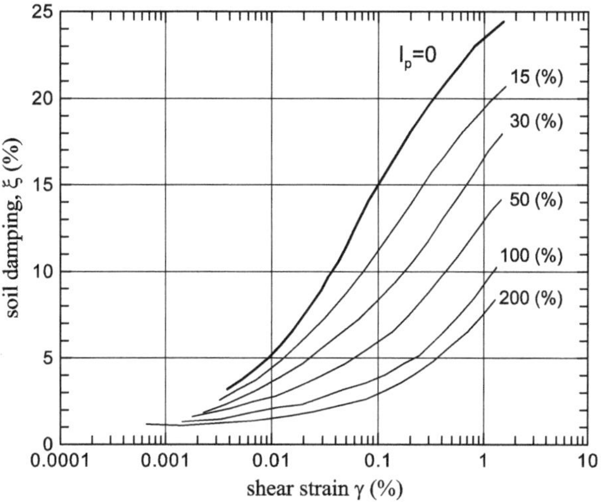

Several damping ratio (ξ_s) – shear strain (γ_s) relationships based on experimental data are available in the literature for fine-grained and granular soils (*e.g*.: Seed et al. 1986; Yasuda and Matsumoto 1993; Ishibashi and Zhang 1993; Vucétic and Dobry 1991). The relationship shown in Fig. 10.14, proposed by Vucétic and Dobry (1991), was used – as a first approximation – to estimate the soil damping coefficients (ξ_{as} , ξ_{ps}) in both active and passive soil wedges (Cecconi et al. 2009).

According to Eq. (10.22a, b) and the adopted $\xi_s - \gamma_s$ relationship, the equivalent damping of the system – calculated through Eq. (10.17) – increases with the maximum expected displacement (Δ_{max}), as shown in Fig. 10.15a, b for the two wall types mentioned above. In these Figures, the value of $\xi_e = 5$ % is imposed at zero displacement in accordance with EC8. For the two retaining walls, it is noted that for $\Delta_{max} > 3$ cm the soil damping reaches values of the order of 15–20 %, much larger than the wall damping.

For cantilever embedded walls, Eq. (10.17) can be usefully simplified by assuming a linear displacement profile leading to a maximum seismic displacement occurring at the top of the wall. With this assumption, it has been found (Model Code 2009) that:

$$\xi_e(\%) \cong 33 + 3.09 \cdot \ln \frac{\Delta_{max}}{H} \qquad \left(for \quad \frac{\Delta_{max}}{H} > 10^{-5} \right) \qquad (10.23)$$

Equation (10.23) noticeably simplifies the DBD calculation steps and it can be usefully adopted for the determination of the reduction factor $V_{as}/V_{as,max}$ introduced in the previous section.

It is worthwhile introducing a last consideration, based on a simple numerical example. It is assumed that, for an anchored retaining wall (for example $H = 8$ m, see Fig. 10.9), a seismic maximum displacement, $\Delta_{max} = 0.03$ m, is set by the designer to ensure an acceptable displacement level. According to Eq. (10.22b), this leads to shear strain within the active (γ_{as}) and passive (γ_{ps}) soil wedges of about $= 7.5 \times 10^{-3}$ and

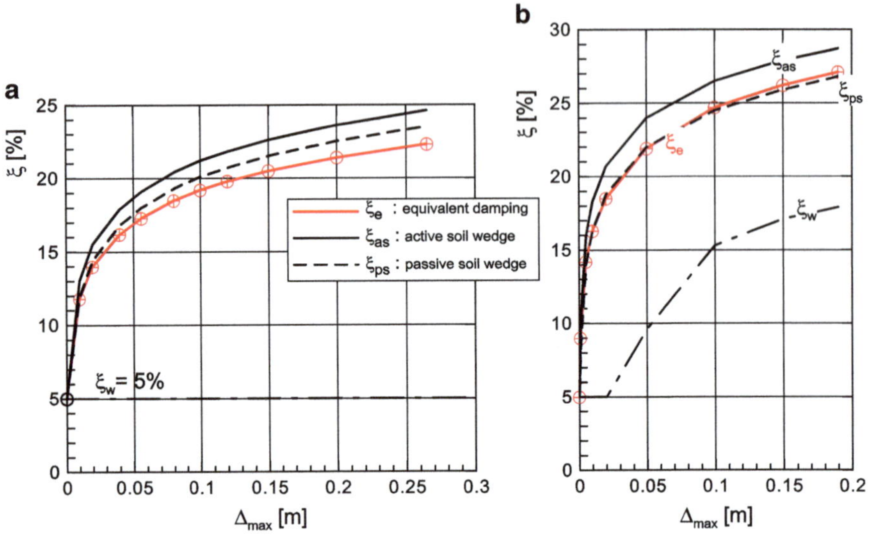

Fig. 10.15 Displacement profile for (**a**) cantilever embedded wall; (**b**) anchored retaining wall (Adapted from Cecconi et al. (2009))

1×10^{-2} respectively. These values shall correspond to a shear modulus reduction G/G_{max} – which can be deduced from the literature for different soils – even lower than 20 % and, hence, the damping ratio could realistically exceed the 20 % (see for example Fig. 10.14 and 10.15b). For a cantilever embedded wall (for example $H = 12$ m, see Fig. 10.10), the same value of displacement $\Delta_{max} = 0.03$ m provides $\gamma_{as} = 2.5 \times 10^{-3}$ within the active soil wedge, leading to slightly lower values of soil damping (see also Fig. 10.15 for a comparison).

Although it is recognized that the proposed formulation (Eq. 10.22) provides some approximation in the evaluation of soil shear strains, similar results can be found from an accurate analysis of admissible strain fields (Bolton and Powrie 1988). Figure 10.16 shows two admissible strain fields for the case of cantilever embedded walls. A wall rotation, $\delta\theta = \Delta_{max}/H$, is consistent with an increment of the soil shear strain $\delta\gamma = 2\ \delta\theta$ in the zone of plastic deformation, independently of whether the wall rotation $\delta\theta$ is inward or outward as well as whether it is around the top or the base of the wall. The shear strain increment could be slightly larger when the wall is propped/anchored at its top (Bolton and Powrie 1988). With regard to the numerical example mentioned above for the cantilever wall, for $\Delta_{max} = 0.03$ m, one obtain $\delta\theta = 0.25$ % and $\gamma_{as} = 2 \cdot \delta\theta = 5 \times 10^{-3}$ for the active soil wedge, leading again to soil damping ξ_{as} of the order of 20 %.

These considerations strengthen the concept that energy dissipation through cyclic deformation of the soil in both passive and active wedges is most probably the dominant mechanism contributing to the equivalent damping ratio of the soil/wall system. For this reason the performance of the retaining wall, under a specific seismic design intensity should be defined by specific strain limits which should be set in relation to an acceptable reduction of the shear modulus (Model Code, DBD09).

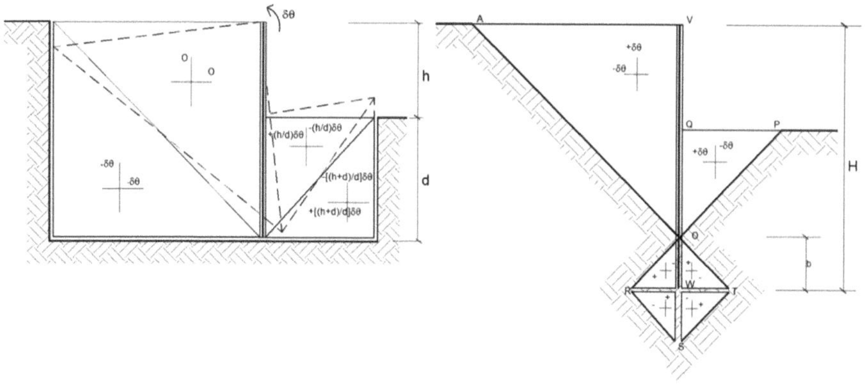

Fig. 10.16 Admissible strain fields for cantilever embedded retaining walls (Adapted from Bolton and Powrie (1988))

10.7 Non-linear Soil-Foundation-Structure Interaction in DDBD

10.7.1 Summary of Experimental and Numerical Progress

In the framework of seismic design according to capacity principles, it is generally recognized that any damage to foundations should be avoided, and the nonlinear capacity of the system is exploited at the superstructure level alone. With the ever increasing interest towards performance based approaches for seismic design, the idea to exploit non-linear energy dissipation at the soil-foundation interface is becoming more and more attractive (ATC-40 1996; Martin and Lam 2000; Pecker 2010; Anastasopoulos et al. 2010; Gajan et al. 2010; Paolucci et al. 2013) and has already led to some outstanding examples of seismic design of foundations allowed to uplift during earthquakes, such as in the case of the Rion-Antirion cable stayed bridge in Greece (Pecker 2006), as well as to increase experimental evidence of the benefits of an integrated foundation-superstructure design, where the controlled share of ductility demand between the superstructure and the foundation may be considered as a key ingredient for a rationale and integrated approach to seismic design of foundations and structures (Liu et al. 2013; Pecker et al. 2012).

In the recent years, substantial progress was achieved to make available well-calibrated experimental tests both on the seismic response of foundations (Shirato et al. 2008; Gajan and Kutter 2008, 2009; Massimino and Maugeri 2013) and on the coupling of nonlinear foundations with nonlinear structures (Chang et al. 2007; Deng et al. 2012; Raychowdhury and Hutchinson 2011; Liu et al. 2013; Drosos et al. 2012; Anastasopoulos et al. 2013). In parallel, the capability to perform reliable numerical simulations of the seismic foundation behaviour with prediction of the residual settlements and rotations under strong seismic shaking was significantly improved, especially based on the application of the non-linear foundation macro-element concept (Nova and Montrasio 1991; Paolucci 1997; Cremer et al. 2001; di Prisco et al. 2006; Muir Wood 2007). In this respect, some key advancements were recently obtained in the development of non-linear macro-element models of the soil-foundation system, by calibration of the numerical models with respect to the previous experimental results (Paolucci et al. 2008a, b; Grange et al. 2008; Chatzigogos et al. 2011; Figini et al. 2012; Correia 2012). A summary of the recent progress achieved in this field is reported by Pecker et al. (2012).

10.7.2 Procedure to Introduce Non-linear Dynamic Soil-Structure Interaction Effects in the DDBD

With the previous achievements both on the experimental and numerical sides, it seems now feasible to close the loop towards an integrated foundation-structure design, by taking into account the non-linear interaction of these elements. In the following, an iterative linear-equivalent procedure is introduced for this purpose, developed in the framework of the displacement based seismic design, which is naturally suited to directly evaluate the system performance in terms of displacements of both the foundation and the superstructure systems.

The procedure is based on the use of empirical curves to evaluate the rotational stiffness degradation (K_F/K_{F0}) and the increase of damping ratio ξ_F as a function of foundation rotation (θ). Iterations are performed to ensure that admissible values of foundation rotations are complied, in addition to the standard checks on structural displacements and drifts. In Fig. 10.17, a set of such curves is presented for shallow foundations on dry sands, as a result of a parametric study, illustrated in more detail by Paolucci et al. (2009). Empirical equations for both K_F/K_{F0} (θ) and ξ_F (θ) as a function of the static safety factor (N_{max}/N) are introduced, for dense ($D_R = 90\ \%$) and medium dense ($D_R = 60\ \%$) soil conditions. An extended set of such curves is under development, based on the availability of new macro-element models, applicable both to cohesive and granular materials and to deep foundations as well (Correia 2012).

The main idea behind the procedure introduced by Paolucci et al. (2013) is the formulation of a linear equivalent oscillator, according to the principles of DDBD, with equivalent period and damping modified according to the non-linear variability of foundation stiffness and damping with foundation rotation, illustrated in Fig. 10.17. For the equivalent period Teq, the classical Eq. (10.22b) is used:

$$T_{eq}^2 = T_S^2 + T_F^2 \tag{10.24}$$

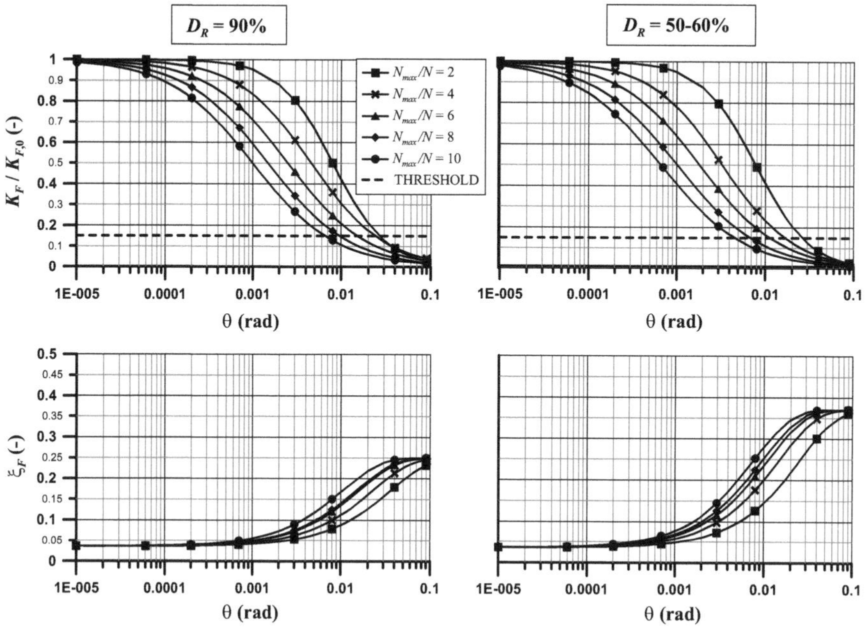

Fig. 10.17 Influence of the loading conditions on the rotational secant stiffness degradation (*top*) and damping increase (*bottom*), for dense (*left*) and medium dense (*right*) sand (Adapted from Paolucci et al. (2009))

where both the fundamental period of the fixed-base structure (T_s) and of the rocking foundation (T_f) are updated within the same iterative procedure. Namely, update of T_f accounts for the amount of rotation calculated at the previous iteration and the consequent variation of foundation rotation according to the graphs in Fig. 10.17. Similarly, the following expression, introduced by Priestley et al. (2007), is used to evaluate the equivalent damping:

$$\xi_{eq} = \frac{\xi_F \Delta_F + \xi_S \Delta_S}{\Delta_F + \Delta_S} \tag{10.25}$$

where $\xi_S = \frac{A_S}{2\pi V_b \Delta_S}$; $\xi_F = \frac{A_F}{2\pi V_b \Delta_F}$ and the meaning of the different symbols is illustrated in Fig. 10.18.

The basic steps of the procedure are summarized as follows, while a flow chart is sketched in Fig. 10.19.

1. Definition of the desired performance in terms of a target design value Δ_d, depending on the limit state to be considered, that may be expressed either in terms of the total lateral displacement Δ_{TOT} (including the foundation rotation contribution) or of the structural displacement alone Δ_S.

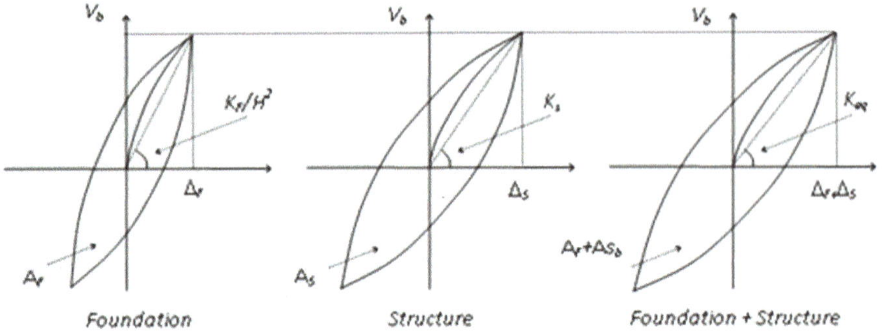

Fig. 10.18 Damping contributions of foundation and structure (Adapted from Priestley et al. (2007))

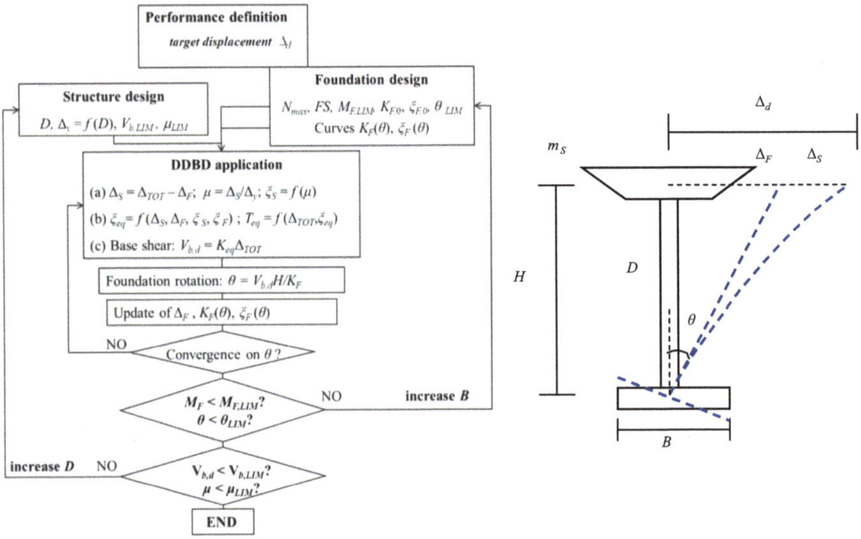

Fig. 10.19 Flow chart summarizing the procedure to account for the effects of non-linear soil-foundation interaction with the superstructure in the framework of DDBD (Adapted from Paolucci et al. (2013))

2. Tentative initial design of the superstructure (e.g., in terms of the pier diameter D and of the reinforcement content).
3. Definition of the initial parameters of the foundation design (width, bearing capacity, stiffness);
4. Evaluation of the 1 DOF substitute structure (1st iteration considers fixed-base), the effective natural period and damping ratio of which are modified to account for the interaction with the foundation;

5. Calculation of base shear and base overturning moment for the prescribed design displacement spectrum;
6. Calculation of the corresponding foundation rotation;
7. Calculation of the contribution of the foundation rotation to the total structural displacement;
8. Update K_F and ξ_F;
9. Test of convergence on foundation rotation (if not, iteration to step 3);
10. Check of the foundation seismic bearing capacity and admissible values of foundation rotation (if check not satisfied, go back to step 1 with an increased value of foundation width).

Details of the procedure are illustrated by Paolucci et al. (2013), together with an application to the seismic design of two bridge piers according to DDBD, both considering fixed base and with consideration of non-linear soil-structure interaction effects.

The design examples clarified that it is possible to achieve a reduction of the ductility demand on the superstructure by reducing the foundation width, at the price of allowing for controlled permanent damage on the foundation itself, both in terms of permanent settlements and rotations. These observations are in agreement with main indications from the balanced foundation-superstructure design, advocated by Liu et al. (2013) with their centrifuge tests on compatible yielding of foundations and structures during seismic loading, who found significant reductions on superstructure demand, particularly with regard to maximum roof drift ratio and maximum base shear.

10.8 Conclusions

This paper is intended to present a state of the art and state of the practice about the potential of the direct displacement based approach to include in design the interaction of structures and soil.

Two cases are explicitly considered and discussed: the design of retaining structures and the inclusion of non-linear soil-foundation interaction in the case of shallow foundations.

For what concerns retaining structures it comes out clearly that the application of DBD principles results in a rational and effective consideration of some allowed displacement, which is hidden in the application of mysterious reduction factors if a forced based approach is adopted. An enlightening example is shown in Fig. 10.9, where the comparison of different codes and approaches shows remarkable differences.

The inclusion of soil-structure interaction, as discussed herein, appears to require some design iteration, which is not needed for structure design alone, but the procedure still appears to be acceptably simple and efficient.

In both cases, the definition of the damping level produced by the soil response, to be associated with the accepted displacement, results to be the crucial and most difficult problem to be faced, requiring further extensive experimental and numerical research, properly focused on this aspect.

It is thus hoped that the very encouraging results presented here will stimulate the geotechnical engineering community towards coordinated research efforts that will turn to be of immediate practical applicability. A specific work frame could be found in the most recent version of a displacement–based Model Code, very recently published (Sullivan et al. 2012).

Acknowledgements The development of the displacement based design method briefly presented in this paper is the summary of a research effort of about 15 years, with the contribution of many Ph.D. students. This effort culminated in the book published in 2007 by Priestley, Calvi and Kowalsky, where a more detailed acknowledgement of contributions is reported.

The application of displacement based concepts to geotechnical structures has been mainly developed within a project funded by the Italian Department of Civil Protection through the Italian Network of University Laboratories of Earthquake Engineering (ReLUIS). The project was coordinated by G.M. Calvi and T. Sullivan. The research units dealing with geotechnical aspects were coordinated by V. Pane and R. Paolucci. Important contributions have been provided by L. Petrini, S. Vecchietti, C. di Prisco, R. Figini, M. Vecchiotti.

References

Akkar S, Boore DM (2009) On baseline corrections and uncertainty in response spectra for baseline variations commonly encountered in digital accelerograph records. Bull Seismol Soc Am 99:1671–1690

Anastasopoulos I, Gazetas G, Loli M, Apostolou M, Gerolymos N (2010) Soil failure can be used for seismic protection of structures. Bull Earthq Eng 8:309–326

Anastasopoulos I, Loli M, Georgarakos T, Drosos V (2013) Shaking table testing of rocking-isolated bridge pier on sand. J Earthq Eng 17:1–32

ATC-40 (1996) Seismic evaluation and retrofit of concrete buildings. Chapter 10: Foundation effects. Technical Rep. SSC 96–01, Seismic Safety Commission, State of California

Bolton MD, Powrie W (1988) Behaviour of diaphragm walls in clay prior to collapse. Geotechnique 38:167–189

Bommer JJ, Mendis R (2005) Scaling of spectral displacement ordinates with damping ratios. Earthq Eng Struct Dyn 34:145–165

Calvi GM, Sullivan TJ (eds) (2009) A model code for the displacement-based seismic design of structures. IUSS Press, Pavia

Calvi GM, Pane V, Paolucci R, (2008) Displacement based seismic design of geotechnical structures, XII Ciclo di Conferenze di Meccanica e Ingegneria delle Rocce, MIR2008, Torino, pp 9–33

Cattoni E, Cecconi M, Pane V, Vecchietti S (2012) Development of DDBD method for retaining walls subjected to seismic loads. In: Proceedings of the II international conference on performance-based design in earthquake geotechnical engineering, Taormina, Italy, 28–30 May, pp 1510–1521

Cauzzi C, Faccioli E (2008) Broadband (0.05 s to 20 s) prediction of displacement response spectra calibrated on worldwide digital records. J Seismol 12:453–475

Cecconi M, Pane V, Vecchietti S (2008) The DDBD method in the a-seismic design of anchored diaphragm walls. In: Mercea 08, international conference on seismic engineering, AIP Conference Proceedings, Reggio Calabria, pp 695–702

Cecconi M, Vecchietti S, Pane V (2009) Il metodo DDBD per la valutazione delle spinte sismiche sulle strutture di sostegno. Progettazione Sismica 2:67–82, IUSS Press

Chang BJ, Raychowdhury P, Hutchinson TC, Thomas JM, Gajan S, Kutter BL (2007) Evaluation of the seismic performance of combined frame-wall-foundation structural systems through centrifuge testing. In: Proceedings of 4th international conference on earthquake geotechnical engineering, Thessaloniki, Greece, Paper no. 1497

Chatzigogos CT, Figini R, Pecker A, Salençon J (2011) A macroelement formulation for shallow foundations on cohesive and frictional soils. Int J Numer Anal Methods Geomech 35:902–931

Correia A (2012) A pile-head macro-element approach to seismic design of monoshaft-supported bridges. PhD thesis, Rose School, Pavia

Cremer C, Pecker A, Davenne L (2001) Cyclic macro-element for soil-structure interaction: material and geometrical non-linearities. Int J Numer Anal Methods Geomech 25:1257–1284

Deng L, Kutter BL, Kunnath S (2012) Centrifuge modeling of bridge systems designed for rocking foundations. J Geotech Geoenviron Eng ASCE 138:335–344

di Prisco C, Massimino M, Maugeri M, Nicolosi M, Nova R (2006) Cyclic numerical analyses of Noto Cathedral: soil-structure interaction modelling. Ital Geotech J 40:49–64

Drosos V, Georgarakos T, Loli M, Anastasopoulos I, Zarzouras O, Gazetas G (2012) Soil–foundation–structure interaction with mobilization of bearing capacity: an experimental study on sand. J Geotech Geoenviron Eng ASCE 138:1369–1386

Dwairi HM, Kowalsky MJ, Nau JM (2007) Equivalent damping in support of direct displacement-based design. J Earthq Eng 11(4):512–530

EC8 (1998) Design provisions for earthquake resistance of structures, Pub. ENV-1998-2. Comité Européen de Normalisation, Brussels

EC8 (2003) Design provisions for earthquake resistance of structures, Pub. ENV-2003-2. Comité Européen de Normalisation, Brussels

Faccioli E, Rovelli A (coordinators) (2007) Definizione dell'input sismico sulla base degli spostamenti attesi, Relazione scientifica finale, Progetto S5. http://progettos5.stru.polimi.it

Faccioli E, Villani M (2009) Seismic hazard mapping for Italy in terms of broadband displacement response spectra. Earthquake Spectra 25:515–539

Faccioli E, Paolucci R, Rey J (2004) Displacement spectra for long periods. Earthquake Spectra 10:347–376

Figini R, Paolucci R, Chatzigogos CT (2012) A macro-element for non-linear soil-structure interaction under seismic loads: development and validation on large scale tests. Earthq Eng Struct Dyn 41:475–493

Gajan S, Kutter BL (2008) Capacity, settlement, and energy dissipation of shallow footings subjected to rocking. J Geotech Geoenviron 134:1129–1141

Gajan S, Kutter BL (2009) Effects of moment-to-shear ratio on combined cyclic load–displacement behavior of shallow foundations from centrifuge experiments. J Geotech Geoenviron 135:1044–1055

Gajan S, Raychowdhury P, Hutchinson TC, Kutter B, Stewart JP (2010) Application and validation of practical tools for nonlinear soil-foundation interaction analysis. Earthquake Spectra 26:111–129

Grange S, Kotronis P, Mazars J (2008) A macro-element for a shallow foundation to simulate soil–structure interaction considering uplift. Comptes Rendus Mécanique 336:856–862

Grant DN, Blandon C, Priestley MJN (2005) Modelling inelastic response in direct displacement-based seismic design. IUSS Press, Pavia

Iervolino I, Galasso C, Paolucci R, Smerzini C (2012) REXEL-DISP – computer-aided real record selection based on displacement spectral compatibility. Beta version. http://www.reluis.it

Ishibashi I, Zhang X (1993) Unified dynamic shear moduli and damping ratios of sand and clay. Soils Found 33:1129–1143

Liu W, Hutchinson T, Kutter B, Hakhamaneshi M, Aschheim M, Kunnath S (2013) Demonstration of compatible yielding between soil-foundation and superstructure components. J Struct Eng 139, SPECIAL ISSUE: NEES 2: Advances in Earthquake Engineering, 1408–1420
Martin GR, Lam IP (2000) Earthquake resistant design of foundations – retrofit of existing foundations. In: Proceedings of GeoEng2000, Melbourne
Massimino MR, Maugeri M (2013) Physical modelling of shaking table tests on dynamic soil–foundation interaction and numerical and analytical simulation. Soil Dyn Earthq Eng 3:1–18
Muir WD (2007) Modelling of dynamic soil problems. In: Earthquake geotechnical engineering, vol 6, Geotechnical, geological, and earthquake engineering. Springer, Dordrecht, pp 131–149
Nova R, Montrasio L (1991) Settlements of shallow foundations on sand. Geotechnique 41:243–256
Pane V, Cecconi M, Vecchietti S (2007) Il Metodo DDBD per il progetto agli spostamenti di strutture di sostegno, XII Convegno Nazionale Anidis – L'Ingegneria Sismica in Italia, Pisa
Paolucci R (1997) Simplified evaluation of earthquake induced permanent displacements of shallow foundations. J Earthq Eng 1:563–579
Paolucci R, Rovelli A, Faccioli E, Cauzzi C, Finazzi D, Vanini M, Di Alessandro C, Calderoni G (2008a) On the reliability of long-period response spectral ordinates from digital accelerograms. Earthq Eng Struct Dyn 37:697–710
Paolucci R, Shirato M, Yilmaz MT (2008b) Seismic behaviour of shallow foundations: shaking table experiments vs. numerical modelling. Earthq Eng Struct Dyn 37:577–595
Paolucci R, di Prisco C, Figini R, Petrini L, Vecchiotti L (2009) Interazione dinamica non lineare terreno-struttura nell'ambito della progettazione sismica agli spostamenti. Progettazione Sismica 2:83–103, IUSS Press
Paolucci R, Figini R, Petrini L (2013) Introducing dynamic nonlinear soil-foundation-structure interaction effects in displacement-based seismic design. Earthquake Spectra 29(2):475–496
Pecker A (2006) Enhanced seismic design of shallow foundations: example of the Rion Antirion bridge. In: Proceedings of 4th Athenian lecture on geotechnical engineering, Athens, Greece
Pecker A (2010) Non-linear soil structure interaction: impact on the seismic response of structures, Keynote lecture at the 14th European conference on earthquake engineering, Ohrid, Macedonia, 30 August – 3 September 2010
Pecker A, Paolucci R, Chatzigogos CT, Correia AA, Figini R (2012) The role of non-linear dynamic soil-foundation interaction on the seismic response of structures – 2nd international conference on performance-based design in earthquake geotechnical engineering, Taormina (Italy), 28–30 May 2012
Priestley MJN (1993) Myths and fallacies in earthquake engineering – conflicts between design and reality. Bull NZSEE 26(3):329–341
Priestley MJN (2000) Performance-based seismic design, keynote address. In: Proceedings of 12th world conference on earthquake engineering, Auckland
Priestley MJN (2003) Myths and fallacies in earthquake engineering, revisited, The Mallet Milne Lecture. IUSS Press, Pavia, 121pp
Priestley MJN, Grant DN (2005) Viscous damping in seismic design and analysis. J Earthq Eng 9(SP2):229–255
Priestley MJN, Calvi MC, Kowalsky MJ (2007) Displacement-based seismic design of structures. IUSS Press, Pavia, 721pp
Raychowdhury P, Hutchinson TC (2011) Performance of seismically loaded shearwalls on nonlinear shallow foundations. Int J Numer Anal Meth Geomech 35:846–858
Seed HB, Wong RT, Idriss IM, Tokimatsu K (1986) Moduli and damping factors for dynamic analyses of cohesionless soils. J Geotech Eng ASCE 112(GT11):1016–1032
Shirato M, Kouno T, Asai R, Nakani N, Fukui J, Paolucci R (2008) Large-scale experiments on nonlinear behavior of shallow foundations subjected to strong earthquakes. Soils Found 48:673–692

Smerzini C, Galasso C, Iervolino I, Paolucci R (2013) Ground motion record selection based on broadband spectral compatibility. Earthq Spectra. Accepted for publication
Sullivan TJ, Priestley MJN, Calvi GM (eds) (2012) A model code for the displacement-based seismic design of structures, DBD12. IUSS Press, Pavia
Vucétic M, Dobry R (1991) Effects of the soil plasticity on cyclic response. J Geotech Eng Div ASCE 117(1)
Yasuda N, Matsumoto N (1993) Dynamic deformation characteristics of sands and rock-fill materials. Can Geotech J 30(5):747–757

Part V
Underground Structures

Chapter 11
Performance and Seismic Design of Underground Structures

Kyriazis Pitilakis and Grigorios Tsinidis

Abstract Underground structures, tunnels, subways, metro stations and parking lots, are crucial components of the build environment and transportation networks. Considering their importance for life save and economy, appropriate seismic design is of prior significance. Their seismic performance during past earthquakes is generally better than aboveground structures. However several cases of severe damage to total collapse have been reported in the literature, with that of the Daikai metro station in Kobe during the Hyogoken-Nambu earthquake (1995) being one of the most characteristic. These recent damages revealed some important weaknesses in the current seismic design practices. The aim of this chapter is not to make another general presentation of the methods used for the seismic design of underground structures, but rather to discuss and highlight the most important needs for an improved seismic performance and design. In that respect it is important to consider that the specific geometric and conceptual features of underground structures make their seismic behavior and performance very distinct from the behavior of aboveground structures, as they are subjected to strong seismic ground deformations and distortions, rather than inertial loads. Several methods are available, from simplified analytical elastic solutions, to sophisticated and in principal more accurate, full dynamic numerical models. Most of them have noticeable weaknesses on the description of the physical phenomenon, the design assumptions and principles and the evaluation of the parameters they need. The chapter presents a short but comprehensive review of the available design methods, denoting the crucial issues and the problems that an engineer could face during the seismic analysis. The main issues discussed herein cover the following topics: (i) force based design against displacement based design, (ii) deformation modes of rectangular underground structures under seismic excitation, (iii) seismic earth pressures on underground

K. Pitilakis (✉) • G. Tsinidis
Research Unit: Soil Dynamics and Geotechnical Earthquake Engineering, Department of Civil Engineering, Aristotle University of Thessaloniki, Thessaloniki, Greece
e-mail: kpitilak@civil.auth.gr; gtsinidi@civil.auth.gr

M. Maugeri and C. Soccodato (eds.), *Earthquake Geotechnical Engineering Design*, Geotechnical, Geological and Earthquake Engineering 28,
DOI 10.1007/978-3-319-03182-8_11,

structures, (iv) seismic shear stresses distribution on the perimeter of the structure, (v) appropriateness of the presently used impedance functions to model the inertial and the kinematic soil-structure interaction effects, (vi) design seismic input motion, accounting of the incoherence effects and the spatial variation of the motion and (vii) effect of the build environment (i.e. city-effects) on the seismic response of underground structures. The discussion is based on detailed numerical analysis of specific cases and recent experimental results in centrifuge tests. Other important issues like the design of submerged tunnels to liquefaction risk, or the complexity to evaluate the response of the joints of submerged tunnels are also shortly addressed. Finally we present the most recent developments on the evaluation of adequate fragility curves for shallow tunnels.

11.1 Introduction

Underground structures such as tunnels, metro stations and underground parking stations are major components of any transportation system. They are more and more frequently constructed especially in densely populated urban areas, to facilitate different needs. Considering their significance in modern societies their seismic design becomes of prior importance in earthquake prone regions.

Their specific geometrical and conceptual features make their seismic behavior very distinct from aboveground structures. The imposed ground deformations during shaking are prevailing, while the inertial forces are generally of secondary importance.

The complexity of the problem and the conscience on the various shortcomings of the available methods lead often to an overdesign of the new structures. The lack of knowledge is attributed, to a certain degree, to the relatively few well-documented cases of important damages during strong earthquakes. Several questions regarding their real seismic behaviour during shaking remain still non-responded. We still cannot evaluate accurately the seismic earth pressures and the seismic shear stresses around a tunnel or on the sidewalls of a deep underground metro station. We don't know how to calculate adequate impedance functions to be used in a SSI analysis and we make several hypotheses on the evaluation of the applied shaking loadings, which may not reflect the real loading conditions during a strong earthquake excitation. These known and accepted weaknesses reflected also to the relative absence of rigorous and detailed seismic code regulations and guidelines for this type of structures.

Several methods have been proposed in the literature and used in practice for the seismic design of tunnels and large underground structures. From simple analytical elastic solutions, to pseudo-static or equivalent static solutions, and from hybrid methods, which try to take into account the relative stiffness of the ground and the structures and the SSI effects, to more sophisticated and a priori accurate models using full dynamic numerical analysis of the soil-structure system. The paper provides a short but comprehensive review of the different design and analysis

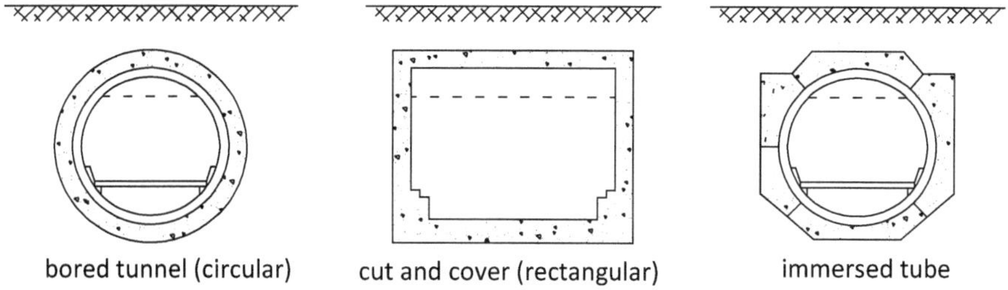

Fig. 11.1 Typical cross sections of underground structures (Modified after Power et al. 1996)

methods for underground structures and tunnels, based on the analysis of some selected case studies and results from experimental testing in centrifuge. Through the presentation, emerging issues that are still open and need further research are highlighted.

Finally we present a new set of fragility curves for the vulnerability and risk assessment of shallow tunnels in alluvial deposits.

11.1.1 Typology – Construction Methods

Underground structures can be grouped into two major categories, namely: (i) multi-story large dimensions structures (e.g. metro stations, parking lots) and (ii) long underground structures (i.e. tunnels). Regarding the construction method, underground structures can be grouped, as follows (Power et al. 1998): (i) bored tunnels, (ii) cut and cover structures and (iii) submerged structures (e.g. immersed tunnels) (Fig. 11.1).

Bored tunnels are constructed by excavating the opening, using boring machines (i.e. TBM) and constructing the support system, underground. They are constructing in rock and stiff soils and they usually refer to circular or semi-circular cross-sections.

Cut and cover structures, typically of rectangular cross-section, are constructed with an open excavation. Commonly, diaphragm walls are used both as temporary support system for the excavation and permanent side-walls of the structure.

Submerged (or immersed) structures are constructed of prefabricated steel or reinforced concrete segments. To avoid buoyancy, they usually have enlarged cross sections to increase mass. The segments are constructed in a dry dock and then moved and placed in shallow trenches under the sea (or river). The connection with each-other is performed through special designed joints (i.e. Kiyomiya 1995). The performance of these joints is of prior importance regarding the longitudinal response of the tunnel, as a possible failure can lead to water leakage inside the tunnel.

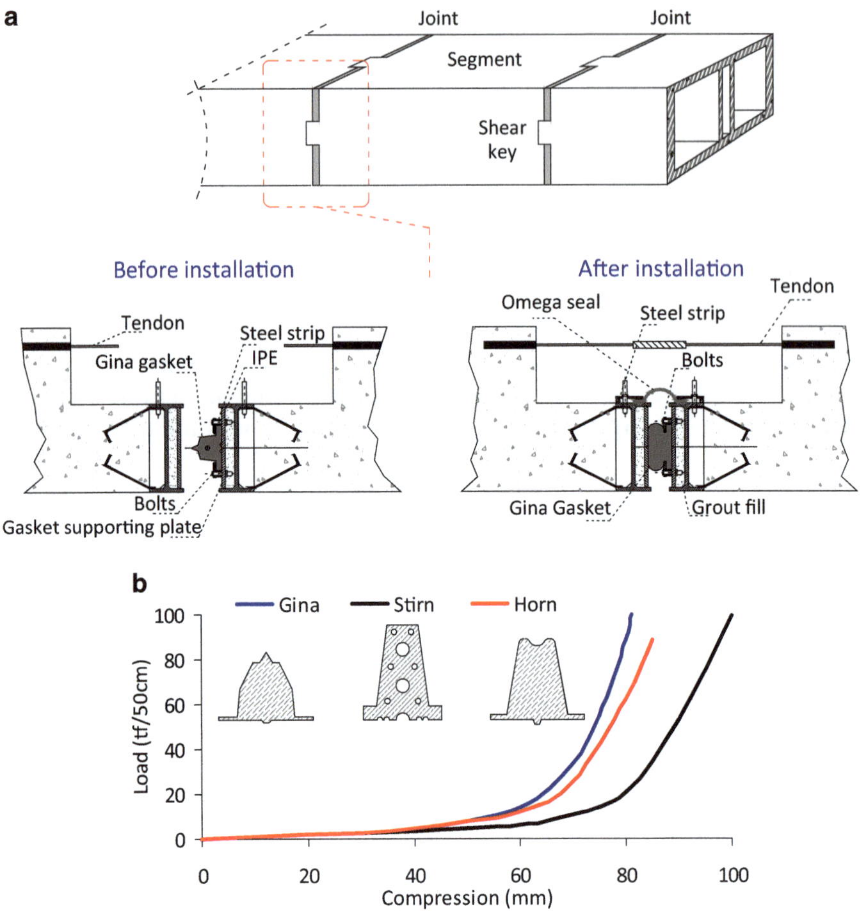

Fig. 11.2 (**a**) Typical joint connection for an immersed tunnel. (**b**) Load-compression relation for several rubber gaskets (Modified after Kiyomiya 1995; Trelleborg 2013)

The main component of the joint is a rubber gasket, functioning only in compression, i.e. Gina type, Horn type or Stirn type (Kiyomiya 1995), while prestressed tendons are used to avoid opening of the joint (Fig. 11.2a). The control of transversal and vertical shear displacements is achieved through specially designed steel or reinforced concrete shear keys. To enhance the waterproof of the joint, usually, a secondary seal is used (i.e. Omega seal). Due to differential pressure inside and outside of the tunnel, the plastic gaskets are pre-compressed to a certain extend. Generally, the gaskets display non-linear behaviour under varying stresses (Fig. 11.2b). Therefore, during an earthquake their stiffness is changing with time due to the transient ground deformations imposed to the structure. This feature makes their modelling difficult and crucial for the tunnel's overall seismic design.

11.2 Seismic Performance

Deep underground structures behave generally better than near surface or above-ground structures, as the imposed ground strains are lower at higher depths. For the same typology the vulnerability is increased for shallow embedment, as the ground strains and velocities are increased when approaching the ground surface. In the following we present and discuss the most important references regarding the seismic performance of tunnels and large underground structures in urban areas.

11.2.1 Tunnels

Seismic performance of tunnels has been reported by several researchers (i.e. Dowding and Rozen 1978; Owen and Scholl 1981; Sharma and Judd 1991; Power et al. 1998; Wang et al. 2001, 2009). Dowding and Rozen (1978), correlated the observed tunnel damages to the estimated (using attenuation relationships) ground surface peak acceleration, using data from 70 case histories. Owen and Scholl (1981) extended the aforementioned database to a total 127 cases. Among others, they concluded that little damage occurred in rock tunnels for peak acceleration below 0.4 g, while cut and cover structures and shallow tunnels tend to be more vulnerable. Sharma and Judd (1991) further extended the database to 192 cases, examining the relationships between observed damages and several factors affecting the behavior, namely the earthquake (magnitude and epicentral distance), the tunnel's supporting system and the geologic conditions. They concluded that deeper structures or rock tunnels seemed to be safer, while the damages are increased with the earthquake magnitude and the decrease of the epicentral distance. Power et al. (1998) used the previous database, adding cases from more recent earthquakes (e.g. the 1995 Kobe earthquake) to further study the performance of bored tunnels. Cases presenting large uncertainties (regarding the performance or the earthquake parameters) and for which the damages were not caused by ground shaking were excluded. Using the most recent at that time ground motion prediction relationships, it was found that for PGAs equal to or less than about 0.2 g, ground shaking caused minor damage. For PGAs in the range of about 0.2–0.5 g, some instances of slight to heavy damages were observed, whereas for PGAs larger than 0.5 g there were many instances of slight to heavy damages.

An indicative example of extensive damage due to ground shaking and permanent displacements is the collapse of the twin Bolu under construction tunnels (Hashash et al. 2001; Kontoe et al. 2008) during the 1999 Kocaeli earthquake. The collapse took place in clay gauge material in the unfinished section of the tunnels. The section was covered with shotcrete and had bold anchors. The tunnel deformed in an oval shape, causing crushing of the shotcrete and bucking of the steel ribs at shoulder and knee location. An important uplift of the invert slab up to 0.5–1.0 m was also reported.

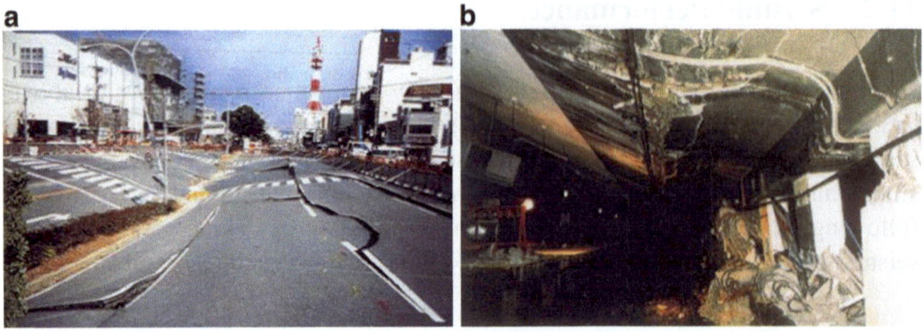

Fig. 11.3 Daikai station. (a) Settlements of the overlaying roadway caused by the subway collapse. (b) Collapse of the central columns of the station (Special Issue of Soils and Foundations 1996)

11.2.2 Long Large Underground Structures

There are few examples of heavy damages of large space underground structures. The most interesting case is that of the Daikai station, that collapsed during the major Hyogoken-Nambu earthquake (1995) (Iida et al. 1996; Kawashima 2000; Hashash et al. 2001) (Fig. 11.3).

This is the first well-reported case of a total collapse of a large underground structure under seismic shaking. During the strong ground shaking the central columns of the station collapsed, leading to collapse of the roof slab. The central columns were designed with very light transversal reinforcement relative to the bending reinforcement. The observed settlements at the surface exceed 2.5 m. The transverse walls at the ends of the station and at areas where the width of the station changed, acted as shear walls preventing the collapse. The walls suffered significant cracking, protecting however the nearby central columns.

The collapse mechanism was extensively studied by several researchers (i.e. EQE 1995; Iida et al. 1996; An et al. 1997; Huo et al. 2005). According to Iida et al. (1996), the relative displacement between the invert slab and the roof slab caused a huge horizontal shear force on the central columns. The overburden soil mass affected also the response, adding inertial force to the structure. The deflection of the roof slab (with respect to the invert slab) could be reduced by the diaphragm action of the roof slab or by the passive earth pressures of the surrounding soils (EQE 1995). Unfortunately, the slab was large enough to act as diaphragm, while a small clearance between the side walls and the sheet pile walls (used during construction), made the compaction of the backfill material difficult, resulting in the structure's inability to mobilize passive earth pressures. According to Huo et al. (2005), the combination of large shear loads with large axial loads (due to the large span) were resulted a brittle shear failure of the columns. Similar conclusions were derived by An et al. (1997). Similar damages were reported for the Kamisawa station (Matsuda et al. 1996; Samata et al. 1997; Kawashima 2000).

Contrary to the above worst cases, the metro stations and the tunnels shafts in Athens performed very well during the Ms 5.9 1999 Athens earthquake, with recorded accelerations of the order of 0.20–0.30 g (Gazetas et al. 2005).

11.2.3 Submerged Tunnels

In 1989 Loma Prieta earthquake (M_s 7.1), the Bay Area Rapid Transit (BART) tunnel system, crossing San Francisco Bay in California, behaved satisfactorily for accelerations in the order of 0.20–0.30 g. Built in the 1960s, BART was one of the first underground structures studied and constructed with seismic design considerations (Kuesel 1969). The design was based on the principle of imposed seismic ground deformations and provision of sufficient ductility. The specially designed joints between the segments behaved quite well suffering small relative displacements. A second example is that of the Osaka South Port immersed tunnel. During the 1995 Hyogoken-Nambu earthquake (M_{JMA} 7.2), the 1 km tunnel behaved very well for accelerations of the order of 0.27 g (Anastasopoulos et al. 2007).

11.3 Seismic Response

Ground shaking due to wave propagation and permanent ground displacements due to lateral spreading, landslides and fault rupture are affecting underground structures during a strong earthquake. The deformation modes dominating the seismic response are somehow different from aboveground structures.

11.3.1 Ground Shaking

The inertia of the surrounding soils is much higher than the inertia of the structure itself; consequently the response of the embedded structure is dominated by the response of the surrounding soil and the transient ground deformations. This feature makes the seismic behavior of underground structures very distinct from the typical aboveground structures, where the response is mainly related to the structure's inertia. An indicative example of the different response of a simple rectangular structure embedded or simply founded on the ground is provided in Fig. 11.4.

When travelling seismic waves joint an underground structure, they "forced" it to deform in various modes, both in the longitudinal and transverse direction (Owen and Scholl 1981). Seismic waves with soil particle movement parallel to the underground structure (e.g. tunnel) axis, result in axial deformation (compression-extension in Fig. 11.5a), whereas soil movement perpendicular to the tunnel axis causes longitudinal bending (Fig. 11.5c). Shear waves propagating upwards from the bedrock may

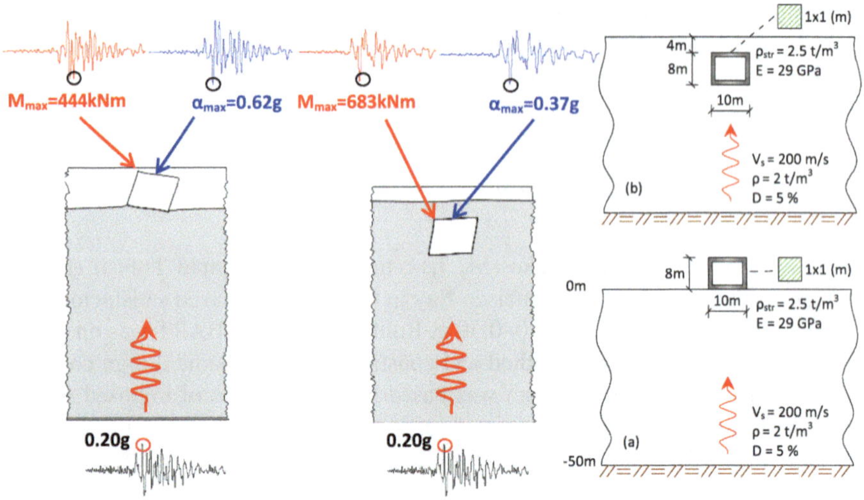

Fig. 11.4 Response of a simple rectangular structure embedded or simply founded on the ground

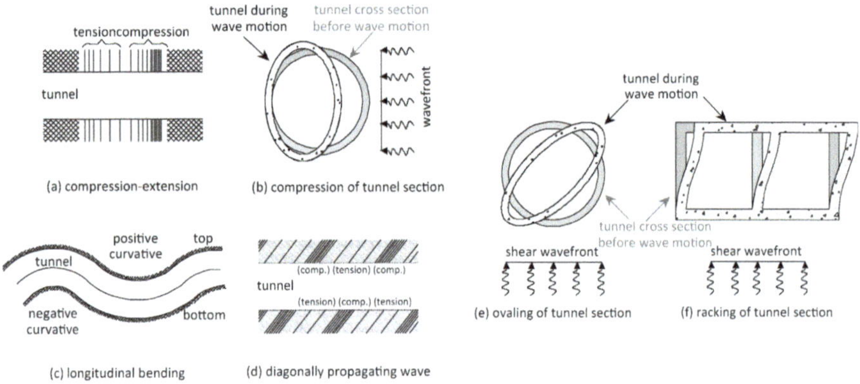

Fig. 11.5 Simplified deformation modes of tunnels due to seismic waves (Modified after Owen and Scholl 1981)

cause the deformation modes presented in Fig. 11.5e and Fig. 11.5f. Ovaling or racking effects for circular and rectangular cross-sections respectively.

Contrary to the pipelines, which are generally very flexible, having small diameters, tunnels and underground structures are quite stiff structures with important dimensions. Hence, while for pipelines the soil-structure effects are generally small if not negligible (Newmark 1968; Mavridis and Pitilakis 1996), for tunnels and other structures considered herein, soil-structure effects cannot be disregarded.

During earthquake shaking strong interaction effects are mobilized between the underground structure and the surrounding soil. This interaction is related to two crucial parameters, namely: (i) the relative flexibility of the structure and the ground and (ii) the interface characteristics between the structure and the surrounding soil. In general, both are changing with the seismic excitation as they depend on the soil shear modulus and strength, which depend on the ground strains attaint and the soil non-linear behavior.

11.3.1.1 Soil-Structure Relative Flexibility

The soil deformation in the proximity of the structure, which in general is not elastic, imposes displacement constraint on the structure's cross-section. Yet, due to the stiffness difference between the two media, the structure does not follow the imposed ground deformation. A relatively rigid structure will resist to the seismic ground deformations, while a flexible structure will follow the ground distortions. For a very flexible structure the structural distortion may be even higher than the free field ground deformation. The overall seismic behavior of the structure depends on the properties of the surrounding soil, which are not constant during strong excitation, and the inertial properties of the structure itself.

The soil to structure relative flexibility is expressed through the so called flexibility ratio F (Owen and Scholl 1981; Wang 1993; Penzien 2000 etc.). The value of the flexibility ratio is closely related to the expected stress level on the structure:

- $F \to 0$: the structure is rigid and will not display any type of deformation.
- $F < 1$: the structure is stiffer than the surrounding soil, thus the structural deformation level will be smaller than the free-field deformation level.
- $F = 1$: the structure and the surrounding soil share the same level of stiffness, so the tunnel will follow the free-field deformation.
- $F > 1$: the racking deformation of the structure is amplified compared to the free-field deformations.

To this end, a crucial point for the seismic evaluation of an underground structure is the proper estimation of the flexibility ratio. For *circular tunnels*, assuming elastic behavior, the flexibility ratio can be computed, using the following analytical formulation (Hoeg 1968; Wang 1993):

$$F = \frac{E_s\left(1 - v_l^2\right)R^3}{6E_l I_l\left(1 + v_s\right)} \tag{11.1}$$

where, E_s is the soil elastic modulus, v_s is the soil Poisson ratio, E_l is the lining elastic modulus, v_l is the lining Poisson ratio, I_l is the lining moment of inertia (per meter) and R is the circular tunnel radius. For *rectangular structures* the flexibility ratio is estimated according to Wang (1993) as:

$$F = \frac{G_m \times W}{S \times H} \tag{11.2}$$

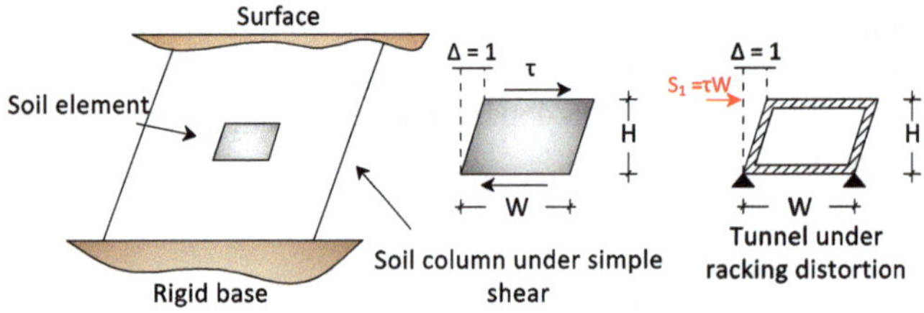

$$\text{Soil: } \gamma = \frac{\Delta}{H} = \frac{\tau}{G_m} \rightarrow \frac{\tau}{\gamma} = \frac{\tau}{\Delta/H} = G_m, \text{ Structure: } \gamma = \frac{\Delta}{H} = \frac{P}{HS_1} = \frac{\tau W}{HS_1} \rightarrow \frac{\tau}{\gamma} = \frac{\tau}{\Delta/H} = \frac{HS_1}{W}, \text{ Flexibitity ratio: } F = \frac{G_m W}{HS_1}$$

Fig. 11.6 Simplified model to evaluate flexibility ratio for rectangular structures (after Wang 1993)

where, G_m is the soil shear modulus, W is the width of the structure, H is the height of the structure and S is the required force to cause a unit racking deflection of the structure, estimated through a simple static elastic frame analysis (Fig. 11.6). The procedure precludes the structure's rocking, assuming pure racking deformation as the prevailing deformation mode. It will be shown that this is not always the case.

Rigid structures can exhibit rocking movement along the structure's vertical line of symmetry, while for flexible structures vertical shear and inward deformations of the sidewalls and the slabs may be observed. To this end, the invert slab and the roof are not always moving and deformed as a simple "shear beam". Rocking or inward deformations of the structure are clearly observed theoretically (Fig. 11.7); rocking modes or inward deformation modes, which prove the emerging need to consider seriously the SSI effects in the seismic design.

Experimental centrifuge results on rectangular model-structures confirm the theoretical observations (Cilingir and Madabhushi 2011a, b, c; Tsinidis et al. 2013b, c, d). As an example, results from a dynamic centrifuge test, performed on a square tunnel model embedded in dry Huston sand, are presented (Tsinidis et al. 2013b, c, d). With flexibility ratio F = 4.69, the tunnel according to the previous criteria is flexible compared to the surrounding soil. Figure 11.8a presents the recorded surface settlements above the tunnel and in free field. Figure 11.8b presents the stress–strain loops, computed directly from the recorded accelerations at different depths from two arrays. The settlements above the tunnel are smaller than those recorded far from the tunnel and at the same time the shear modulus is much higher in the soil mass above the tunnel. The observations indicate that actually the structure is rather rigid with respect to the surrounding soil, contrary to the initial hypothesis. These remarks and other that will be discussed later, stress the need for an improved and rational evaluation of the flexibility ratio.

11.3.1.2 Soil-Structure Interface

The interface characteristics between the structure lining and the surrounding soil affect considerably the seismic behavior of the system (Huo et al. 2005; Sedarat et al. 2009). Actually the development of the shear stresses is supposed to reduce to

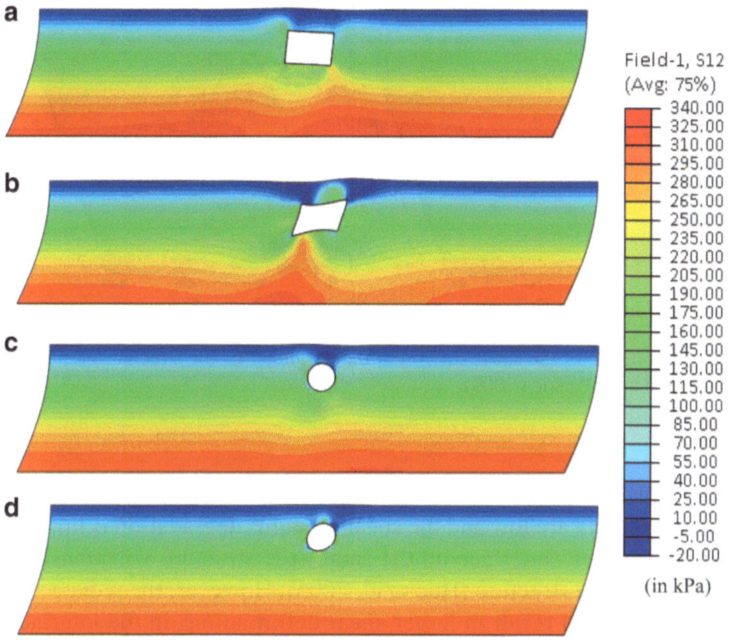

Fig. 11.7 Deformation modes of rectangular and circular tunnels in sand ($V_s = 200$ m/s) subjected to sine wavelets – Shear stresses contours at maximum distortion: (**a**) rigid rectangular tunnel (F = 0.18), (**b**) flexible rectangular tunnel (F = 11.9), (**c**) rigid circular tunnel (F = 0.07), (**d**) flexible circular tunnel (F = 2.07)

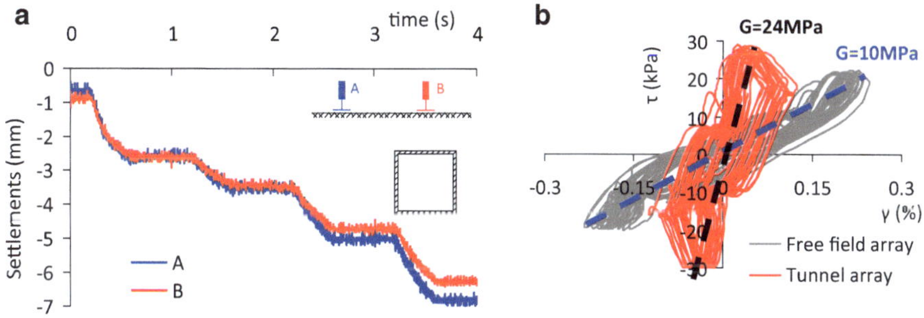

Fig. 11.8 (**a**) Dynamic settlements as recorded during a centrifuge test at two points of the soil surface, (**b**) Stress–strain loops computed by the recorded accelerations at free field and near the tunnel (Tsinidis et al. 2013b, c, d)

a minimum the in-plane vertical shear deformation of the structure and it's rocking, forcing it to deform predominantly in "horizontal shear" (Fig. 11.9). However, this is not observed both numerically and experimentally.

Rough, "strong" interface, capable to accommodate high shear stresses with limited deformations, results in high stresses on the lining. If the structure is stiff

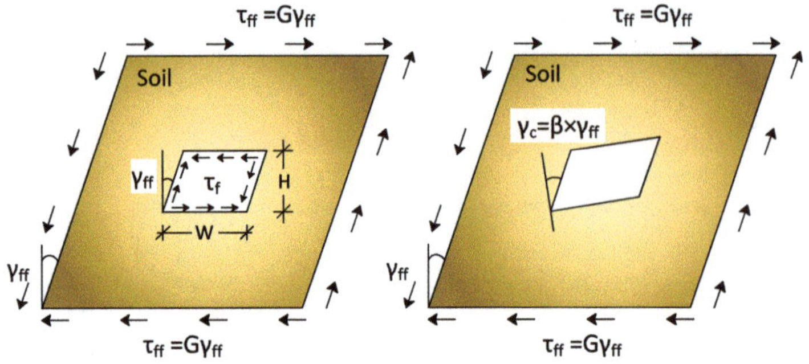

Fig. 11.9 Importance of the shear stresses at the perimeter of the structure on the deformation mode (Modified after Penzien 2000)

the displacements of the surrounding soil are constrained, the soil deformations around the structure are low and the shear modulus degradation is limited. On the contrary, a "weaker" interface on the same stiff structure will transmit smaller shear stresses to the structure (Huo et al. 2005). However, the reduced constraint on the soil mass will result in larger shear deformations in the interface, which may even produce a slip along the interface. The slip will produce higher concentration of stresses, higher strains and consequently a further degradation of the soil shear modulus, causing larger deformations in the surrounding soil and high normal stresses at the interface. Hence the potential to develop considerable soil displacements will be increased, although the capability to transmit them in shear to the structure will be small. To this end, the soil-structure relative stiffness and the interface characteristics are interrelated and may have opposite effects on the structure's response. Figure 11.7 depicts the total soil shear stresses (static and dynamic part) around tunnels of different rigidity, at the time of maximum structural distortion. Shear stresses are more uniformly distributed around the rigid tunnels, whereas for the flexible tunnels sharp stress concentrations are observed at some locations (i.e. roof slab to side walls corners for the rectangular tunnel).

In conclusion the complex soil to structure relative flexibility, usually expressed in terms of the flexibility ratio, and the interface seismic behavior have a paramount effect on the dynamic response of an underground structure. To this end, they affect not only the soil-structure interaction effects but also the earthquake loads applied on the structure, which are normally expressed in terms of (i) seismic earth pressures on the side walls, (ii) seismic shear stresses around the perimeter and (iii) inertial loads. All these parameters are poorly known resulting to simplified assumptions. The present chapter aims in contributing to a better understanding and quantification of these complex phenomena in order to improve the seismic design.

11.3.2 Ground Failure

Ground failure is related to large permanent displacements and ground deformations, caused by liquefaction, slope failure or fault movements. Once the amount of permanent total and differential displacements is known, the design of an underground structure to resist these deformations is a classical static problem. The main issue is the proper and rational estimation of the expected deformations and permanent displacements. Usually, existing guidelines for buried pipelines (i.e. ASCE 1984) or other relevant references are used for this estimation (i.e. for fault crossing, Newmark and Hall 1975; Kennedy et al. 1977). A detailed numerical approach is probably the best solution, when the parameters of the problem are somehow known (ASCE 1984; Anastasopoulos and Gazetas 2010). Having an idea of the ground deformations and the permanent displacements, several techniques can be applied to improve the design and mitigate the risk. In case of seismically induced landslides among the most common techniques is strengthening and stabilization of the ground using piles and anchors or/and improving drainage conditions. Similar techniques are used in case of liquefaction induced lateral spreading or upward buoyancy effects. For the case of fault displacements, there are some strategies to minimize the effects, namely enlarging the cross-sections of the structure, construction of ductile joints etc. (Power et al. 1998). Sometimes it is proposed to make contingency plans to facilitate the rapidly repair of the structure, instead of trying to design the structure to withstand the fault displacements.

11.4 Design Principles

It is common in engineering practice, to evaluate the static and the seismic behavior separately. The final detailing is performed for a combination of the static and the seismic loads. The analysis and design of underground structures consists of four major steps:

- Seismic hazard assessment and selection of the design earthquake.
- Evaluation of transient ground response and induced phenomena i.e. liquefaction, ground failure.
- Evaluation of structure seismic response.
- Synthesis of static and seismic loads.

11.4.1 Seismic Hazard Assessment and Selection of the Design Earthquake

The seismic design of an underground structure needs a site-specific evaluation of the ground motion characteristics along it's axis for different return periods.

The design ground motion for ordinary structures is usually proposed in national seismic codes. However tunnels and large space underground structures are not ordinary structures, which makes necessary the performance of site-specific hazard analysis. There are two methods of analysis:

- A deterministic seismic hazard analysis (DSHA) that is normally applied when the seismic catalogue is poor and there are well-identified active seismic faults affecting directly the structure.
- A probabilistic seismic hazard analysis (PSHA) that quantifies the uncertainties in the analysis and develops a range of expected ground motions with their probabilities of occurrence.

Once the level of seismicity in a region established, the design earthquake has to be defined. Two design earthquake levels are commonly defined, the Maximum Design Earthquake (MDE) and the Operational Design Earthquake (ODE) (Wang 1993; Hashash et al. 2001). The first level refers to the maximum level of ground motion expected in the region, with low probability of occurrence, and the design goal in this case is the public safety during and after the earthquake. The second level is expected to occur once during the life of the underground facility and the design goal is to continue operating during and after an earthquake. This means that the damages must be kept to a minimum. The aforementioned procedures provide the necessary ground motion parameters for analysis, such as ground acceleration, velocity and displacement amplitudes for outcrop conditions or at the ground surface if appropriate GMPE will be applied considering local soil conditions.

11.4.2 Evaluation of Transient Ground Response and Induced Phenomena

Regardless of the method of analysis, an important element in any design process is the adequate evaluation of the seismic ground motion imposed on the underground structure, expressed in terms of acceleration or ground deformations according to the method to be applied. The use of seismic code prescriptions should be used with caution for a preliminary analysis of the ODE case. As mentioned, a detailed site-specific analysis for different return periods of MDE and ODE is mandatory for the final design. This should be done for all segments of the long span structure in order to specifically address the variations of the local site conditions. Once the outcrop motion is defined then there are several methods (linear, equivalent linear, non-linear) to estimate the ground motion characteristics at the depth of the underground structure. For high structures (i.e. multi-story metro stations or parking lots) the ground motion characteristics should be defined at different depths. Peak amplitudes and time histories are both necessary. Traditionally a 1D analysis of upward propagating SH and SV waves is applied. Oblique incidence

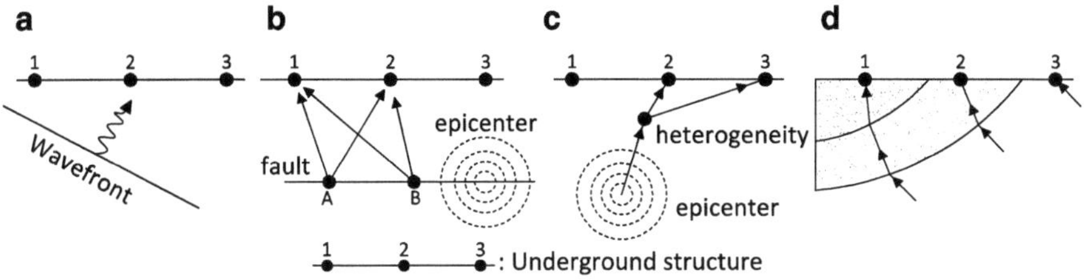

Fig. 11.10 Sources of asynchronous shaking and spatial variation of ground motion (**a**) inclined wavefront, (**b**) fault mechanism, (**c**) heterogeneity of the soil deposits, (**d**) soil local conditions

may be also used for near field conditions. In case of complex ground conditions and geometries (i.e. valleys, lateral geologic discontinuities), 2D or even a 3D analysis should be useful. The analysis is normally performed in total stresses. Effective stress analysis is applied when necessary.

For the longitudinal seismic analysis, spatial variability of ground motion must be properly estimated and accounted (Kramer 1996; Zerva et al. 1988; Zerva 1993, 1994; Zerva and Zervas 2002; Zerva and Beck 2003). Ground shaking is varying in space and time in terms of wave amplitude, phase, frequency characteristics and duration. A simple phase difference, due to incident angle of traveling waves is not always sufficient. These aspects can significantly affect the response of long structures (L > 200–300 m). In example, differential movements may be very important even in small distances. A schematic illustration of the potential sources of ground motion incoherency is presented in Fig. 11.10.

Commonly the input motion is estimated assuming free field conditions. This is not obviously the case for underground structures constructed in urban areas, where complex structural systems of aboveground and/or other underground structures exist. In these cases, the interactions between the structures may affect the input motion on the under consideration underground structure. A discussion is made on this issue in an ensuing section.

11.4.3 Structure Seismic Response

The evaluation of seismic response of underground structures is usually performed separately in the transversal and longitudinal direction and then the computed internal forces are combined (Hashash et al. 2001; ISO 23469 2005; FWHA 2009 etc.). Several methods are available in the literature for the seismic design in both directions. A short but comprehensive review is presented in the ensuing. The application of these methods in two representative real cases reveals the differences between them. This important remark is actually the starting point of the discussion in the present chapter. The discussion is enriched with recent experimental data from centrifuge testing (Tsinidis et al. 2013b, c, d).

11.4.3.1 Static and Seismic Loads

Once the seismic analysis of an underground structure is done, the results, in terms of internal forces must be combined with the results of the static analysis. It should be noted, as it will be discussed in detail later, that usually the static loads are prevailing, especially in case of large buoyancy loads. The combination is normally performed on the basis of load factors, which are not always available for underground structures. For cut and cover tunnels, Wang (1993) suggests the following combinations for MDE and ODE respectively:

$$U = D + L + E_1 + E_2 + EQ\ \{MDE\} \tag{11.3}$$

$$U = 1.05 \times D + 1.3 \times L + \beta_1(E_1 + E_2) + 1.3 \times EQ\ \{ODE\} \tag{11.4}$$

while for bored and mined tunnels the following expressions are proposed:

$$U = D + L + EX + H + EQ\ \{MDE\} \tag{11.5}$$

$$U = 1.05 \times D + 1.3 \times L + \beta_2(EX + H) + 1.3 \times EQ\ \{ODE\} \tag{11.6}$$

U is the required structural strength capacity, D is the effects due to dead loads of structural components, L is the effects due to live loads, E_1 is the effects due to vertical loads of earth and water, E_2 is the effects due to horizontal loads of earth and water, EX is the effects of static loads due to excavation, H is the effects due to hydrostatic loads, EQ is the effects due to design earthquake motion, β_1 is a factor taking values 1.05 if extreme loads are assumed for E_1 and E_2 with little uncertainty, otherwise 1.3 and β_2 is a factor taking values 1.05 if extreme loads are assumed for EX and H with little uncertainty, otherwise 1.3.

Considering the design goals for Maximum and Operational Design Earthquake, structures must provide strength and sufficient ductility, through special structural considerations. Unfortunately, there is not a clear answer on whether an embedded structure should be designed to remain in the elastic range or not, as there are not any strict code regulations regarding this issue. In some guidelines it is implied that embedded structures could be designed to undergo damages to a certain degree (i.e. Wang 1993; Hashash et al. 2001; FWHA 2009), especially for the case of MDE. In this case, special design arrangements must be considered to provide the required ductility to the structure. AFPS/AFTES (2001) give some very general guidelines on how to evaluate the post-elastic behavior of an underground structure. Introduction of plastic hinges for the critical sections is proposed. Some references are given for the estimation of their ductility, based on some general assumptions. On the other hand, considering the importance of embedded structures, their special features and the difficulty in retrofitting after serious damages, new structures are usually designed to remain in the elastic range (behavior factor q = 1). In any case, as the uncertainties on many crucial issues affecting the seismic behavior of the structure are quite large, sufficient ductility should be given to the structure, as a secondary defense mechanism.

11.5 Design Methods for Shaking

Several methods are available in the literature for the seismic design of tunnels and large underground structures (i.e. Wang 1993; Penzien 2000; Kawashima 2000; Hashash et al. 2001; AFPS/AFTES 2001; ISO 23469 2005; FWHA 2009 among others). Based on the way that the seismic loading is described and modeled, they can be classified in three general categories:

- Force-based methods.
- Displacement-based methods.
- Numerical methods where the structure and the soil are analyzed as a coupled system using different numerical techniques (finite element, finite difference, etc.)

With the exception of the third one, where the soil-structure interaction effects are inherently considered in the model, methods belonging in the other two general categories can be also classified accounting or not for soil-structure interaction. For the methods where the soil-structure interaction is not considered (free field deformation methods), it is assumed that the structure will undergo the free field ground deformations, whereas for the soil-structure interaction methods, the input motion (in terms of ground displacements or equivalent forces) is modified to account of the existence of the structure. Methods for the transversal and the longitudinal analysis are shortly presented in the following paragraphs.

11.5.1 Transversal Seismic Analysis

The transversal response is crucial for the evaluation of the maximum internal loads and the lining detailing. Several methods may be found in the literature both for circular cross-section structures (i.e. tunnels) and rectangular structures.

11.5.1.1 Forced-Based Methods

According to these methods, the seismic load is introduced in terms of equivalent forces acting on the structure in a static way. The structure is commonly simulated as a frame model using beam elements. The main differences between the methods

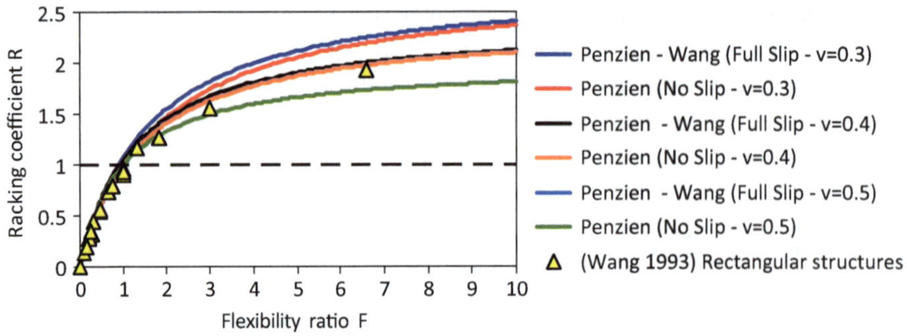

Fig. 11.11 F-R correlation for circular and rectangular structures (After Wang 1993 and Penzien 2000)

are related to the way that the equivalent forces are estimated and the way the soil-structure interaction is modeled, if accounted.

F-R Method for Rectangular Structures (Wang 1993; Penzien 2000)

Generally the soil-structure interaction effects are more significant in case of rectangular structures, as they usually have larger and thicker dimensions, to withstand the static loads. This feature may lead to a strong modification of the structural racking deformation with respect to the ground racking distortion. Simplified approaches have been proposed in the literature (Wang 1993; Penzien 2000), accounting of this modification, through the so called racking ratio (structural/ground distortion). According to these methods, the racking ratio is related to the flexibility ratio. Wang (1993) related the racking ratio (R) to the flexibility ratio (F) using results from full dynamic analyses of soil-tunnel systems, while Penzien (2000) correlated the two ratios based on a pseudo-static approach. Indicative F-R relations are presented in Fig. 11.11. Wang (1993) showed that for a given flexibility ratio, the normalized structural distortion (racking ratio) of a rectangular structure tends to be reduced by 10 % with respect to a circular tunnel. To this end, results for circular tunnels are proposed to be used as an upper limit.

Having an estimation of the racking ratio R, using diagrams like those in Fig. 11.11, the structural deflection can be computed as:

$$\Delta_{\mathrm{structure}} = \mathrm{R} \times \Delta_{\mathrm{ff}} \tag{11.7}$$

where, Δ_{ff} is the free field soil distortion.

An equivalent, to the structural deflection, force (force producing the same structural drift displacement as computed earlier) can be estimated and imposed on the structure, as a concentrated force or as distributed force (Fig. 11.12) in a simple elastic static frame analysis. The internal forces computed by this procedure, are combined with the static internal forces, for the final structure detailing.

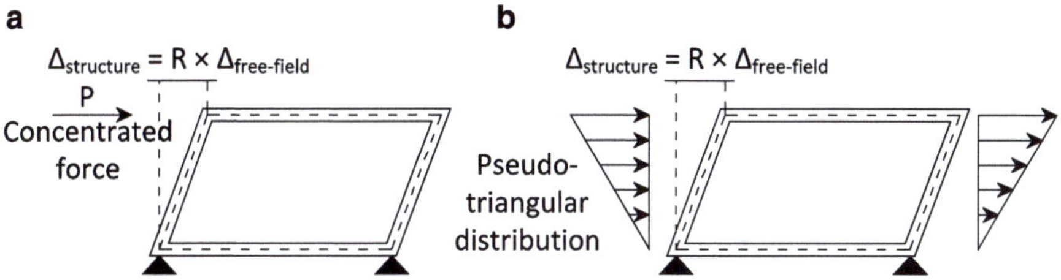

Fig. 11.12 Simplified frame analysis models (**a**) concentrated load at the corner, (**b**) distributed load on the side walls (Modified after Wang 1993)

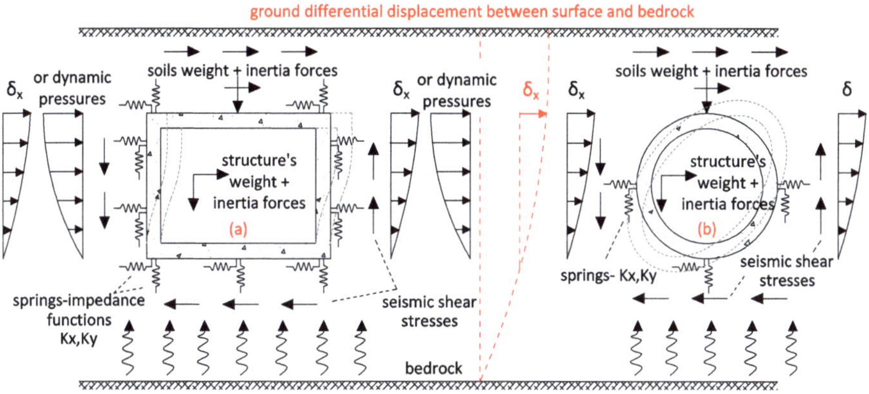

Fig. 11.13 Simplified equivalent static method – equivalent forces. (**a**) rectangular structure, (**b**) circular tunnel

The methodology is based on the hypothesis of pure racking deformation of the structure, both for the flexibility ratio determination (as mentioned above) and for the final analysis procedure. As mentioned, for rigid structures rocking movement may be observed, whereas for relatively flexible structures inward deformations of the slabs and the side walls may be occurred. The exclusion of these deformation patterns may modify considerably the design internal loads.

Simplified Equivalent Static Analysis

It is quite common in engineering practice to simplify the seismic analysis of a structure using an equivalent static procedure. According to ISO 23469 (2005), an underground structure can be analyzed in the transversal direction using a frame-spring model (Fig. 11.13). The structure in this case, is modeled with beam elements while the soil-structure interaction is modeled through appropriate springs (impedance functions). In the simplest case, the soil-structure interaction effects are precluded and the analysis is done with the frame model omitting the springs. The seismic loading is statically introduced in terms of: (i) equivalent inertial static loads (caused by the structure and the overburden soils mass), (ii) seismic shear stresses at the perimeter of the structure and (iii) seismic earth pressures imposed on the side walls of the structure. The equivalent inertial forces are estimated based on an average acceleration computed at the depth of the structure, while the shear stresses can be estimated through a 1D soil response analysis, accounting probably of the non-linear soil behavior through a non-linear or an equivalent linear procedure (Schnabel et al. 1972). Seismic earth pressures (for the case of rectangular structures) are usually estimated, using seismic code regulations for retaining walls, i.e. Mononobe Okabe approach (Okabe 1926; Mononobe and Matsuo 1929), regulations for rigid walls (Greek Seismic Code 2003; CEN 2004b) or other guidelines (i.e. Seed and Whitman 1970).

The method can easily be applied, but it has some important shortcomings. The accurate magnitude and distribution of the seismic earth pressures for the case of fully embedded structures are not well known. Estimations according to regulations referring to semi-embedded structures (i.e. retaining walls) may overestimate or underestimate the seismic earth pressures, leading to erroneous results. Similarly, the accurate estimation of the seismic shear stresses around the tunnel is still an open issue. Complex sliding and/or slip and gap between the soil and the structure may lead to redistributions of the stresses on the interface, phenomena that can not be captured with this method.

Moreover, the determination of impedance functions (e.g. springs and dashpots) for underground structures, is a quite delicate problem, as very few suggestions can be found in the literature for this type of structures. For example St. John and Zahrah (1987) proposed the following formulation for the evaluation of the longitudinal and transversal soil springs, assuming that the structure is subjected to sinusoidal input motion:

$$K = \frac{16 \times \pi \times G \times (1 - \nu)}{(3 - 4\nu)} \times \frac{d}{L} \tag{11.8}$$

where G is the soil shear modulus, v is the Poison ratio of the soil, d is the structure width and L is the sinusoidal wavelength. According to AFPS/AFTES (2001) guidelines, the soil springs may be estimated using the following formulation:

$$K = 0.5 \times \frac{G}{H} \tag{11.9}$$

where H is the structure's height. Other analytical expressions, referring to deep or surface foundations or even retaining walls, may be found in the literature (Scott 1973; O'Rourke and Dobry 1978; Gazetas and Dobry 1984; Gazetas 1991; Kavvadas and Gazetas 1993; Veletsos and Younan 1994; Mylonakis 1995; Building Seismic Safety Council 2003; Gerolymos and Gazetas 2006). Considering the conceptual features of underground structures that are very distinct compare to foundations and retaining walls, the utilization of these formulations may lead to large discrepancies (Pitilakis et al. 2007a; Pitilakis et al. 2009). Acknowledging these uncertainties AFPS/AFTES (2001), rather than offering an analytical expression, recomends a prior finite element analysis to determine the reactions of the surrounding ground on the structure for virtual unit displacement in the horizontal and vertical directions. In general the issue of the evaluation of appropriate impedance functions for underground structures is still open and further study deemed to be necessary.

11.5.1.2 Displacement-Based Methods

Contrary to the force-based methods, in the displacement-based methods the seismic load is introduced in terms of seismic displacements, an assumption, which is closer to the problem's physics. Once again, the main differences between the methods are related to the way that the seismic displacements are computed and the way in which the soil-structure interaction is modeled, if accounted.

Simplified Solutions – Free Field Approach

The main assumption of this approach is that the structure will undergo the free field ground deformations. For circular tunnels, simplified closed form solutions exist, whereas for the case of rectangular structures, a simplified static frame analysis is proposed. More specifically, circular tunnels under transversal wave passage will undergo ovaling deformations that can be calculated from the free field shear strains (γ_{max}), assuming non-perforated or perforated elastic ground, as:

$$\frac{\Delta d}{\mathrm{d}} = \pm \frac{\gamma_{max}}{2}, \text{ for non perforated ground} \tag{11.10}$$

$$\frac{\Delta d}{\mathrm{d}} = \pm 2\gamma_{max}(1 - \mathrm{v_m}), \text{ for perforated ground} \tag{11.11}$$

For rectangular structures a simplified static frame analysis is proposed to estimate the response. The structure is modeled as a frame model subjected to the free field soil distortions.

The soil shear distortions may be obtained with simplified methods (i.e. Newmark 1968; FWHA 2009; Gingery 2007) or through a 1D soil response analysis (i.e. Schnabel et al. 1972). The method may overestimate or underestimate the structure's response, depending on the rigidity of the structure (Hashash et al. 2001). Generally it is considered acceptable for relatively flexible structures in rigid soils or rock.

Simplified Solutions for Circular Tunnels Including Soil-Structure Interaction

Several analytical solutions, to determine the diametric deflection and the internal forces of a circular tunnel subjected to seismic shear waves, can be found in the literature (Hoeg 1968; Wang 1993; Penzien and Wu 1998; Penzien 2000; Bobet 2003; Park et al. 2009). The most known and widely used methods are those proposed by Wang (1993) and Penzien (2000). The methods consider two soil-tunnel interface conditions, namely: (i) full slip, where no bonding in the tangential direction between the two media is considered and (ii) no slip, where the soil and tunnel are considered to be in perfect bonding. The true interface-contact conditions are often between these two limit states. The effect of the existence of the structure and the soil-structure interaction effects are expressed through the flexibility ratio and the compressibility ratio, which are measures of the flexural and extensional stiffness of the structure with respect to the surrounding soils. The solutions are summarized in Table 11.A1, of Appendix A. The approach seems to be fair for relative rigid tunnels embedded in soft soils, where the soil-structure interaction effects are expected to be increased. A more complex analytical solution is presented by Huo et al. (2006) for rectangular tunnels.

Simplified Equivalent Static Analysis

As mentioned before the structure is modeled as a frame subjected to equivalent static loads. The only difference with the forced-based case is that instead of imposing seismic earth pressures on the side-walls of the structure, seismic ground deformations are imposed. The ground deformations can be computed from a simple 1D soil response analysis, usually applying an equivalent linear approach.

11.5.1.3 Numerical Methods

To avoid the problem of proper determination of impedance functions for underground structures or proper estimation of earth pressures, seismic shear stresses or even predefinition of the deformation mode, the complete soil-structure system can be modelled and analysed, using 2D numerical models (FE, FD, spectral elements, etc.) (ISO 23469 2005; FWHA 2009). Commonly, when a FE model is used, the structure is modelled with beam elements and the surrounding soil with plane strain elements. To this end, the soil-structure interaction is explicitly reproduced, while layered soil deposits can be accurately simulated. The non-linear behavior of the soil, the structure or the interface can be also taken into account, using appropriate constitutive laws. The main difference between the numerical methods is the way in which the seismic load is introduced.

Detailed Equivalent Static Analysis

According to the method, the seismic load is introduced statically, in terms of "free-field" seismic ground deformations, imposed on the side boundaries of the 2D numerical model. They are separately computed using a 1D soil response analysis of the free-field soil conditions. The imposed static ground displacements may be the maximum at each depth independently of the time or those corresponding to the time step when the soil shear distortions at the structure's depth are maximized. Normally several time-histories should be used (more than ten) taking the average at each depth. FHWA (2009) guidelines describe a stepping procedure, where instead of imposing ground displacements, the model is statically enforced by displacement time histories, computed for each depth from a 1D soil response analysis.

An important issue is the selection of the appropriate lateral extends of the model. The distance of the side-boundaries to the structure cross section, can substantially affect the results. When the seismic ground deformation is imposed at a long distance from the structure, as proposed in ISO 23469 (2005), a great amount of induced ground strain might be artificially absorbed by the soil response, thus "relieving" the structure and altering the analysis results. If, on the other hand, the model is laterally reduced bringing the side boundaries very close to the structure, it is doubtful whether the soil-structure interaction mechanisms will deploy to their full extent.

Full Dynamic Time History Analysis

Full dynamic time history analysis is considered to be the most sophisticated and accurate method for the seismic analysis of underground structures (ISO 23469 2005; FWHA 2009). The method can efficiently describe the kinematic and inertial aspects of the soil-structure interaction and the complex geometry of the soil deposit. Moreover, the non linear behavior of both the structure and the soils can be efficiently simulated using appropriate constitutive laws. In the simplest case, an equivalent linear approximation may be employed in order to simulate the soil behaviour under seismic excitation (Bardet et al. 2000). The soil shear modulus and damping properties are modified in a multi-step procedure at every analysis cycle, according to the shear deformation (γ) of the soil deposit, estimated in the previous time step, and this at each depth. The degradation of soil shear modulus [G(γ)] and material damping increase [D%(γ)] with increasing shear strain (γ), are described by selected G-γ-D curves corresponding to each soil type.

The material damping for soils is normally of hysteretic type and frequency independent. Nevertheless, employing the frequency dependent Rayleigh-type damping facilitates the dynamic analysis. Indeed, since Rayleigh damping is a linear combination of mass and stiffness matrices, it is efficiently incorporated into the analysis procedure. Yet, the selection of the damping parameters and the resulting damping curve should be carefully inspected, in order to achieve constant damping properties at the frequency range of interest.

The size of the plane strain elements, simulating the soil (when the FEM is incorporated), should be carefully selected, in order to efficiently reproduce propagation of seismic wave frequencies up to 10–15Hz, an upper frequency bound considered adequate for civil engineering purposes. The analysis is performed imposing the input motion at the base of the model (i.e. bedrock level). If outcrop time histories will be selected for the analysis, it is proposed to use elastic basement instead of a rigid one.

11.5.2 Discussion on the Transversal Analysis

11.5.2.1 Force-Based Against Displacement-Based Design

The real seismic loading on tunnels and underground structures is applied through the seismic wave propagation and the associated ground distortions imposed on the underground structure. So, displacement-based design is more appropriate and consistent with the physics of the problem. However force-based design, despite several drawbacks, remains always familiar to the engineering community and simpler to apply. We are discussing in the following the main issues associated with the current force-based and displacement-based design methods for underground structures. Further discussion is made in the following paragraphs on specific issues like seismic earth pressures and shear stress distribution around the tunnel.

Much research has been done the last years regarding the use of these two general approaches in the seismic design of aboveground structures. Several arguments, in favor of the inconsistency of force-based design, as discussed by Priestley (1993) and recently by Calvi and Sullivan (2009), are also valid in the case of large space underground structures and tunnels. We shortly discuss some of these issues in the following.

- The use of ductility capacity dependent force-reduction factors is inappropriate because the actual ductility demand for a structural component is typically smaller than the ductility capacity of the structures'. Moreover it is not clear whether an embedded underground structure should be designed to remain in the elastic range or not.
- The elastic force distributions are computed using the elastic stiffness for analysis. The inelastic design forces are then computed by reducing uniformly the elastic forces by a behavior factor, neglecting the fact that within the structure do not all the elements yield at the same level of deformation (e.g. see the failure of the Daikai station).
- It is very difficult to define the system ductility for mixed structural systems often present in underground structures like metro stations, parking lots and roadway or metro tunnels. Using the same behavior factor to the lower of the different components is inappropriate, as we don't know how the ductility demands and forces will develop for the complex system.
- There is a strong inter-dependency of strength and stiffness in RC structures. Their stiffness depends on the strength each component is assigned and it is not purely a function of the section geometry. Therefore, in order to know the cracked elastic period of vibration, the flexural strength is required. However the force-based design procedure, which relies on the period of vibration in order to determine the required strength, is not capable to provide this information.
- The force-based design approach estimates the inelastic displacement based on the elastic displacement response. In Eurocode 8 (CEN 2004a) the inelastic displacement is assumed to be equal to the elastic displacement, contrary to the United States and Japan regulations where different procedures are adopted.

For underground structures, we are demonstrating herein, the development of inelastic displacements (see the following paragraphs in this chapter), which depend on the hysteretic properties of the structural members and the structure as a whole. These inelastic displacements cannot be estimated using a force-based design procedure.

All these problems are emerged with force-based design approach. The displacement-based design is proved to be not only more consistent with the physics of the problem, but also more appropriate for the efficient seismic design of RC underground structures with ductility higher that unity.

11.5.2.2 Seismic Earth Pressures

The distribution and the magnitude of the seismic earth pressures on an embedded structure are quite complicate and generally not well known. Consequently, designers are obliged to use traditional approaches inspired or directly related to Mononobe-Okabe approach for retaining walls (Okabe 1926; Mononobe and Matsuo 1929; Seed and Whitman 1970; Greek Seismic Code 2003; CEN 2004b). Obviously, they are applied in case of force-based pseudo-static analysis mentioned earlier. These regulations and guidelines refer to partially embedded structures, or above ground retaining walls, whose behavior is quite different from deep large underground structures. Therefore, their computed internal forces might differ considerably from the real ones. We discuss in the ensuing, several issues related to the seismic earth pressures, using as examples two real structures (Pitilakis and Tsinidis 2010) and results from preliminary theoretical and experimental studies.

The first example is the case of an immersed tunnel planning to be constructed in Thessaloniki, Greece (Fig. 11.14a). In Fig. 11.14b we compare the dynamic part of the seismic earth pressures estimated according to Mononobe-Okabe method, with the maximum dynamic earth pressures (regardless the time step), computed with a full dynamic numerical time-history analysis. Two different input motions have been used both scaled at the same PGA. The M-O average dynamic earth pressures on the tunnel's side-walls are poorly capturing the pattern and the amplitudes of the developing maximum dynamic earth pressures, which are clearly depending on the relative flexibility of the tunnel and the soil, an issue that is not considered in the M-O approach. The assumption of non-deformable wall in M-O deviates considerably from the numerical results, where the actual flexibility of the tunnel is properly modelled. Actually, due to the tunnel's complex seismic response and deformation, the seismic earth pressures are varying between the passive (close to the slabs) and active (between the two slabs) limit states. To this end, the M-O approach assuming "active conditions" and the similar Greek seismic code (EAK2000 2003) one, are clearly underestimating the earth pressures. On the contrary assuming "passive conditions" the earth pressures are overestimated.

A major problem of the M-O methods, in particular in case of deep and high underground structures is the estimation of the actual PGA, which will be used in

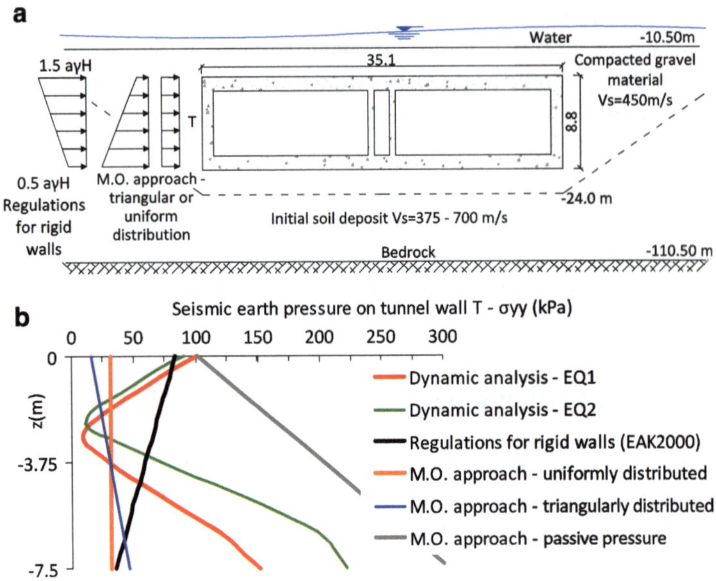

Fig. 11.14 (**a**) Tunnel simplified cross section and (**b**) comparison of the computed dynamic part of seismic earth pressures on the sidewalls through a full dynamic numerical analysis with the Mononobe-Okabe limit state approach and the Greek seismic code regulations (EAK2000 2003) for rigid walls

the analysis. The average value over the high of the structure, often used in practice, may be inaccurate for large dimension structures (i.e. multi story metro stations or parking lots).

In a strong earthquake, soil plastic deformations around the tunnel, may cause redistributions of the stresses affecting also the dynamic earth pressures. Figure 11.15 presents the computed dynamic earth pressures time histories at two characteristic points of the sidewall of a simple rectangular culvert (2m×2m) embedded in sand (F = 11.6). The culvert was subjected to excitations of increasing amplitude, causing soil plastic deformations. These deformations resulted in residual earth pressures on the structure, especially at the middle height of the wall, where the soil deformations were increased due to the inward deformation of the tunnel.

Similar behaviour was also observed during centrifuge tests on square model tunnels embedded in dry sand (Cilingir and Madabhushi 2011a, b, c; Tsinidis et al. 2013b, c, d). Figure 11.16 presents the recorded earth pressures at two locations, as measured during a centrifuge test on a square tunnel embedded in dry sand (Tsinidis et al. 2013b, c, d). Both the experimental and the theoretical data indicate three phases for the seismic earth pressures, namely a transient stage, a steady state stage and finally a post-earthquake residual stage. During the first few cycles of loading, seismic earth pressures were building up. Then in the steady state stage the earth pressures were fluctuating around a mean value. Finally, in the post-earthquake stage

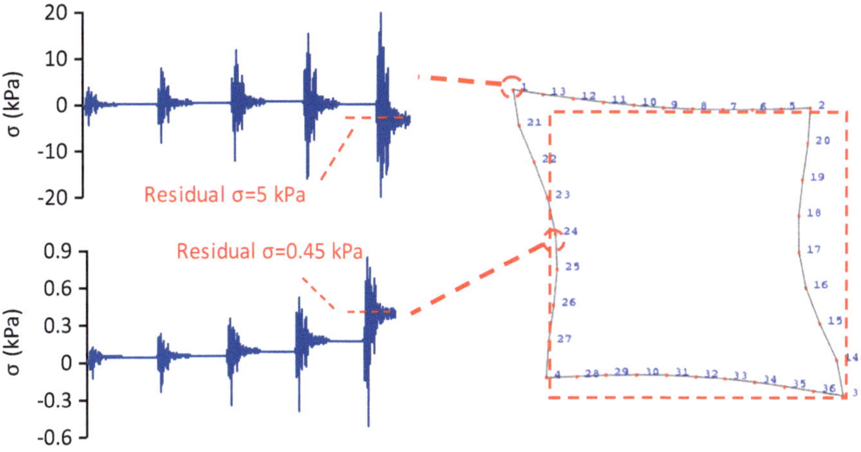

Fig. 11.15 Dynamic earth pressures time histories at two points of a rectangular culvert (F = 11.6) embedded in sand (V_s = 230 m/s, γ = 16.3kN/m^3, φ = 32°, ψ = 4°)

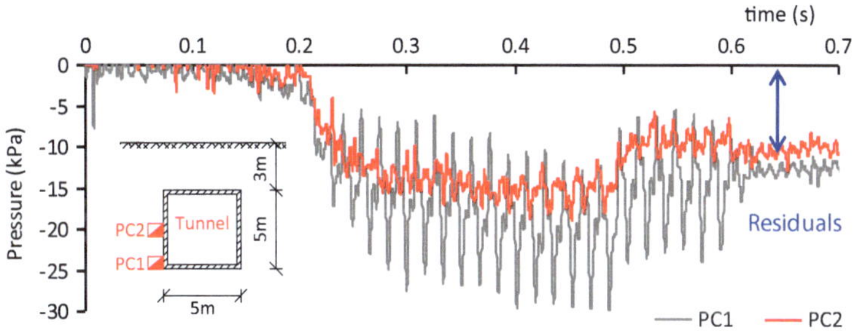

Fig. 11.16 Dynamic earth pressures time histories as recorded at two points of a rectangular tunnel embedded in dry sand during a dynamic centrifuge test (Tsinidis et al. 2013b, c, d)

some residual stresses were obtained on the tunnel lining exactly as in the numerical example.

The second case study is a multi-storey deep underground structure (e.g. metro station, Fig. 11.17a). More specifically, the station consists of 3 levels; each 7 m high and it is constructed in stiff clay. The dynamic earth pressure distribution is now even more complicated. The "blind" use of the M-O method is certainly erroneous as it does not respect the fundamental assumptions of a limit state approach as M-O. An approach close to that of braced diaphragm walls should be more adequate. The main question is again the amplitude of the ground inertial forces (peak ground acceleration). In Fig. 11.17b we compare the maximum dynamic earth pressures developed on the side-walls of the structure. The comparison is globally good. The earth pressures calculated for non-deformable walls seem to be an upper limit of the dynamic earth pressures. However stress concentrations

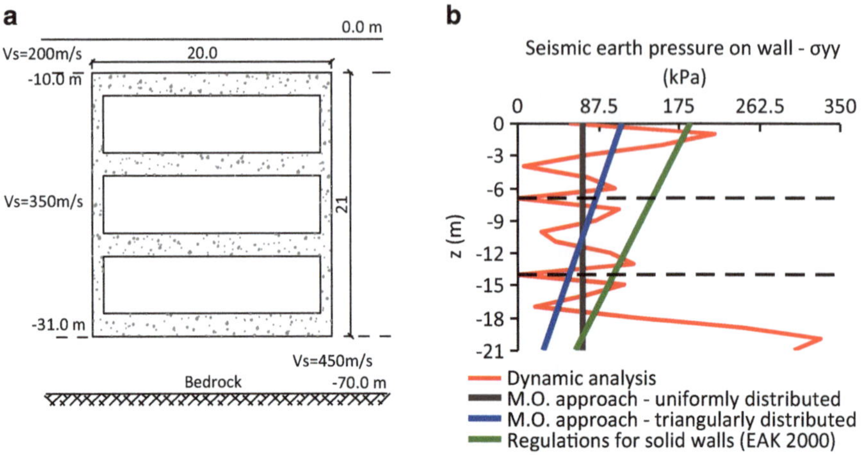

Fig. 11.17 (**a**) Station cross section, (**b**) seismic earth pressures on the sidewalls

at the slabs levels cannot be captured, neither the important earth-pressures at the base of the structure.

In conclusion we demonstrated, through few representative examples, that the available analytical methods to estimate the seismic earth pressures on underground structures, for the force-based approach, are not accurate enough. Their main weakness is the fact that they neglect the relative soil-structure flexibility, which depends on a number of parameters namely, the structure geometry and deformation modes, the soil properties, as well as the amplitude and frequency content of the input motion; hence this issue remains an open research subject.

11.5.2.3 Seismic Shear Stresses

The seismic shear stresses distribution mobilized between the lining and the soil can significantly affect the seismic behavior of the structure. However, the proper estimation of these stresses is quite difficult, as their values depend on several factors, i.e. the deformation pattern of the tunnel, which is not always in simple shear, the intensity of the input motion and the rugosity of the soil-lining interface. Moreover a detailed measurement of the shear stress distribution along the perimeter of the structure is very difficult to our knowledge. Some recent researches (Tohda et al. 2010; Chen et al. 2010) revealed that the values of the mobilized shear stresses are quite small and slippage at the soil-structure interface is happening even for very low amplitudes of input motion. As the measurement of local shear stresses is quite difficult, further study deemed to be necessary to verify these observations.

A real structure, i.e. the immersed tunnel presented earlier, and some preliminary results from theoretical and experimental studies in the centrifuge are used as examples, to discuss herein some crucial issues related to seismic shear stresses.

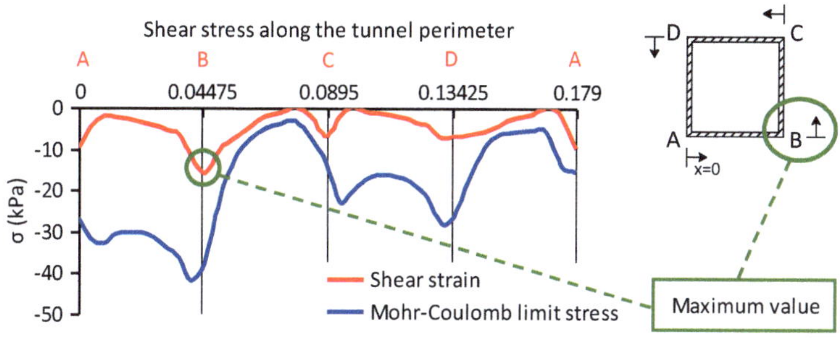

Fig. 11.18 Seismic shear stresses developed around a square culvert (for the time step of maximum racking distortion) compared to Mohr – Coulomb limit stress

Commonly, for simplification, a uniform distribution for the circumferential shear stresses is adopted. As mentioned above, this is not always true. For flexible structures sharp stress concentrations may be observed at some locations (i.e. slab-side walls corners for the rectangular structures). Figure 11.18 presents the computed shear stresses distribution around the perimeter of the culvert presented before (Fig. 11.15), for the time step of maximum racking distortion. Full bonding between the soil and the culvert was assumed. The computed stresses are compared with the Mohr-Coulomb limit values, which are in general much higher. It can be seen that shear stresses are not distributed uniformly around the perimeter of the culvert getting their maximum value close to the corners (e.g. point B in Fig. 11.18). In some regions they tend to zero indicating the presence of a gap or slippage.

Similar observations are made in the case of the immersed tunnel (Fig. 11.19a). The seismic shear stresses are computed through full dynamic analysis and different simplified design approaches, where the input motion is introduced in terms of ground displacements (maximum values with depth, independently of the time step) on the lateral boundaries of the model. Solid connection of the tunnel nodes to the adjacent soil elements was assumed (i.e. very rough interface). The method proposed by Wang (1993) is also presented, together with the shear stresses calculated at the same depths for free field conditions, without the presence of the tunnel, using a classical 1D EQL model.

Different methods and approaches, with different working assumptions, lead to different results, compared to the full dynamic numerical analysis. For some cases the differences are too important, so as they certainly affect the whole design, including earth pressures and internal forces. As mentioned, shear stresses are affected by the characteristics of the soil-structure interface (i.e. rough or smooth surface) and the input motion (duration, amplitude etc.). Moreover, as for the seismic earth pressures, soil plastic deformations may affect the shear stresses around the structure, leading to residual stresses after shaking. A very crucial parameter that can significantly affect the distribution and the magnitude of the seismic shear stresses around the perimeter of an underground structure is its deformation pattern during the earthquake. The latter is not rationally captured by

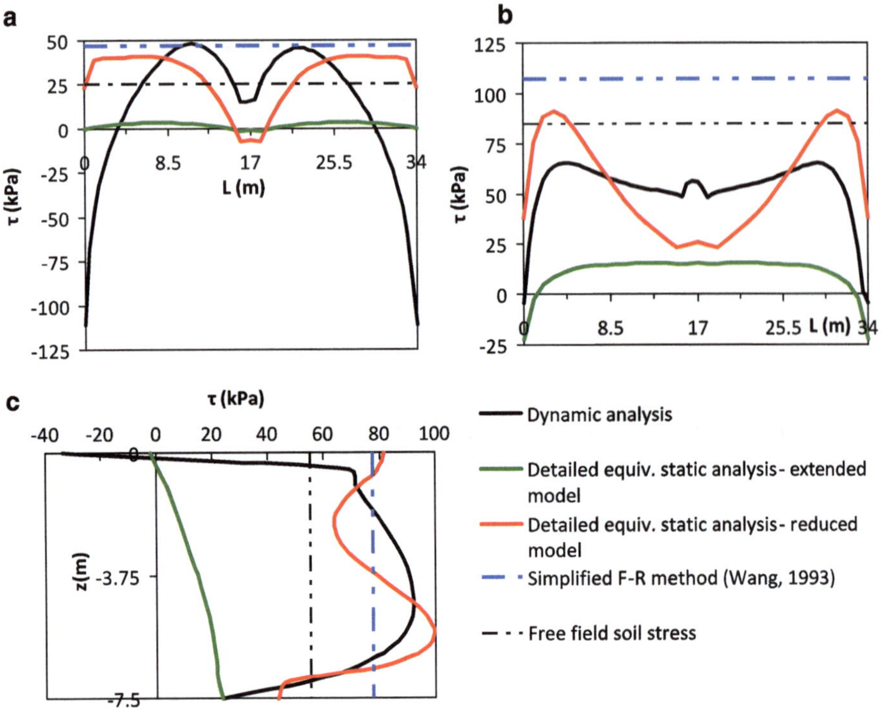

Fig. 11.19 Comparison of the computed seismic shear stresses with different methods at (**a**) the roof slab, (**b**) the inverted slab and (**c**) the side- walls of the Thessaloniki immersed tunnel

any simplified method, leading probably to the observed discrepancies. The strong correlation between the relative flexibility and the tangential behavior of the soil-structure interface need to be further studied.

A very interesting point has been recently observed from the analysis of experimental data of a centrifuge test (Tsinidis et al. 2013b, c, d). We compare the experimentally derived, using the Zeghal and Elgamal (1994) method, average shear stresses, with the shear stresses estimated in free field conditions applying the Seed and Idriss (1971) approach for shear waves propagating upward (Fig. 11.20). The experimental shear stresses are computed at two locations, namely in free field far from the tunnel and above the tunnel. We remark an excellent comparison in free field conditions for increased input excitations (0.21–0.36 g). On the contrary, above the tunnel the average shear stress magnitude depends on the level of excitation. For small intensities it is slightly larger than the free field and progressively, with increasing excitation, it decreases. In very strong excitation, the stresses above the tunnel are clearly lower than the free field. It is believed that this is not related to the non-linear soil behaviour, as it does not be observed at the free field. It should be related to the complex modes of deformations of the tunnel deviating from the simple shear beam hypothesis. It is an important result proving that the mobilization of the seismic shear stresses around an embedded structure

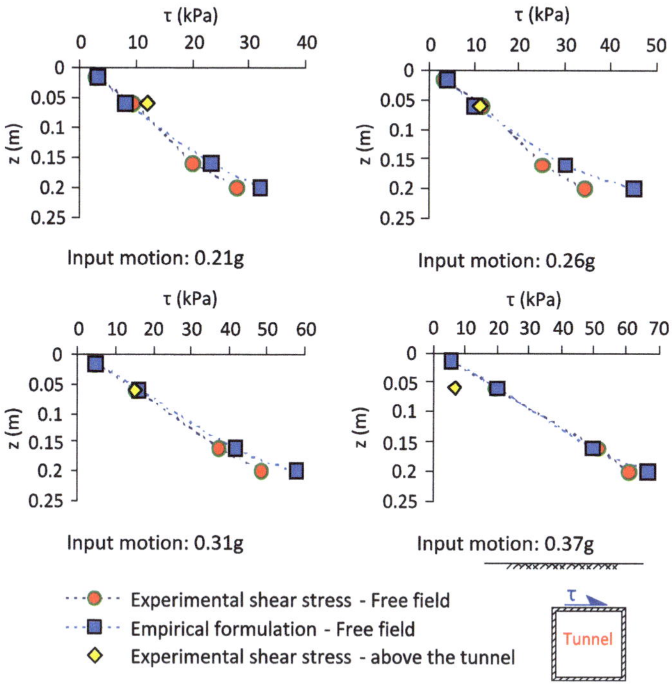

Fig. 11.20 Experimental derived shear stresses estimated in free field conditions and above the tunnel. Comparison with analytical solutions (Seed and Idriss 1971)

plays an important role in the seismic response, a fact that we didn't pay much attention until now.

11.5.2.4 Internal Forces

Internal forces are of prior importance for detailing the structure. Underestimating them is unsafe, while an overdesign is not only needlessly expensive but may lead to the stiffening of the structure that may change the whole response pattern in a detrimental way. To discuss the differences in internal forces applying different force-based and displacement-based methods, we compare, in the ensuing, their results with the full dynamic analysis for two case studies, namely, a multi storey metro station and the immersed tunnel.

Figure 11.21b presents the comparison of the computed bending moments for the typical cross-section of the "Venizelos" metro station in Thessaloniki (Pitilakis et al. 2007b; Pitilakis et al. 2008). In general, the simplified equivalent static analysis gives more conservative results (up to 100 %) in most locations, compared to the full dynamic analysis. This conservatism is probably necessary to cover various uncertainties of simplified equivalent static analysis method, such as the

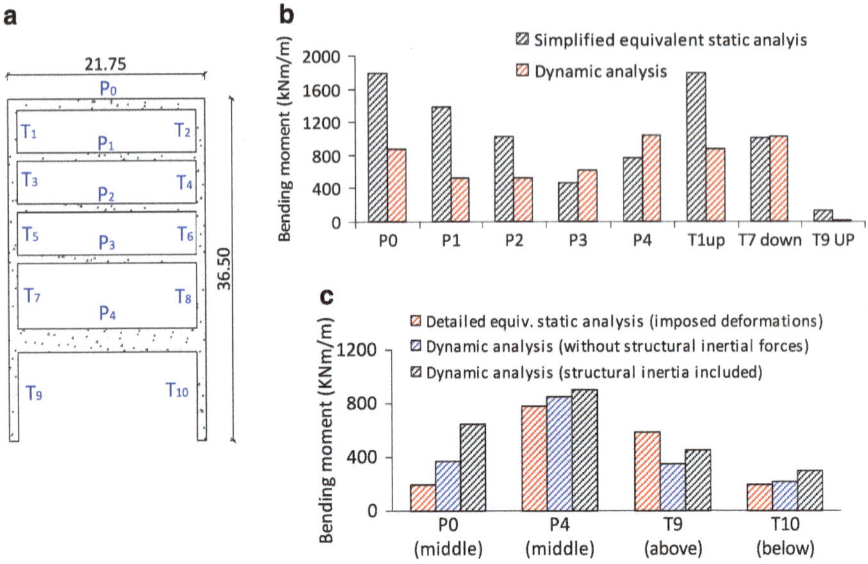

Fig. 11.21 (**a**) "Venizelos" metro station cross section, (Thessaloniki, Greece) (**b**) and (**c**) Comparisons of bending moments at crucial sections as computed by several methods

estimation of the impedance functions, or the estimation of the equivalent static loads (i.e. seismic earth pressures and inertia forces). In Fig. 11.21c we compare the bending moments of the detailed equivalent static analysis with the dynamic analysis. The equivalent seismic loading is introduced in terms of seismic ground deformations at the model side boundaries. The comparison is quite good for most segments with the exception of the roof slab, where the simplified method gives much lower bending moments.

In Fig. 11.22 we present comparisons of the computed bending moments (M) and axial forces (N) at six characteristic sections of the immersed tunnel. Ignoring soil-structure interaction effects is leading to large differences on the internal forces with respect to the dynamic analysis results. The improved ranking method proposed by Wang (1993) presents lower values compared to the full dynamic analysis, for both loading procedures, i.e. concentrated force at the roof slab-wall corner or triangular pressure distribution on the sidewalls (the latter is presented here). The simplified equivalent static analysis underestimates in the majority of the sections the seismic internal forces. In the case where the racking distortions are imposed in terms of equivalent pressures (estimated according to the Greek Seismic Code – EAK2000 2003 regulations for rigid walls), the results (not presented here) deviate even more with respect to the dynamic analysis. The response calculated with the detailed equivalent static analysis is generally closer to the dynamic analysis. The general remark is that the results using different methods present important differences.

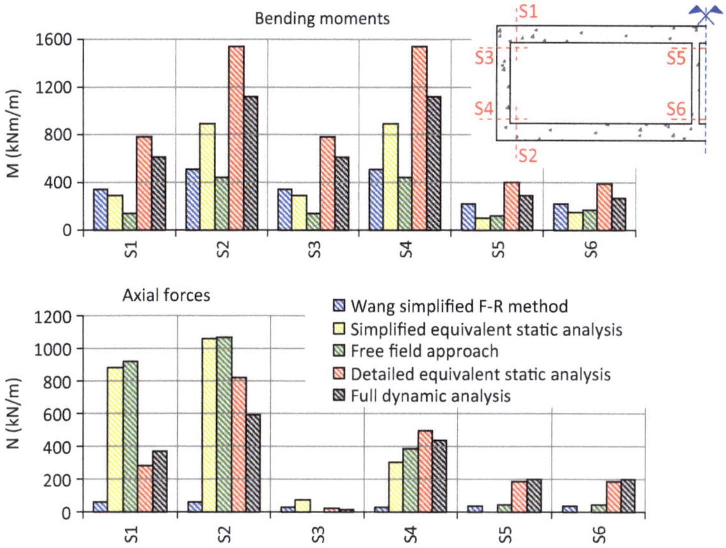

Fig. 11.22 Internal forces at six crucial sections of the Thessaloniki immersed tunnel as computed by several methods

Figure 11.23 presents comparisons of the computed internal forces of a circular tunnel, in terms of bending moments and axial forces, using a 2D full dynamic analysis and the simplified solutions proposed by Wang (1993) and Penzien (2000). Full-slip and no-slip assumptions were studied for the tunnel-soil interface, while the soil non-linear behavior was taken into account through an equivalent linear approximation. The comparisons indicate a good agreement between the analytical predictions and the numerical analyses, with an exception for thrust in case of no-slip conditions, where Penzien (2000) model underestimates the internal forces, a fact already stated by Hashash et al. (2005). To this end, the method summarized by Wang (1993) could be used.

An important issue, which has been disregarded so far, is the development of large soil plastic deformations that may cause stress redistributions around the underground structure, producing residual internal forces. Figure 11.24 is a good example of this phenomenon. Dynamic centrifuge tests were carried out on a circular flexible aluminum model tunnel embedded in dry sand under centrifuge acceleration of 80 g (Bilotta et al. 2009; Lanzano et al. 2010, 2012; Tsinidis and Pitilakis 2012; Tsinidis et al. 2013a). Four earthquakes (sine waves of increasing amplitude and frequency) were fired in a row. The three distinctive stages that were described before for the earth pressures, are now clearly observed for the bending moments of the tunnel lining. A single event may develop extra residual bending moments of about 20 % of the initial value. For a sequence of earthquakes the extra residual moments are much higher.

The development of excessive residual stresses and internal forces is explicitly related with the intensity of the ground motion and consequently with the non-linear

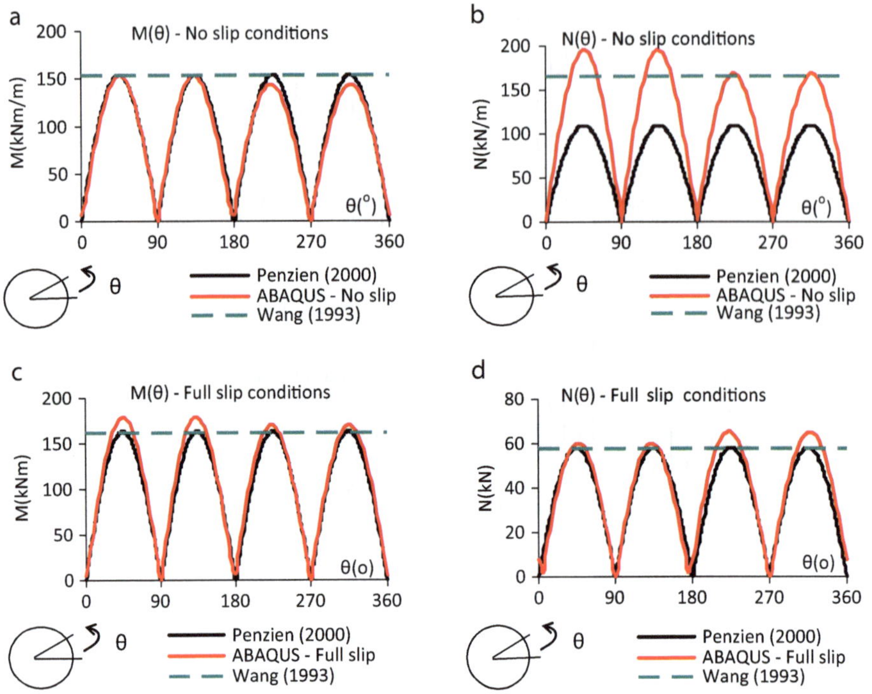

Fig. 11.23 Internal forces of a circular tunnel embedded in sand (**a**) bending moment for no slip conditions, (**b**) axial force for no slip conditions, (**c**) bending moment for full slip conditions, (**d**) axial force for full slip conditions (burial depth = 15 m, diameter: d = 6 m, lining thickness: t = 0.4 m, E_s = 36GPa, V_{so} = 250 m/s, G/G_o = 0.76, γ_{eff} = 0.059 %)

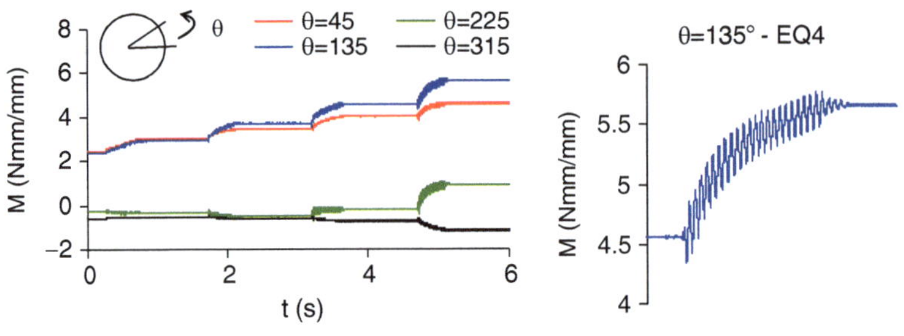

Fig. 11.24 Bending moment records from a dynamic centrifuge test on a circular model tunnel (after Lanzano et al. 2012)

soil behaviour. This behaviour cannot be efficiently captured with an equivalent linear approach, as it is stated by Amorosi and Boldini (2009) and observed in the example presented in Fig. 11.25. It is a flexible tunnel (F = 13) embedded in a soil deposit consisting of a soft layer on the top 15 m (V_s = 100 m/s, φ = 32°, c = 0 MPa), overlaying a stiffer soil (V_s = 350 m/s). The soil-tunnel system was analyzed, using a

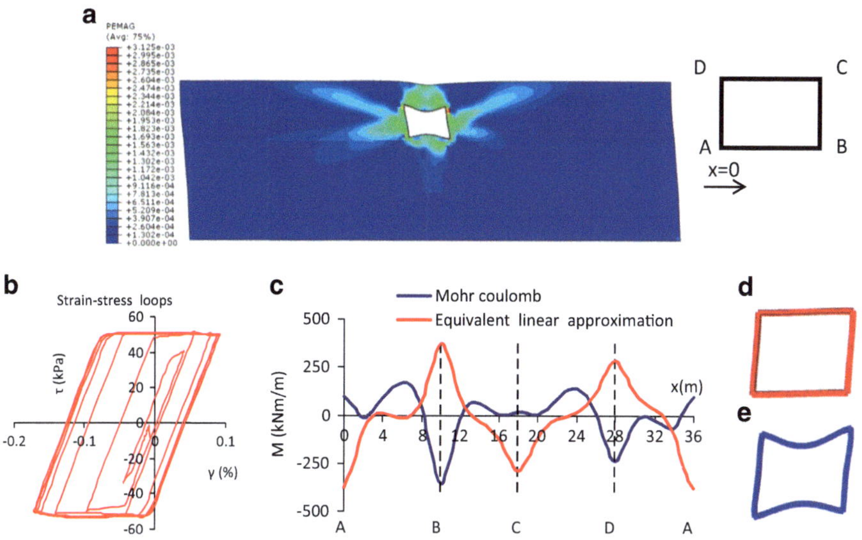

Fig. 11.25 (**a**) Plastic deformations around the tunnel, (**b**) soil shear strain–stress loops at the middle of the roof slab, (**c**) bending moments along the perimeter at maximum racking distortion, (**d**) tunnel deformed shape for the equivalent linear approximation analysis, (**e**) tunnel deformed shaped for the elastoplastic analysis

2D plane strain model in ABAQUS (ABAQUS 2009). Equivalent linear and elasto-plastic Mohr-Coulomb models were used. The plastic deformations around the tunnel are depicted at Fig. 11.25a, while the bending moments in the tunnel lining are presented in Fig. 11.25c. The soil plastic deformations are causing inward deformations of the lining (Fig. 11.25e), changing significantly the bending moment distribution around the tunnel, causing also residual bending moments.

11.5.3 Longitudinal Seismic Analysis

Longitudinal seismic analysis of long tunnels is equally important to the transversal analysis. Features, such as incoherence of seismic motion and seismic performance of joints (construction joints, joints between segments etc.) can substantially affect the seismic behavior of the structure. As for the transversal analysis, several methods have been proposed for the longitudinal seismic analysis (i.e. St. John and Zahrah 1987; ISO 23469 2005; FWHA 2009; Kawashima 2000 etc.). They are classified into two main categories, namely displacement-based methods and full numerical methods using either the finite element or difference method or other

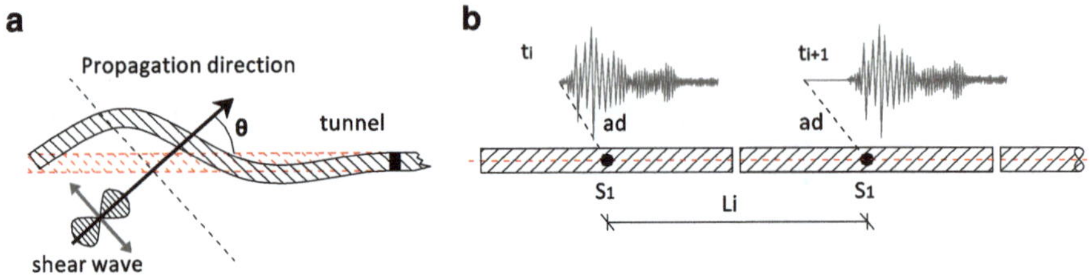

Fig. 11.26 (**a**) Wave propagation angle effect on the structure's response, (**b**) Time lag phenomenon for long underground structures

numerical techniques. There are also some analytical simplified solutions, which are also classified as displacement-based design approaches.

An important issue for the longitudinal seismic design is the selection of the input motion, which is controlled by the spatial incoherency of the ground motion, including local site effects (Fig. 11.10). Ground shaking is continuously varying in space, in terms of wave amplitude, frequency characteristics, time of arrival and duration. These aspects can significantly affect the response of long structures. Shaking table tests on rectangular RC tunnel-models embedded in unsaturated clay (Chen et al. 2010) revealed that non-uniform earthquake excitations produce higher intensities compared to the uniform ones, while they can lead to important differential displacements and rotation at the joints in case of segmented structures.

The simplest source of asynchronous shaking lies in the angle of the wave-front to the tunnel axis. The apparent velocity C_s of propagating waves, causing different arrival times between two separate points, can be calculated as:

$$C_s = {}^{V}\!/_{\sin\theta} \tag{11.12}$$

where, V is the velocity of the P, S and R waves and θ the angle, in which the seismic wave reaches the structure. Figure 11.26a schematically explains this effect.

In Fig. 11.26b the time-lag effect is presented, albeit in a crude application of the asynchronous shaking theory, with no alteration to the signal. The time lag can be computed using the following expression and for an apparent velocity ranging from 1,000 to 2,500 m/s (i.e. Power et al. 1998).

$$t_{i+1} - t_i = {}^{L_i}\!/_{C_s} \tag{11.13}$$

A detailed description of the different aspects for the evaluation of the ground motion spatial variability and its effects on long structures may be found in Zerva et al. 1988 and other relevant publications (Zerva et al. 1988; Zerva 1993, 1994; Zerva and Beck 2003 etc.).

11.5.3.1 Seismic Design Methods in the Longitudinal Axis

Displacement-Based Design Methods

All methods belonging in this category are using in one or another way the analogue of a beam on soil springs and dashpots to model the tunnel and the soil compliance. The input is always the ground deformation from the passage of seismic waves.

Analytical Solutions

Analytical solutions, precluding or accounting of the soil-structure interaction effects, can be found in the literature for the computation of the structural response. In the simplest case, analytical solutions provide the axial and the curvature ground strains for S, P and Rayleigh waves (i.e. St. John and Zahrah 1987 – Table 11.A2 of Appendix A). The strains are enforced directly on the structure that is modelled as an equivalent beam. Moreover, St. John and Zahrah (1987), have proposed analytical solutions for the calculation of the internal forces of the structure under sinusoidal excitation. The solutions are tabulated in Table 11.A3 of Appendix A. The same researchers have also proposed analytical solutions for the calculation of the internal forces accounting of the soil-structure interaction effects, for several types of seismic waves, using approximate expressions for the soil springs supporting the elastic beam. The appropriateness of the presently available impedance functions to account of SSI effects is a major problem of the seismic design of tunnels and underground structures. It has been discussed in several parts of this work and it will be further elaborated in this paragraph.

A similar method is proposed by JRA (1992), referring to common utility tunnels (Kawashima 2000). It is again based on a Winker model to describe the soil-structure interaction; the structure is modelled as a beam subjected to a predefined seismic deformation profile (Fig. 11.27). The ground deformation profile takes into account, in a simplified way, the variation of soil properties, the thickness of the soil deposits and the ground motion incoherency, introducing the so called modified equivalent wavelength (L'). For the given seismic ground displacement (Table 11.A4), the internal forces of the structure can be obtained, using the theory of an elastic beam founded on an elastic foundation (Table 11.A4).

Mass-Spring Model Method

The method has been proposed for immersed tunnels, (Kiyomiya 1995), but it can generalised for all long underground structures. According to this method, the surface layer is divided into a number of slices, perpendicular to the structure axis. Each slice is represented by an equivalent mass-spring system; the mass represents the soil slide mass, while a spring and a dashpot connect the mass to the rigid base. The spring constant is estimated, so as the system response to coincide to the natural period of the first mode of soil deposit shear vibration.

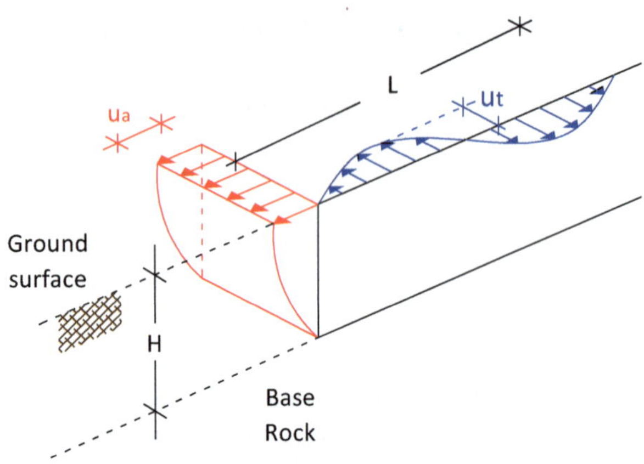

Fig. 11.27 Seismic ground deformations according to JRA displacement-based method (Kawashima 2000)

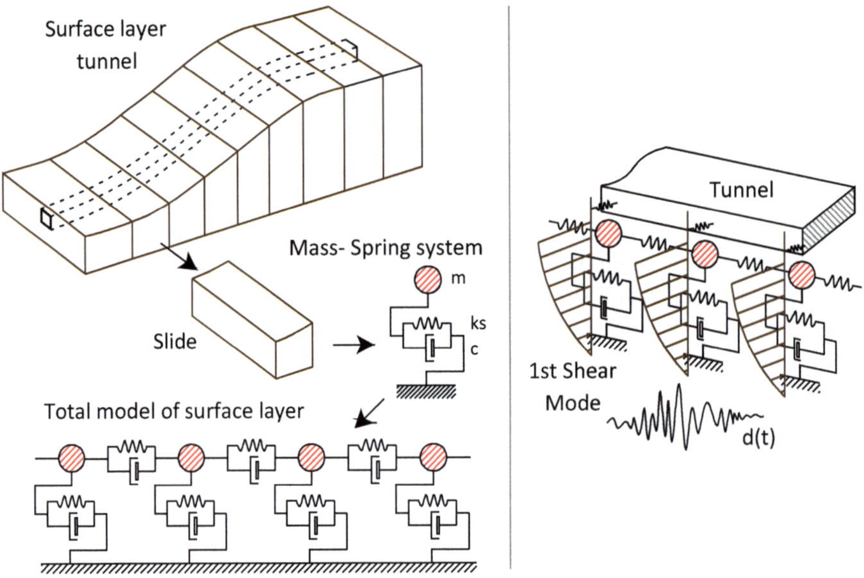

Fig. 11.28 Mass spring model method (Modified after Kiyomiya 1995)

The neighbouring masses are connected to each another, along the structure axis, with springs and dashpots, to simulate the connection between the adjacent ground slices. Using the general dynamic equilibrium equation of the system, the response of the soil deposit can be computed. Then, the structural response may be estimated, assuming the structure as a beam supported by springs, subjected to the ground displacement computed by the aforementioned soil model (Fig. 11.28).

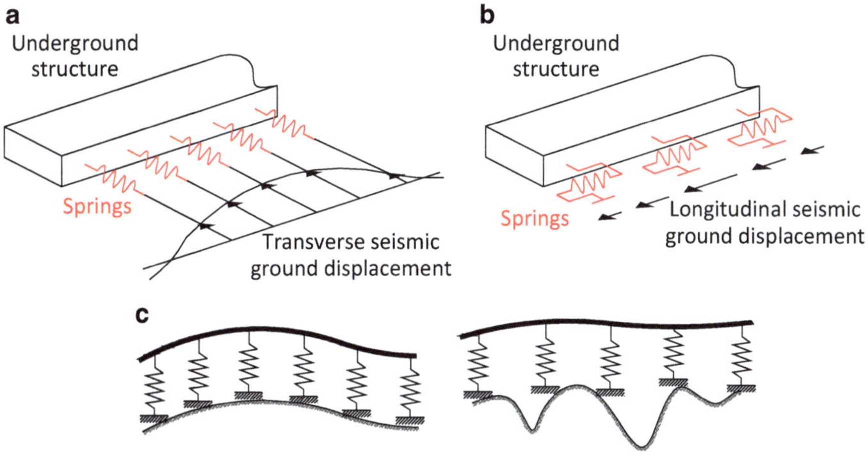

Fig. 11.29 Simplified equivalent static analysis (Modified after ISO 23469 2005), (**a**) transversal analysis, (**b**) longitudinal analysis, (**c**) discretization of soil-springs

Simplified Equivalent Static and Dynamic Analysis

The underground structure is again modeled as a beam on elastic foundation, i.e. springs (ISO 23469 2005). The analysis is performed using an appropriate finite element or finite difference code. The seismic load is applied on the springs as equivalent static ground deformation that can account of the spatial variation (Fig. 11.29). A significant factor is the distance between the springs, which is depending on the frequency range of interest (1–15 Hz).

Similar to the simplified equivalent static analysis, is the dynamic analysis. The only difference is that instead of applying statically the maximum seismic ground deformations, a dynamic time history analysis is performed, where the model is subjected to displacement time histories that account of the ground motion spatial variation. An example is presented by Anastasopoulos et al. (2007).

When Winkler-type models are used, the estimation of the impedance functions (springs and dashpots) to model the soil compliance is of prior importance. Unfortunately, as for the transversal direction, no plausible solutions exist in the literature for longitudinal analysis of underground structures, expect some general ideas inspired from the elastic impedance functions of surface foundations (i.e. Clough and Penzien 1993; AFPS/AFTES 2001; Vrettos 2005). As an example, AFTES/AFPS (2001) propose to use springs for the analysis equal to the soil stiffness. In that respect, the impedance functions might substantially vary, affecting seriously the computation of the internal loads and the design of the structure.

Full Numerical 3D Dynamic Analysis

Full dynamic time-history analysis, using 3D finite element, finite difference or lumped mass models (ISO 23469 2005), is by far the most adequate method to design large and long underground structures. With this method, both the transversal and the longitudinal directions of the structure can be modeled and analyzed simultaneously, taking into account the complicated soil deposit geometry, the non-linear behavior of the soil and the structure, using appropriate constitutive relationships, and the behavior of the soil-structure interface. While the input motion definition remains a decisive and constantly debated parameter, the main shortcoming of the method is the high computational cost, which makes its use difficult for parametric analysis, usually needed during the design phases of underground structures. To this end, full dynamic 3D analysis is practically used in cases of structures of significant importance (i.e. nuclear station underground ducts). Moreover, several issues, such as the appropriate simulation of the soil-tunnel interface or of the incoherence and spatial variation of the seismic input motion are still open.

11.5.3.2 Seismic Design of Joints for Tunnels

Long underground structures, often, have abrupt changes in structural stiffness (connection to ventilation buildings and stations, junctions etc.), while they are crossing different ground conditions. Important stress concentrations might happen in these locations; to reduce them flexible joints are commonly used. For long underground structures, and in particular for immersed tunnels, the proper design of the joints is of paramount importance (Hashash et al. 2001; FWHA 2009). The differential movements and rotation can be estimated using analytical solutions or numerical methods. As an example, we illustrate herein, the evaluation of the joints (Gina gaskets) deformations of the immersed tunnel in Thessaloniki. We applied two different methods: an analytical closed form solution (St. John and Zahrah 1987 and Power et al. 1996) and a simplified dynamic analysis method. Assuming that the tunnel segments are rigid with respect to the joints, the joint deformation can be obtained as:

$$\Delta_u = \varepsilon_{axial} \times l_e \tag{11.14}$$

where l_e is the segment length (equal to 153 m in this case) and ε_{axial} is the axial deformation estimated according to Power et al. (1996). The computed axial deformations (in compression) are functions of the apparent velocity and incidence angle (Fig. 11.30). It is observed that these two parameters are affecting significantly the joints deformations. It must be also noticed that this method does not separate the deformations to tension or compression deformations.

The dynamic analysis was performed with the FE code ADINA (ADINA 2005), using an equivalent elastic beam on dynamic Winkler foundation (simplified dynamic analysis). The surrounding soil was modeled with constant springs calculated according to Mylonakis et al. (2006), taking into consideration the level of the ground strains developed at the middle depth of the tunnel (i.e. equivalent linear approach) (Tsinidis and Chalatis 2008). The model is schematically depicted in

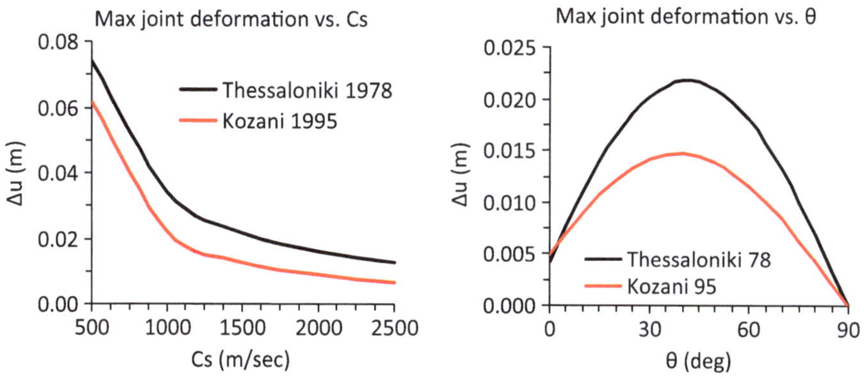

Fig. 11.30 Variation of joints axial deformations (**a**) with the apparent velocity C_s and (**b**) with the angle of incidence θ

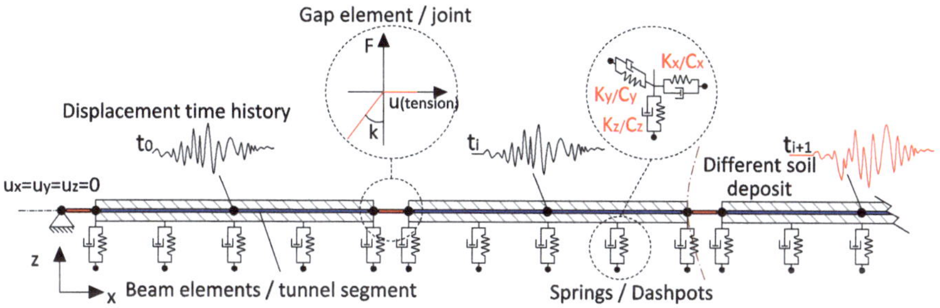

Fig. 11.31 Schematic representation of the numerical model for the longitudinal analysis of the immersed tunnel

Fig. 11.31. The non-linear joints (Gina-gaskets) were modeled with special gap elements, functioning only in compression (compression springs). An initial joint stiffness was calculated, assuming that the joint is deformed (pre-compressed) by approximately 10 cm, due to the hydrostatic pressure at the tunnel's depth. The input motion was displacement time-histories travelling with pre-determined time lag along the tunnel axis, taking into consideration the spatial variability of the soil layers. Two models for the joint behavior were adopted, a linear one (cases A and B) and a bi-linear one (cases C and D), while two assumptions for the soil conditions along the tunnel were made, a uniform one (cases A and C) and a diversified in axis (cases B, D).

Some indicative results are given in Fig. 11.32 and in Table 11.1. Model D gives the worst results; however the initial joint compression of 10 cm is not exceeded. In any case, the utilization of prestressed tendons at the joints is needed to reduce any extensional deformation of the joints.

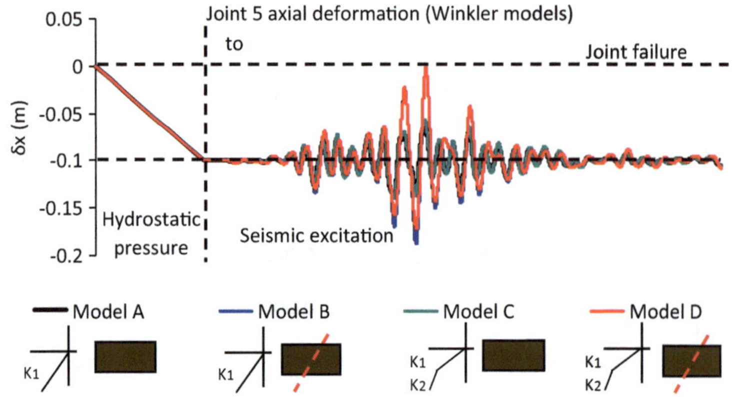

Fig. 11.32 Typical joint axial deformation time-histories for the examined numerical models

Table 11.1 Maximum joint axial deformations for each model (Joints 1 and 9 at exits, Joint 5 at the middle)

	max Δx (m)		
Numerical model	Joint 1	Joint 5	Joint 9
A	0.070	0.044	0.060
B	0.070	0.099	0.088
C	0.070	0.043	0.059
D	0.070	0.099	0.078
Analytical solution ($\varphi = 0°$, $C_s = 1{,}000$ m/s)	0.0095		

11.6 Seismic Design Against Ground Failure

Underground structures are quite vulnerable to ground failures associated with large permanent deformations, caused by ground liquefaction, slope instability and fault movements.

11.6.1 Liquefaction

Liquefaction can produce important longitudinal or transversal deformations on underground structures, due to lateral spreading and settlements of the liquefiable soils. These deformations can act as differential quasi-static deformations on the structure. Liquefaction can also cause uplift phenomena to an immersed structure and buoyancy loads that should not be disregarded (Travasarou and Chacko 2008; Kutter 2008; Travasarou 2010). The main mechanisms of uplift of light immersed

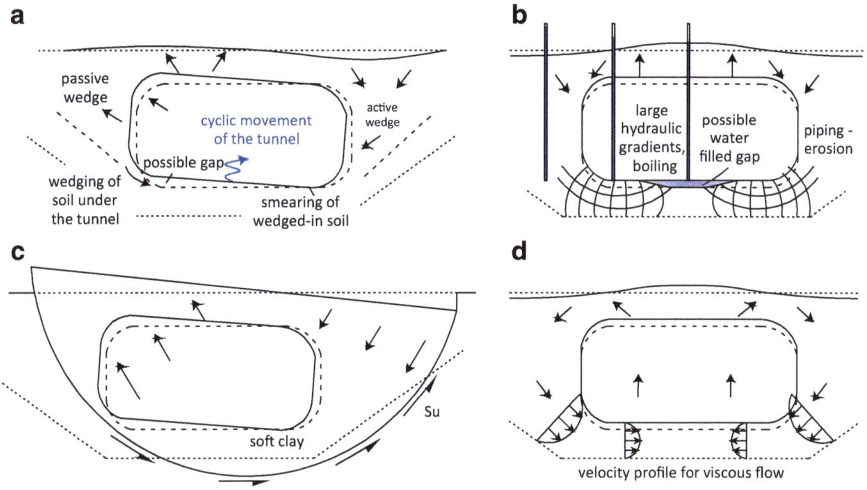

Fig. 11.33 Mechanisms of uplift of an immersed structure during liquefaction; (**a**) the ratcheting mechanism, (**b**) the pore water migration mechanism, (**c**) the bottom heave mechanism and (**d**) the viscous flow mechanism (after Chou et al. 2010)

tunnels in liquefiable soils, as described by Chou et al. (2010) and schematically presented in Fig. 11.33, can be summarized as follows:

- Ratcheting mechanism (Fig. 11.33a): During the shaking the tunnel moves back and forth relative to the backfill. When the tunnel moves to the one side, an active wedge of soil is been developed on the other side, funneling sand toward the invert slab of the tunnel. When the motion reserve, some of this sand wedges under the tunnel and it is "ratcheted" up with each cycle of motion. If there is large lateral displacement between the tunnel and the soil beneath it, this soil that is trapped under the tunnel, may be spread and smoothed along the tunnel's base.
- Pore water migration mechanism (Fig. 11.33b): During the shaking excess pore pressure is builded up. The excess pore pressure beside the tunnel and under the tunnel, are differing, as the overburden pressure differs between these two positions (greater pore pressure beside the tunnel). This difference is causing flow of the pore water under the tunnel and finally uplift of the tunnel.
- Bottom heave mechanism (Fig. 11.33c): This mechanism is the same as in the case of a braced excavation or a slope instability problem. The shearing of the soft ground under the liquefiable soil may case more upward movement of the tunnel.
- Viscous flow mechanism (Fig. 11.33d): After liquefaction, the tunnel may undergo further uplift due to the viscous flow of the liquefied soil. If viscous flow is significant, parabolic velocity profiles in the soil and vertical movement of the tunnel (uplift movement) are expected after the shaking end.

In order to avoid the liquefaction phenomena, several mitigation measures can be taken. Ground improvement, increase of the dead load of the structure and/or drainage or soil replacement, are among the most popular countermeasures to reduce liquefaction risk (Power et al. 1998).

11.6.2 Fault Crossing – Slope Failure

Tunnels and long underground tunnels are very vulnerable to fault displacements and generalized large-scale massive slope failures. The main problem is the accurate identification of the fault zone and the angle of thrust. With the exception of crossing some well identified and described fault zones, in most cases the exact location and angle of the fault trace are practically unknown. Existing methods for buried pipelines might be used, for the preliminary design in case of relatively flexible structures (Newmark and Hall 1975; Kennedy et al. 1977; ASCE 1984). A detailed numerical approach is certainly the best solution (ASCE 1984; Anastasopoulos et al. 2008; Anastasopoulos and Gazetas 2010). The fault displacements can be estimated, using empirical relationships that correlate the displacement with the earthquake magnitude and the fault type (i.e. Wells and Coppersmith 1994) and applied statically, as an imposed dislocation of the basement of the soil-structure model. It is obvious that the design depends on the magnitude of the expected displacement and the width of zone over which the displacement is distributed (Power et al. 1998). If the displacements are concentrated in a narrow zone, usually a local enlarge of the structure section is proposed across and beyond the fault zone, while, if the displacements are small and distributed over a relatively wide zone, it is possible to design the structure to accommodate the permanent ground displacements, providing sufficient ductility (i.e. ductile joints etc.) In case of slope instabilities, special mitigation measures to stabilize the ground should be used.

11.7 Complex "City Effects" on Underground Structures

In densely populated urban areas, tunnels and other underground structures are often passing beneath high-rise buildings or they are located close to them. The existence of these structures may create complex interaction effects with the underground structures usually referred as "city effects". These effects can affect the seismic wave propagation field, altering the seismic input motion with respect to the free field case (i.e. Semblat et al. 2008; Ghergu and Ionescu 2009). In this sense, they may modify considerably the seismic response of the underground structure, while at the same time the existence of the embedded structure, close to the surface and buildings foundations may alter the response of the buildings themselfs. A first attempt to identify and understand these effects is shortly presented here (Tzarmados 2011). More specifically, the interaction of aboveground structures with a shallow rectangular tunnel is discussed, through few representative results illustrated in Fig. 11.35. Full dynamic analyses was performed with ABAQUS (Fig. 11.34) assuming elastic behavior for both the soil and the structures. Perfect bonding between soil and the structures was assumed, while the aboveground structures were modeled in a simplified way, as equivalent SDOF systems. Three different soil types were studied, namely soil type B, C and D according to EC8 (CEN 2004a). The lining thickness was properly selected in order to model both flexible and rigid underground structures with respect to the surrounding soil.

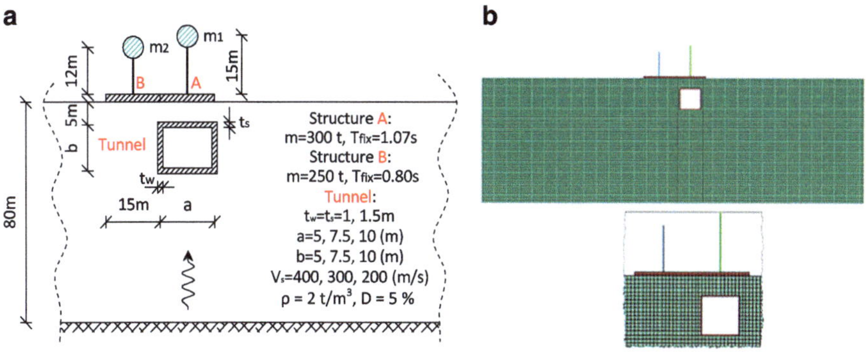

Fig. 11.34 Interaction between above ground structures and shallow tunnel: (**a**) Case studies, (**b**) Finite Element model in ABAQUS – Two aboveground structures case

The analyses performed using two input motions, a simplified Ricker wavelet and the Takatori record from the 1995 Great Hansin earthquake in Kobe, Japan.

The main conclusions of this study can be summarized as follows:

- The existence of the underground structure, with dimensions much smaller compared to the seismic wave lengths, has a limited effect on the aboveground buildings.
- On the contrary, the existence of aboveground structures affect the underground structure response, increasing, in the majority of the cases, the seismic loading expressed in terms of racking deflections.
- The presence of above ground structures has a strong effect on the flexibility to racking ratio (Fig. 11.35), as a consequence of the complex wave propagation effects, i.e. reflections of the waves between underground structure and above-ground structures. In particular, in case of a rigid tunnel ($F < 1.0$), the seismic waves are trapped between the roof slab of the tunnel and the foundation of the building resulting in an increase of the racking distortion of the tunnel. This phenomenon is less important in case of flexible tunnels.

The soil nonlinearity, expected during strong earthquake shaking, and the introduction of interfaces to better model the contact of the lining with the soil, might affect these first conclusions.

11.8 Seismic Retrofit of Damaged Tunnels

Several retrofit schemes for tunnels are proposed in the literature (i.e. Power et al. 1998; Hashash et al. 2001). For circular tunnels, the most common damage typologies are: (i) the lose of contact with the surrounding soil and (ii) the reduction of the lining strength. For the first problem, contact strengthening using grounding techniques may be used, whereas for the latter, strengthening of the lining, using

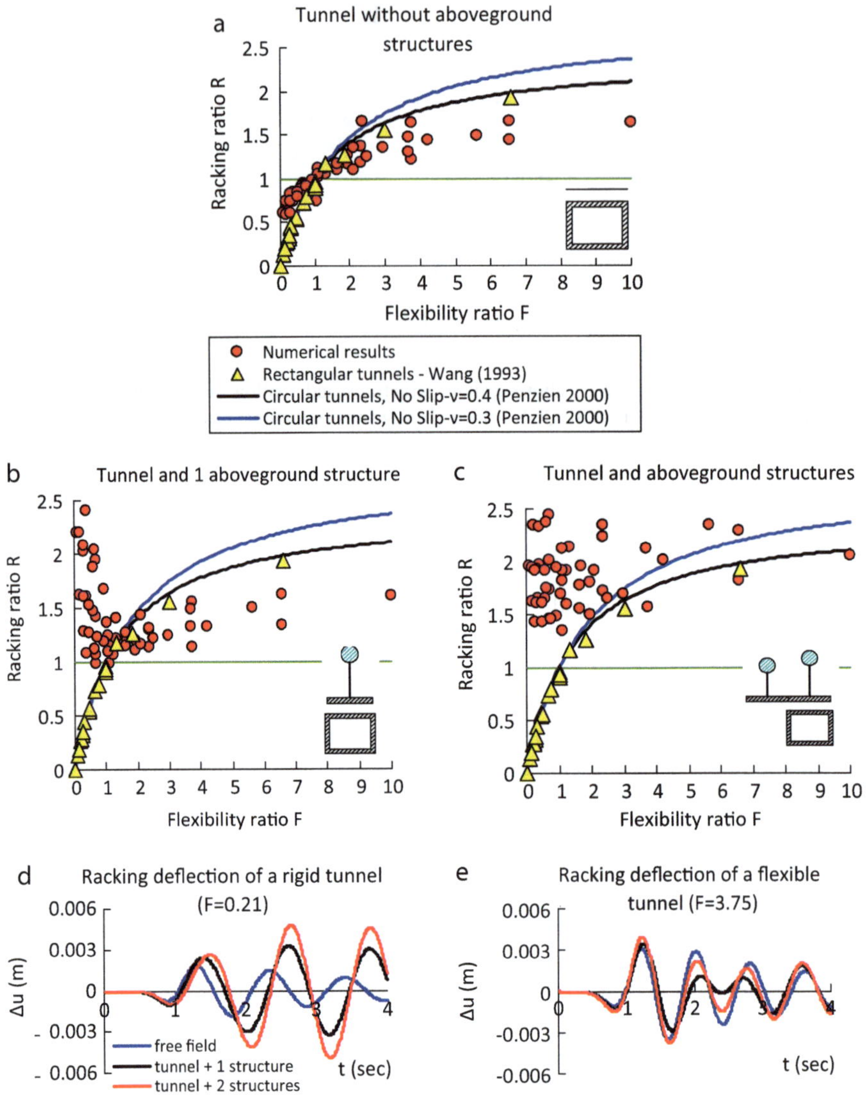

Fig. 11.35 Representative results for the tunnel-soil-aboveground structures interaction effects – response of the tunnel. (**a**) F-R relationship in case of absence of the aboveground structures, (**b**) F-R relationship in case of one building, (**c**) F-R relationship for two buildings, (**d**) effect of the existence of buildings on the racking deflection of a rigid tunnel, (**e**) effect of the existence of buildings on the racking deflection of a flexible tunnel

reinforced concrete or steel members is commonly proposed. It is noted that increasing the structural stiffness is not always the best solution, as stress concentrations due to the increased stiffness, may occur. On the contrary, in these cases, a design scheme that provides sufficient ductility may be the solution (Hashash et al. 2001). Similar measures can be taken for cut and cover structures.

11.9 Importance of Seismic Design Compared to the Static

11.9.1 Static Analysis of Underground Structures-Tunnels

Usually, the static analysis of underground structures in the transversal or the longitudinal direction, is using linear frame-type elements for the structure and constant linear springs to simulate the soil support. More sophisticated 2D or 3D FE models may be also used, to account of the heterogeneous stratigraphy and the soil non-linear behavior. Generally, the static loads are the following (Fig. 11.36):

- Dead loads of the structure (g1 + g2).
- Live loads of the structure (q).
- Hydrostatic pressures and uplift force at the structure (E1).
- Geostatic pressures at the structure (E2).

The uplift force is of utmost importance for the static analysis of underground structures. The combination that maximizes the structural response at each section must be used for the structure detailing.

11.9.2 Comparison Between Seismic and Static Internal Forces

Seismic loads and the associated internal forces in large underground structures are in most cases lower than the static loads. A metro station in Thessaloniki ("Venizelos station") is used again, as an example, to discuss the importance of the static loads and in particular the role of the uplift force to the final design. The station is constructed in a location with very high water table, which produces significant uplift forces. In order to reduce the calculated high stresses at the inverted slab of the station, caused by uplift forces, four different possible solutions have been examined (Matsoukas and Fletzouris 2007), as described in Fig. 11.37a.

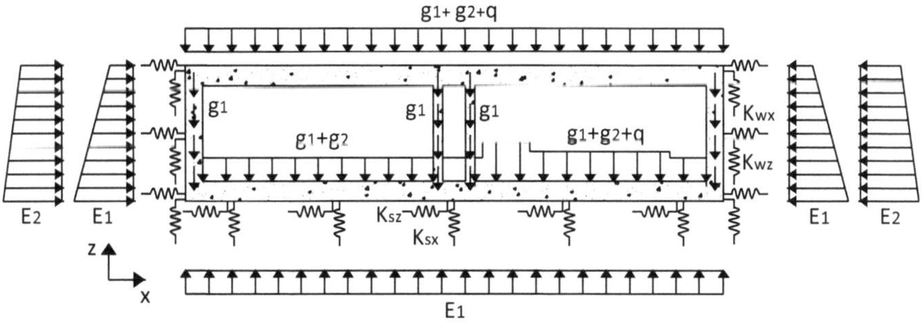

Fig. 11.36 Static loads on an underground structure (g_1, g_2: dead loads, q; live loads, E_1: hydrostatic loads – uplift load, E_2: geostatic pressures)

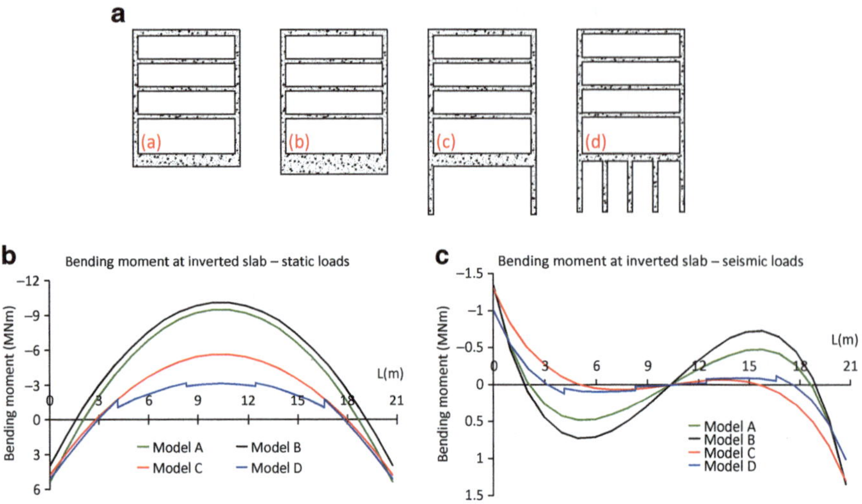

Fig. 11.37 (**a**) Different models for the station foundation, (**b**) bending moment at the inverted slab of the station for static loads and different models for the foundation, (**c**) bending moment at the inverted slab of the station for seismic loads and different models for the foundation

The long side diaphragm walls play an important role for the whole stability of the station and the control of underwater flow. The friction piles beneath the foundation were proposed to reduce the uplift forces mobilizing extra friction. Figure 11.37b highlight the favorable effect of the friction piles solution to the bending moments of the inverted slab, for both static and seismic conditions.

In Fig. 11.38 we estimate the participation of the static, uplift and seismic loads to the internal forces in some critical locations of the structure. The static loads are actually dominating the design values of the station (Fig. 11.38). The uplift force has the biggest contribution to the final design forces of the station's foundation slab, whereas the other static loads seem to dominate the station's upper levels response. The use of friction piles has a very beneficial effect on the foundation slab internal forces, reducing considerably (almost 3 times) the bending moments. In general, the contribution of the seismic forces is about 20 % of the total forces. This is due to the very important contribution of the buoyancy (uplift) forces reaching almost the 100 % of the total loads in some locations. In other cases the seismic forces are of the order of 30 % of the total forces.

11.10 Vulnerability Assessment of Underground Structures

So far, the vulnerability assessment of tunnels has been mainly based on expert judgment (ATC-13 1985; NIBS 2004) or empirical fragility curves (ALA 2001), derived from the statistical analysis of observed damages in past earthquakes

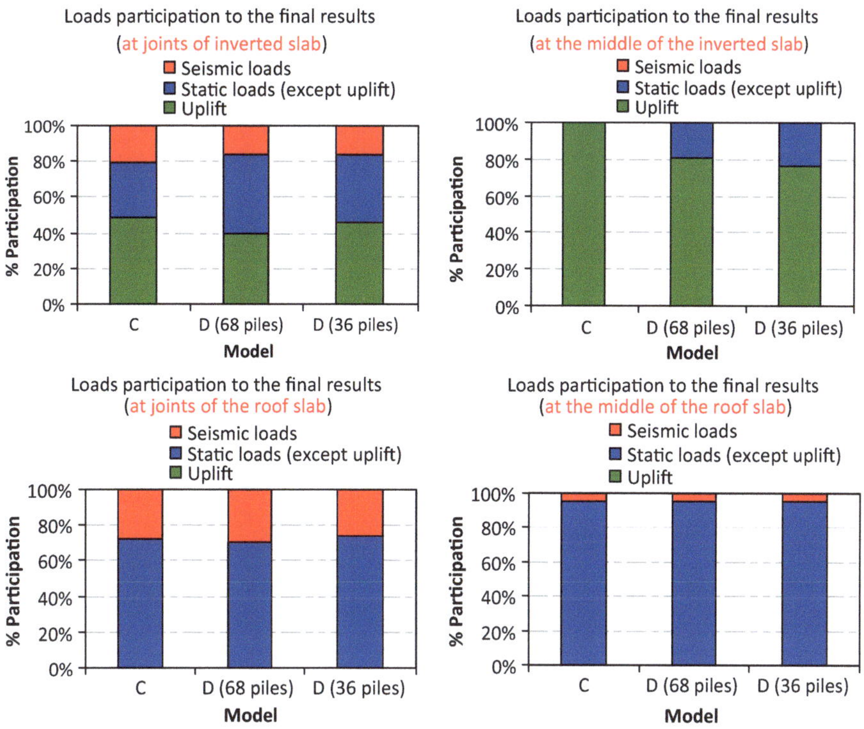

Fig. 11.38 Loads participation to the final design loads in the middle and the corners of the roof and the inverted slab of the "Venizelos" Metro station

(Dowding and Rozen 1978; Owen and Scholl 1981; Wang 1985; Sharma and Judd 1991). American Lifelines Alliance (2001) produced empirical fragility curves for peak ground acceleration, for bored and cut and cover tunnels with poor-to-average and good construction assumptions, based on regression analysis of a worldwide damage database. Corigliano (2007) proposed empirical fragility curves for peak ground velocity, for deep tunnels using similar empirical data, without considering the type of support or the stiffness and strength of the soil.

Recently, Argyroudis and Pitilakis (2012) proposed fragility curves for shallow circular and cut and cover tunnels, considering different tunnel geometries and strength and different soil properties. The transversal seismic response of the tunnel, for upward travelling SH or SV waves, was evaluated under quasi-static conditions, applying the free field ground deformations at a certain distance from the tunnel cross section, assuming that the surrounding soil is deformed uniformly in shear and the embedded tunnels are forced to follow this movement (detailed equivalent static analysis). In that respect, soil-structure interaction effects were considered explicitly. The free field seismic ground deformations were calculated through 1D equivalent linear analysis (EQL). The analysis was contacted with the finite element code PLAXIS 2D (Plaxis 2002).

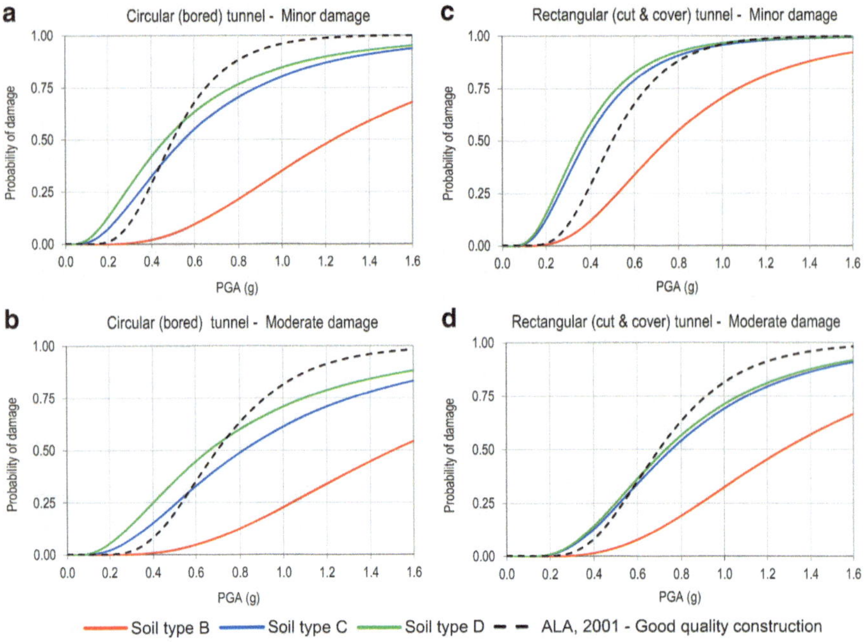

Fig. 11.39 Fragility curves for circular (bored) (**a**, **b**) and rectangular cut and cover (**c**, **d**) tunnels embedded in soil type B, C or D according to Eurocode 8, compared with empirical fragility curves for minor and moderate damage states, after ALA 2001 (Argyroudis and Pitilakis 2012)

Three damage levels were defined (minor, moderate, extensive) according to the exceedance of strength capacity of the most critical sections of the tunnel. The analyses were carried out for different soil profiles corresponding to soil types B, C and D of Eurocode 8 (CEN 2004a). Two typical shallow tunnel sections were considered, a circular (bored) tunnel with a 10 m diameter and a rectangular (cut and cover) with dimensions 16 × 10 m. The corresponding flexibility ratios varied from 0.2 to 55 covering a wide range of soil and tunnel stiffness.

The proposed sets of fragility curves, for soil type B, C and D according to EC8 (CEN 2004a) are presented in Fig. 11.39 for the circular and the rectangular tunnel cross-sections. As it should be expected, soil conditions modify drastically the vulnerability of the tunnels, which was disregarded so far (see curves proposed by ALA 2001). The most important difference is observed in case of stiff soils (class B) where ALA (2001) suggests much higher probability of damages.

11.11 Conclusions

Considering the importance of underground structures and tunnels, their seismic design is of paramount significance. The chapter provides a short review of the available seismic analysis and design methods, while it presents a comprehensive discussion of several issues that are still open and can significantly affect the seismic performance and design of underground structures. The main conclusions are summarized below:

- Underground structures are in general less vulnerable than aboveground structures for the same intensity level, at least for the case of ground shaking. However, there are several well-reported cases, from recent strong earthquakes, of severe damages and even collapses (i.e. Daikai station collapse).
- The seismic behavior of tunnels and underground structures is different from that of the above ground structures, as the kinematic loading is prevailing. The inertial effects are of secondary importance. Therefore, underground structures should be designed for imposed seismic ground deformations rather than inertial forces, as in the case of above ground structures.
- Two parameters are highly affecting the seismic behavior of underground structures, namely the surrounding soil to structure relative flexibility, usually expressed in terms of flexibility ratio and the characteristics of the soil-structure interface. These parameters might be competitive to each other.
- The proper and rational estimation of the soil to structure relative flexibility for rectangular structures is still an open issue. Theoretical and experimental studies prove that the structure exhibits more complex deformation modes during dynamic loading, compare to the simple horizontal shear deformation mode, commonly assumed so far. The side-walls and the slabs of flexible structures may undergo inward deformations, whereas for rigid structures, a rocking mode of vibration may be observed. Vertical shear may be also observed. The actual shear stresses, developed during the shaking, around the perimeter of the structure, are affecting the deformation modes of the structure. Unfortunately, these phenomena are not accounted by existing methodologies for the evaluation of the soil-structure relative flexibility and further study deemed to be necessary.
- The input motion and the seismic design loads must be estimated through a detailed seismic hazard analysis, considering seriously the specific site effects due to the local soil conditions.
- Transversal seismic analysis: There are several methods in the literature classified as force-based, displacement-based and numerical methods. In the later case, the structure and the soil are analyzed as a coupled system, using different numerical techniques (finite, element, finite difference, etc.). The displacement-based methods are closer to the physics of the problem and present several advantages including the proper evaluation of the inelastic response of the structural components. However, forced-based methods are still used in engineering practice, mainly because they can be easier implemented and engineers are still more familiar with this approach. The numerical methods and in particular the full dynamic analysis of the coupled soil-structure system are certainly more appropriate for the seismic analysis, as the complex soil and structure geometries and the non-linear behavior of both the soil and the structure can be accurately described. The main shortcoming of the method is the high computational cost.
- We discussed in detail several important issues for the transversal design namely the seismic earth pressures on the side-walls of the structures, the seismic shear stresses at the perimeter of the structures and the internal forces on the lining. It is observed that different approaches lead to significantly different results. The proper

estimation of impedance functions for underground structures and tunnels, the rational estimation of the seismic earth pressures on the side-walls and the estimation of the seismic shear stresses around the perimeter of the structure are still important open issues that can affect considerably the seismic design.

- Longitudinal seismic analysis of long segmented structures is equally important to the transversal analysis. As for the transversal analysis, several methods are available, classified as displacement-based and numerical methods using full 3D FE or FD models. In the first case, usually a Winkler type model is assumed to model the soil compliance and the soil-structure interaction effects, whereas for the structure beam elements are used. Poor knowledge on the evaluation of soil impedances is the major weakness of these methods. The full dynamic analysis using 3D numerical models is usually used in cases of structures of high importance (i.e. nuclear power plant ducts), as it is the most accurate but at the same time the most time-consuming method. Asynchronous motion must incorporate in the analysis procedure. Simple phase difference introducing a simple time lag may not be accurate enough. Special seismic design provisions must be taken in case of joints between segments, as these joints can be the most vulnerable parts of an underground structure.
- Ground failures due to liquefaction, slope instability or fault movements affect seriously the overall design of the structure. A static approach is usually used.
- The complex structural environment of urban areas may cause complex interaction phenomena, altering the input motion of an underground structure with respect to the free field case. Preliminary results, from elastic dynamic analyses, show that the existence of aboveground structures may affect the seismic response of an underground structure. On the contrary, the existence of underground structures, with small dimensions compared to the predominant wave lengths, seems not to affect the response characteristics of aboveground structures.
- Generally, static gravity loads and especially uplift are dominating the response of an underground structure. Seismic loads are more important when uplift loads are minimized.
- Improved fragility curves, for circular bored tunnels and rectangular cut and cover tunnels, are finally proposed revealing the importance of the soil conditions.

Acknowledgements The research leading to the presented experimental results has received funding from the European Community's Seventh Framework Programme [FP7/2007 – 2013] for access to the Schofield Centre of the University of Cambridge under grant agreement n° 227887 (SERIES project). The authors would like to acknowledge Prof. Gopal Madabhushi for the fruitful discussions they had with during the study. The contribution of Dr. Sotiris Argyroudis to the vulnerability assessment section of the chapter is also kindly acknowledged.

Appendix A

Closed form solutions presented in the literature for both the transversal and the longitudinal analysis are summarized in the following Tables (Table 11.A5).

Table 11.A1 Closed form solutions for internal forces of circular tunnels (After Wang 1993 and Penzien 2000)

Wang solutions (1993)	Penzien solutions (2000)
Full slip:	
$C = \frac{E_s(1-\nu_l^2)R}{E_l t(1+\nu_s)(1-2\nu_s)}, F = \frac{E_s(1-\nu_l^2)R^3}{6E_l I_l(1+\nu_s)}$	$\pm\Delta d^n_{stru} = \pm R^n \Delta d_{ff}, \quad R^n = \frac{\pm\Delta d^n_{stru}}{\Delta d_{ff}} = \pm\frac{4(1-v_s)}{(1+a^n_s)}$
$\frac{\Delta d}{d} = \pm\frac{1}{3}K_1 F\gamma_{max}, K_1 = \frac{12(1-\nu_s)}{2F+5-6\nu_s}$	$a^n_s = \frac{12(5-6v_s)E_l I_l}{D^3 G_s(1-v_l^2)}$
$T_{max} = \pm\frac{1}{6}K_1\frac{E_s}{(1+\nu_s)}R\gamma_{max}, M_{max} = \pm\frac{1}{6}K_1\frac{E_s}{(1+\nu_s)}R^2\gamma_{max}$	$T(\theta) = -\frac{12E_l I_l \Delta d^n_{stru}}{D^3(1-v_l^2)}\cos 2\left(\theta+\frac{\pi}{4}\right)$
	$M(\theta) = -\frac{6E_l I_l \Delta d^n_{stru}}{D^2(1-v_l^2)}\cos 2\left(\theta+\frac{\pi}{4}\right)$
	$V(\theta) = -\frac{24E_l I_l \Delta d^n_{stru}}{D^3(1-v_l^2)}\sin 2\left(\theta+\frac{\pi}{4}\right)$
No slip:	
$T_{max} = \pm K_2\tau_{max} r = \pm K_2\frac{E_s}{2(1+\nu_s)}R\gamma_{max}$	$\pm\Delta d_{stru} = \pm R\Delta d_{ff}, \quad R = \frac{\Delta d_{stru}}{\Delta d_{ff}} = \pm\frac{4(1-v_s)}{(1+a_s)}$
$K_2 = 1 + \frac{F[(1-2\nu_s)-(1-2\nu_s)C]-\frac{1}{2}(1-2\nu_s)^2+2}{F[(3-2\nu_s)+(1-2\nu_s)C]+C\left[\frac{5}{2}-8\nu_s+6\nu_s^2\right]+6-8\nu_s}$	$a_s = \frac{24(3-4v_s)E_l I_l}{D^3 G_s(1-v_l^2)}$
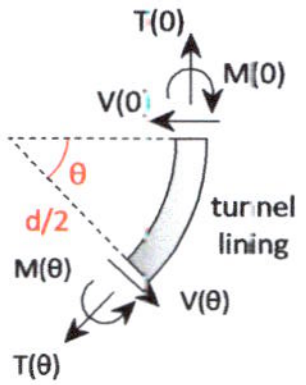	$T(\theta) = -\frac{24E_l I_l \Delta d_{stru}}{D^3(1-v_l^2)}\cos 2\left(\theta+\frac{\pi}{4}\right)$
	$M(\theta) = -\frac{6E_l I_l \Delta d_{stru}}{D^2(1-v_l^2)}\cos 2\left(\theta+\frac{\pi}{4}\right)$
	$V(\theta) = -\frac{24E_l I_l \Delta d_{stru}}{D^3(1-v_l^2)}\sin 2\left(\theta+\frac{\pi}{4}\right)$

Table 11.A2 Strain and curvature due to body and surface waves (After St. John and Zahrah 1987)

Wave type		Longitudinal strain	Normal strain	Shear strain	Curvature
P waves		$\varepsilon_l = \frac{V_P}{c_P} \cos^2\varphi$	$\varepsilon_n = \frac{V_P}{c_P} \sin^2\varphi$	$\gamma = \frac{V_P}{c_P} \sin\varphi \cos\varphi$	$\frac{1}{\rho} = \frac{\alpha_P}{c_P^2} \sin\varphi \cos^2\varphi$
		$\varepsilon_{lm} = \frac{V_P}{c_P}$ for $\varphi = 0^o$	$\varepsilon_{nm} = \frac{V_P}{c_P}$ for $\varphi = 90^o$	$\gamma_m = \frac{V_P}{2c_P}$ for $\varphi = 45^o$	$\frac{1}{\rho_m} = 0.385 \frac{\alpha_P}{c_P^2}$ for $\varphi = 35^o16'$
S waves		$\varepsilon_l = \frac{V_s}{c_s} \sin\varphi \cos\varphi$	$\varepsilon_n = \frac{V_s}{c_s} \sin\varphi \cos\varphi$	$\gamma = \frac{V_s}{c_s} \cos^2\varphi$	$K = \frac{\alpha_s}{c_s^2} \cos^3\varphi$
		$\varepsilon_{lm} = \frac{V_s}{2c_s}$ for $\varphi = 45^o$	$\varepsilon_{nm} = \frac{V_s}{2c_s}$ for $\varphi = 45^o$	$\gamma_m = \frac{V_s}{c_s}$ for $\varphi = 0^o$	$K_m = \frac{\alpha_s}{c_s^2}$ for $\varphi = 0^o$
Rayleigh waves	Compressional component	$\varepsilon_l = \frac{V_{RP}}{c_R} \cos^2\varphi$	$\varepsilon_n = \frac{V_{RP}}{c_R} \sin^2\varphi$	$\gamma = \frac{V_{RP}}{c_R} \sin\varphi \cos\varphi$	$K = \frac{\alpha_{RP}}{c_R^2} \sin\varphi \cos^2\varphi$
		$\varepsilon_{lm} = \frac{V_{RP}}{c_R}$ for $\varphi = 0^o$	$\varepsilon_{nm} = \frac{V_{RP}}{c_R}$ for $\varphi = 90^o$	$\gamma_m = \frac{V_{RP}}{2c_R}$ for $\varphi = 45^o$	$K_m = 0.385 \frac{\alpha_{RP}}{c_R^2}$ for $\varphi = 35^o16'$
	Shear component	–	$\varepsilon_n = \frac{V_{RS}}{c_R} \sin\varphi$	$\gamma = \frac{V_{RS}}{c_R} \cos\varphi$	$K = \frac{\alpha_{RS}}{c_R^2} \cos^2\varphi$
			$\varepsilon_{nm} = \frac{V_{RS}}{c_R}$ for $\varphi = 90^o$	$\gamma_m = \frac{V_{RS}}{c_R}$ for $\varphi = 0^o$	$K_m = \frac{\alpha_{RS}}{c_R^2}$ for $\varphi = 0^o$

Table 11.A3 Closed form solutions for the determination of bending moment and trust of the lining under sinusoidal input motion in the longitudinal direction (After St. John and Zahrah 1987)

Free field approach	Soil-structure interaction approach
$M=\left(\frac{2\pi}{L}\right)^2\cos^3\varphi E_l I_l A\sin\left(\frac{2\pi x}{L/\cos\varphi}\right)$	$M=\dfrac{\left(\frac{2\pi}{L}\right)^2\cos^3\varphi}{1+\frac{E_l I_l}{K_h}\left(\frac{2\pi}{L}\right)^4\cos^4\varphi}E_l I_l A\sin\left(\frac{2\pi x}{L/\cos\varphi}\right)$
$T=\left(\frac{2\pi}{L}\right)\sin\varphi\cos\varphi E_l A_l A\cos\left(\frac{2\pi x}{L/\cos\varphi}\right)$	$T=\dfrac{\left(\frac{2\pi}{L}\right)\sin\varphi\cos\varphi}{1+\frac{E_l A_l}{K_a}\left(\frac{2\pi}{L}\right)^2\cos^2\varphi}E_l A_l A\cos\left(\frac{2\pi x}{L/\cos\varphi}\right)$
	where : $K_h=K_a=\frac{16\pi G_s(1-v_s)}{(3-4v_s)}\frac{d}{L}$

Table 11.A4 Closed form solutions for the determination of bending moment and trust of the lining (After Kawashima 2000)

Design seismic deformation	Internal forces on the lining
$u_g(x,z) = u_{ga}(x,z) = u_{gt}(x,z)$	$P_h = c_{ta} \times c_{ja} \times {}^{\pi EA}/_{L} \times \tilde{u_h}$
$u_g(x,z) = u_h \times \cos\left(\frac{\pi z}{2H}\right) \times \sin\left(\frac{2\pi x}{L'}\right)$	$P_v = c_{ta} \times c_{ja} \times {}^{\pi EA}/_{L} \times \frac{\tilde{u_h} + \tilde{u_v}}{2}$
$u_h = \frac{2}{\pi^2} S_V T_S$	$M_h = c_{tt} \times c_{jt} \times {}^{4\pi^2 EI_h}/_{L^2} \times \tilde{u_h}$
$T_S = 1.25 \times \sum_i \frac{4H_i}{V_{Si}}$	$M_v = c_{tv} \times c_{jv} \times {}^{4\pi^2 EI_v}/_{L^2} \times \tilde{u_v}$
$L' = \frac{2L_1L_2}{L_1 + L_2}$, $L_1 = V_S \times T_S$, $L_2 = V_{SB} \times T_S$	$c_{ta} = \frac{1}{1 + \left(\frac{2\pi}{\lambda_a L''}\right)^2}$, $c_{tt} = \frac{1}{1 + \left(\frac{2\pi}{\lambda_h L'}\right)^4}$, $c_{tv} = \frac{1}{1 + \left(\frac{2\pi}{\lambda_v L'}\right)^4}$
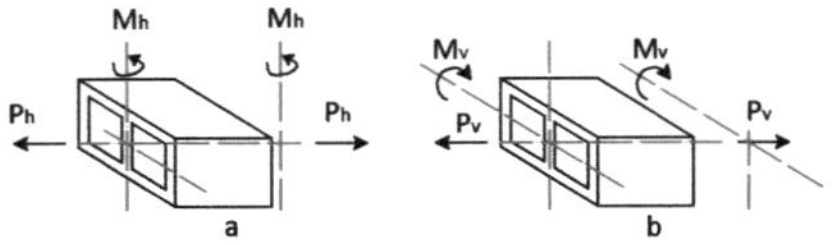	$\lambda_\alpha = \sqrt{\frac{k_a}{EA}}$, $\lambda_t = \sqrt[4]{\frac{k_t}{EI_h}}$, $\lambda_v = \sqrt[4]{\frac{k_v}{EI_v}}$ $L'' = L\sqrt{2}$

Table 11.A5 Nomenclature for the previous tables

Symbol	Definition
C	Compressibility ratio
F	Flexibility ratio
D	Tunnel diameter
d	Width of structure
γ_{max}	Maximum soil shear strain
v_s	Soil medium Poison ratio
E_s	Soil medium elastic modulus
G_s	Soil medium shear modulus
v_l	Structure lining Poison ratio
E_l	Structure lining elastic modulus
c	Wave velocity
V	Maximum particle velocity
a	Maximum particle acceleration
φ	Angle of incidence (with respect to the structure axis)
I_l	Moment of inertia of the structure (per meter)
A_l	Cross-sectional area of the structure lining
K_h	Foundation soil modulus for transverse-horizontal load
K_a	Foundation soil modulus for axial load
L	Design sinusoidal wavelength
A	Design sinusoidal amplitude
V_s	Averaged shear wave velocity of the subsurface ground
V_{SB}	Bedrock shear wave velocity
H_i	Thickness of the i-th layer
V_{si}	Shear wave velocity of the i-th layer
T_s	Natural period of the subsurface ground
S_V	Design velocity response spectrum at the bedrock
EI_h, EI_v	Flexural rigidities of the structure in the horizontal and in the vertical plane
$\tilde{u}_h$	Design horizontal ground displacement at structure's depth
$\tilde{u}_v$	Design vertical ground displacement at structure's depth
c_{ja}, c_{jt}, c_{jv}	Modification factors depending on the boundary conditions at the ends of the structure in the longitudinal, transversal and vertical direction
k_a, k_h, k_v	Stiffnesses of soil springs in the longitudinal, transversal and vertical direction

References

ABAQUS (2009) Analysis user's manual – vols I – IV – v6.9. Dassault Systèmes, SIMULIA Inc. {Computer program}

ADINA (2005) Automatic dynamic incremental nonlinear analysis. Theory and modelling guide. ADINA R&D, Inc., Watertown, USA {Computer program}

AFPS/AFTES (2001) Guidelines on earthquake design and protection of underground structures. Working group of the French association for seismic engineering (AFPS) and French tunneling association (AFTES) Version 1

American Lifelines Alliance (ALA) (2001) Seismic fragility formulations for water systems, part 1-guideline. ASCE-FEMA, p 104, California, USA

Amorosi A, Boldini D (2009) Numerical modelling of the transverse dynamic behaviour of circular tunnels in clayey soils. Soil Dyn Earthq Eng 59:1059–1072

An X, Shawky A, Maekawa K (1997) The collapse mechanism of a subway station during the great Hanshin earthquake. Cem Concr Compos 19:241–257

Anastasopoulos I, Gazetas G (2010) Analysis of cut-and-cover tunnels against large tectonic deformation. Bull Earthq Eng 8(2):283–307

Anastasopoulos I, Gerolymos N, Drosos V, Kourkoulis R, Georgarakos T, Gazetas G (2007) Nonlinear response of deep immersed tunnel to strong seismic shaking. J Geotech Geoenviron 133(9):1067–1090

Anastasopoulos I, Gerolymos N, Drosos V, Georgarakos T, Kourkoulis R, Gazetas G (2008) Behaviour of deep immersed tunnel under combined normal fault rupture deformation and subsequent seismic shaking. Bull Earthq Eng 6(2):213–239

Argyroudis S, Pitilakis K (2012) Seismic fragility curves of shallow tunnels in alluvial deposits. Soil Dyn Earthq Eng 35:1–12

ASCE Committee in Gas and Liquid Fuel Lifelines (1984) Guidelines for the seismic design of oil and gas pipeline systems. Technical Council on Lifeline Earthquake Engineering, ASCE, New York

ATC-13 (1985) Earthquake damage evaluation data for California. Applied Technology Council, Redwood City

Bardet JB, Ichii K, Lin CH (2000) EERA: a computer program for equivalent-linear earthquake site response analyses of layered soil deposits. University of Southern California, Department of Civil Engineering, Los Angeles, 40 p

Bobet A (2003) Effect of pore water pressure on tunnel support during static and seismic loading. Tunn Undergr Space Technol 18:377–393

Bilotta E, Lanzano G, Russo G, Silvestri F, Madabhushi SPG (2009) Seismic analyses of shallow tunnels by dynamic centrifuge tests and finite elements. In: Proceedings of the 17th international conference on soil mechanics and geotechnical engineering, Alexandria, Egypt

Building Seismic Safety Council (2003) FEMA 450-NEHRP recommended provisions for seismic regulations for new buildings and other structures. Federal Emergency Management Agency, Washington, DC

Calvi GM, Sullivan TJ (eds) (2009) A model code for the displacement-based seismic design of structures. IUSS Press, Pavia

CEN (2004a) Eurocode 8: design of structures for earthquake resistance – Part 1: general rules, seismic actions and rules for buildings, EN 1998–1:2004. Comité Européen de Normalisation, Brussels

CEN (2004b) Eurocode 8: design of structures for earthquake resistance – Part 5: Foundations, retaining structures and geotechnical aspects, EN 1998–5: 2004 Comitee Europeen de Normalisation, Brussels

Chen J, Shi X, Li J (2010) Shaking table test of utility tunnel under non-uniform earthquake wave excitation. Soil Dyn Earthq Eng 30:1400–1416

Chou JC, Kutter BL, Travasarou T, Chacko JM (2010) Centrifuge modeling of seismically induced uplift for the BART transbay tube. J Geotech Geoenviron 137(8):754–765

Cilingir U, Madabhushi SPG (2011a) A model study on the effects of input motion on the seismic behaviour of tunnels. Soil Dyn Earthq Eng 31:452–462

Cilingir U, Madabhushi SPG (2011b) Effect of depth on the seismic behaviour of circular tunnels. Can Geotech J 48:117–127

Cilingir U, Madabhushi SPG (2011c) Effect of depth on the seismic response of square tunnels. Soils Found 51(3):449–457

Clough RW, Penzien J (1993) Dynamics of structures, 2nd edn. McGraw-Hill, New York

Corigliano M (2007) Seismic response of deep tunnels in near-fault conditions. PhD dissertation. Politecnico di Torino, Italy

Dowding CH, Rozen A (1978) Damage to rock tunnels from earthquake shaking. J Geotech Eng Div 104:175–191

EQE International (1995) The January 17, 1995 Kobe earthquake: an EQE summary report. EQE International, Oakland, CA USA

FWHA (2009) Technical manual for design and construction of road tunnels—civil elements. U.S. Department of transportation. Federal Highway Administration. Publication No. FHWA-NHI-10-034, 702 p

Gazetas G (1991) Formulas and charts for impedances of surface and embedded foundations. J Geotech Eng ASCE 117(9):1363–1381

Gazetas G, Dobry R (1984) Horizontal response of piles in layered soils. J Geotech Eng ASCE 110 (1):20–40

Gazetas G, Gerolymos N, Anastasopoulos I (2005) Response of three Athens metro underground structures in the 1999 Parnitha earthquake. Soil Dyn Earthq Eng 25:617–633

Gerolymos N, Gazetas G (2006) Winkler model for lateral response of rigid caisson foundations in linear soil. Soil Dyn Earthq Eng 26(5):347–361

Ghergu M, Ionescu IR (2009) Structure–soil–structure coupling in seismic excitation and "city effect". Int J Eng Sci 47:342–354

Gingery J (2007) A simplified method for estimating shear strains for ovaling and racking analysis of tunnels. In: Proceedings of the 4th international conference on earthquake geotechnical engineering, Thessaloniki, Greece, 25–28 June 2007

Greek Seismic Code – EAK2000 (2003) Ministry of public works, Athens, Greece (in Greek)

Hashash YMA, Hook JJ, Schmidt B, Yao JI-C (2001) Seismic design and analysis of underground structures. Tunn Undergr Space Technol 16(2):247–293

Hashash YMA, Park D, Yao JIC (2005) Ovaling deformations of circular tunnels under seismic loading, an update on seismic design and analysis of underground structures. Tunn Undergr Space Technol 20:435–441

Hoeg K (1968) Stresses against underground structural cylinders. J Soil Mech Found Div 94:833–858

Huo H, Bodet A, Fernández G, Ramírez J (2005) Load transfer mechanisms between underground structure and surrounding ground: evaluation of the failure of the Daikai station. J Geotech Geoenviron 131(12):1522–1533

Huo H, Bodet A, Fernández G, Ramírez J (2006) Analytical solution for deep rectangular structures subjected to far-field shear stresses. Tunn Undergr Space Technol 21:613–625

Iida H, Hiroto T, Yoshida N, Iwafuji M (1996) Damage to Daikai subway station. Soils Found, Special Issue on Geotechnical Aspects of the January 17 1995, Hyogoken-Nambu Earthquake. Japanese Geotechnical Society, Japan, pp 283–300

ISO 23469 (2005) Bases for design of structures – seismic actions for designing geotechnical works. ISO International Standard. ISO TC 98/SC3/WG10

Japan Road Association (1992) Guide specifications of design and construction of underground parking lots. Japan Road Association, Tokyo

Kavvadas M, Gazetas G (1993) Kinematic seismic response and bending of free – head piles in layered soil. Geotechnique 43(2):207–222

Kawashima K (2000) Seismic design of underground structures in soft ground: a review. In: Kusakabe, Fujita, Miyazaki (eds) Geotechnical aspects of underground construction in soft ground. Balkema, Rotterdam. ISBN 90 5809 1 066

Kennedy RP, Chow AW, Williamson RA (1977) Fault movement effects on buried oil pipeline. J Transp Eng Div ASCE 103(TE5):617–633

Kiyomiya O (1995) Earthquake-resistant design features of immersed tunnels in Japan. Tunn Undergr Space Technol 10(4):463–475

Kontoe S, Zdravkovic L, Potts D, Mentiki C (2008) Case study on seismic tunnel response. Can Geotech J 45:1743–1764

Kramer SL (1996) Geotechnical earthquake engineering. Prentice Hall Inc., Upper Saddle River

Kuesel TR (1969) Earthquake design criteria for subways. J Struct Div ASCE, USA ST6: 1213–1231

Kutter BL, Chou JC, Travasarou T (2008) Centrifuge testing of the seismic performance of a submerged cut and cover tunnel in liquefiable soils. In: 4th geotechnical earthquake engineering and soils dynamics conference; Proceeding of 4th GEESDC, Sacramento, May 2008

Lanzano G, Bilotta E, Russo G, Silvestri F, Madabhushi SPG (2010) Dynamic centrifuge tests on shallow tunnel models in dry sand. In: Proceedings of the VII international conference on physical modelling in geotechnics (ICPMG 2010). Taylor & Francis, Zurich, pp 561–567
Lanzano G, Bilotta E, Russo G, Silvestri F, Madabhushi SPG (2012) Centrifuge modelling of seismic loading on tunnels in sand. Geotech Test J 35(6):854–869. doi:10.1520/GTJ104348
Matsoukas P, Fletzouris V (2007) Seismic design of metro stations – application at the Thessaloniki Venizelos metro station. MSc thesis, Civil Engineering Department, Aristotle University of Thessaloniki, Greece (in Greek with English abstract)
Matsuda T, Samata S, Iwatate T (1996) Seismic response analysis for a collapsed underground subway structure with intermediate columns. The 1995 Hyogoken-Nambu Earthquake: investigation into damage to civil engineering structures. Japan Society of Civil Engineers Committee of Earthquake Engineering, pp 277–285
Mavridis GA, Pitilakis KD (1996) Axial and transverse seismic analysis of buried pipelines. In: Proceedings of the 11th world conference on earthquake engineering, Acapulco, Mexico
Mononobe N, Matsuo H (1929) On the determination of earth pressure during earthquakes. In: Proceedings of the world engineering conference, vol 9, Tokyo, pp 177–185
Mylonakis G (1995) Contribution to static and seismic analysis of piles and pile – supported bridge piers. PhD thesis, State University of New York, Buffalo
Mylonakis G, Nikolaou S, Gazetas G (2006) Footings under seismic loading: analysis and design issues with emphasis on bridge foundations. Soil Dyn Earthq Eng 26(9):824–853
Newmark NM (1968) Problems in wave propagation in soil and rock. In: Proceedings of the international symposium on wave propagation and dynamic properties of earth materials, Albuquerque, August 1968
Newmark NM, Hall WJ (1975) Pipeline design to resist large fault displacement. In: Proceedings U.S. national conference on earthquake engineering, Ann Arbor, MI
NIBS (2004) HAZUS-MH: technical manuals. Federal Emergency Management Agency and National Institute of Building Science, Washington, DC
O'Rourke MJ, Dobry R (1978) Spring and dashpot coefficients for machine foundations on piles. American Concrete Institute, Publication SP – 10, Detroit, pp 177–198
Okabe S (1926) General theory of earth pressure. J Jpn Soc Civil Eng 12(1)
Owen GN, Scholl RE (1981) Earthquake engineering of large underground structures. Report no. FHWA/RD-80/195, Federal Highway Administration and National Science Foundation, p 279
Park KH, Tantayopin K, Tontavanich B, Owatsiriwong A (2009) Analytical solution for seismic-induced ovaling of circular tunnel lining under no-slip interface conditions: a revisit. Tunn Undergr Space Technol 24:231–235
Penzien J (2000) Seismically induced racking of tunnel linings. Earthq Eng Struct Dyn 29:683–691
Penzien J, Wu C (1998) Stresses in linings of bored tunnels. Earthq Eng Struct Dyn 27:283–300
Pitilakis K, Tsinidis G (2010) Seismic design of large, long underground structures: metro and parking stations, highway tunnels. In: Proceedings of international geotechnical conference "Geotechnical Challenges in Megacities" (GEOMOS2010), Moscow, Russia, 7–10 June 2010
Pitilakis K & Associates (2007a) Design principles of underground metro stations under earthquake loading according to the equivalent static force method. Technical report no. 2 (in Greek with English summary), Thessaloniki, Greece
Pitilakis K & Associates (2007b) Seismic analysis of the underground metro station "VENIZELOS" according to (a) Imposed seismic ground deformations method and (b) Full dynamic analysis. Technical report no. 3 (in Greek with English summary), Thessaloniki, Greece
Pitilakis K, Kirtas E, Riga E, Matsoukas P, Flentzouris V (2008) Seismic design of underground structures with large dimensions. In: Proceedings of 3rd HCEEES conference, Athens, Greece, 5–7 Nov 2008 (in Greek with English abstract)

Pitilakis K, Chalatis A, Tsinidis G, Kirtas E (2009) Numerical analysis and seismic design of shallow tunnels in soft alluvial deposits. In: Proceedings of computational methods in structural dynamics and earthquake engineering (COMPDYN 2009), Rhodes, Greece, 22–24 June 2009

Plaxis 2D (2002) Reference manual. Plaxis bv, Netherlands. {Computer program}

Power MS, Rosidi D, Kaneshiro J (1996) Vol. III Strawman: screening, evaluation, and retrofit design of tunnels. Report draft. National Center for Earthquake Engineering Research, Buffalo, NY

Power M, Rosidi D, Kaneshiro J, Gilstrap S, Chiou SJ (1998) Summary and evaluation of procedures for the seismic design of tunnels. Final report for task 112-d-5.3(c). National Center for Earthquake Engineering Research, Buffalo

Priestley MJN (1993) Myths and fallacies in earthquake engineering – conflicts between design and reality. Bull Natl Soc Earthq Eng NZSEE 26(3):329–341

Samata S, Ohuchi H, Matsuda T (1997) A study of the damage of subway structures during the 1995 Hanshin-Awaji earthquake. Cem Concr Compos 19:223–239

Schnabel PB, Lysmer J, Seed HB (1972) SHAKE: a computer program for earthquake response analysis of horizontally layered sites. Report no. UCB/EERC-72/12, Earthquake Engineering Research Center, University of California, Berkeley

Scott RF (1973) Earthquake induced pressures on retaining walls. In: Proceedings of the 5th world conference on earthquake engineering, Rome, vol 2, pp 1611–1620

Sedarat H, Kozak A, Hashash YMA, Shamsabadi A, Krimotat A (2009) Contact interface in seismic analysis of circular tunnels. Tunn Undergr Space Technol 24(4):482–490

Seed HB, Idriss IM (1971) Simplified procedure for evaluating soil liquefaction potential. J Soil Mech Found Div 97(SM9):1249–1273

Seed HB, Whitman RV (1970) Design of earth retaining structure for dynamic loads. ASCE specialty conference on lateral stresses in the ground and design of earth retaining structures

Semblat JF, Kham M, Bard PY (2008) Seismic-wave propagation in alluvial basins and influence of site-city interaction. Bull Seismol Soc Am 98(6):2665–2678

Sharma S, Judd WR (1991) Underground opening damage from earthquakes. Eng Geol 30:263–276

Special Issue of Soils and Foundations (1996) Selected photographs on the 1995 Hyogoken – Nambu earthquake. Special issue on geotechnical aspects of the 17 January 1995 Hyogoken–Nanbu earthquake. Japanese Geotechnical Society, Japan

St. John CM, Zahrah TF (1987) Aseismic design of underground structures. Tunn Undergr Space Technol 2(2):165–197

Tohda J, Yoshimura H, Ohsugi A, Nakanishi K, Inoue Y, Ko HY, Wallen RB (2010) Centrifuge model tests on the dynamic response of sewer trunk culverts. In: Laue J, Seward L, Springman SM (eds) Modelling in geotechnics. Taylor & Francis, London. ISBN 978-0-415-59288-8

Travasarou T (2010) Insights from the vulnerability studies of liquefaction-induced uplift of an immersed tunnel. In: Proceedings of the 6th Greek geotechnical and geo-environmental engineering conference, Volos, Greece, September 2010 (in Greek)

Travasarou T, Chacko JM (2008) Liquefaction-induced uplift mechanics of immersed tunnel. In: Proceedings of the 3rd Greek conference of earthquake mechanics and engineering seismology, Athens, November 2008 (in Greek)

Tsinidis G, Chalatis A (2008) Seismic design of immersed tunnels – application at the Thessaloniki immersed roadway tunnel. MSc thesis, Civil Engineering Department, Aristotle University of Thessaloniki, Greece (in Greek with English abstract)

Tsinidis G, Pitilakis K (2012) Seismic performance of circular tunnels: centrifuge testing versus numerical analysis. In: II international conference on performance based design in earthquake geotechnical engineering, Taormina, Italy, May 2012

Tsinidis G, Pitilakis K, Trikalioti AD (2013a) Numerical simulation of round robin numerical test on tunnels using a simplified kinematic hardening model. Acta Geotechnica. doi:10.1007/s11440-013-0293-9 (accepted – in print)

Tsinidis G, Pitilakis K, Heron C, Madabhushi SPG (2013b) Experimental and numerical investigation of the seismic behavior of rectangular tunnels in soft soils. In: Computational methods in structural dynamics and earthquake engineering conference (COMPDYN 2013), Kos, Greece, June 2013

Tsinidis G, Heron C, Pitilakis K, Madabhushi G (2013c) Physical modeling for the evaluation of the seismic behavior of square tunnels. In: Ilki A, Fardis MN (eds) Seismic evaluation and rehabilitation of structures. Geotechnical geological and earthquake engineering, vol 26. Springer, Cham, pp 389–406. doi:10.1007/978-3-319-00458-7_22

Tsinidis G, Pitilakis K, Anastasiadis A, Pitilakis D, Heron C, Madabhushi SPG, Stringer M, Paolucci R (2013d) TUNNELSEIS: investigation of several aspects affecting the seismic behaviour of shallow rectangular underground structures in soft soils. Final report. Series: Seismic engineering research infrastructures for European synergies. http://www.series.upatras.gr/TUNNELSEIS

Tzarmados D (2011) Dynamic interaction of shallow metro tunnels with aboveground structures at the surface. MSc thesis, Civil Engineering Department, Aristotle University of Thessaloniki, Greece (in Greek with English abstract)

Trelleborg (2013) Gina Gasket. Brochure available on internet: http://www.trelleborg.com/upload/Infrastructure/Files/Gina%20Gasket.pdf. Last accessed 1 Dec 2013

Veletsos AS, Younan AH (1994) Dynamic soil pressures on rigid vertical walls. Earthq Eng Struct Dyn 23:275–301

Vrettos (2005) Design issues for immersed tunnel foundations in seismic areas. In: Proceedings of the 1st Greece-Japan workshop: seismic design, observation, and retrofit of foundations, Athens, 11–12 October 2005

Wang JM (1985) The distribution of earthquake damage to underground facilities during the 1976 Tang-Shan earthquake. Earthquake Spectra 1:741–757

Wang JN (1993) Seismic design of tunnels: a simple state of the art design approach. Parsons Brinckerhoff Inc., New York

Wang WL, Wang TT, Su JJ, Lin CH, Seng CR, Huang TH (2001) Assessment of damage in mountain tunnels due to the Taiwan Chi-Chi earthquake. Tunn Undergr Space Technol 16:133–150

Wang ZZ, Gao B, Jiang YJ, Yuan S (2009) Investigation and assessment on mountain tunnels and geotechnical damage after the Wenchuan earthquake. Sci China Ser E-Technol Sci 52(2):546–558

Wells DL, Coppersmith KJ (1994) New empirical relationships among magnitude, rupture length, rapture width, rapture area and surface displacement. Bull Seismol Soc Am 84(4):974–1002

Zeghal M, Elgamal AW (1994) Analysis of site liquefaction using earthquake records. J Geotech Eng ASCE 120(6):996–1017

Zerva A (1993) Pipeline response to directionally and spatially correlated seismic ground motions. J Press Vessel Tech ASME 115:53–58

Zerva A (1994) On spatial variation of seismic ground motions and its effects on lifelines. Eng Struct 16:534–546

Zerva A, Beck JL (2003) Identification of parametric ground motion random fields from spatially recorded seismic data. Earthq Eng Struct Dyn 32:771–791

Zerva A, Zervas V (2002) Spatial variation of seismic ground motions: an overview. ASME Appl Mech Rev 55(3):271–297

Zerva A, Ang AH, Wen YK (1988) Lifeline response to spatially variable ground motions. Earthq Eng Struct Dyn 7:361–379

Part VI
Special Topics

Chapter 12
Reinforced Soil Walls During Recent Great Earthquakes in Japan and Geo-Risk-Based Design

Yoshihisa Miyata

Abstract As a country affected by frequent earthquakes, Japan has accumulated experience on using reinforced soil wall technology as a seismic countermeasure. This paper briefly reviews the recent development of this technology in Japan. Approximately 1,600 case histories on the seismic performance of such walls during the 2011 Great East Japan Earthquake were collected and analyzed. Statistical data on the seismic damage revealed that all types of reinforced soil walls performed well during the 2011 earthquake. The case study where the damage was closely investigated was also particularly focused upon and discussed.

In current infrastructure projects, the most economical solution with the required performance is selected from several candidate solutions. In general, the term "most economical" implies that the initial cost is the lowest for that particular solution. Risk-induced seismic events are not usually considered in many projects. In this paper, a new design concept that considers the lifecycle cost including the geo-risk for a reinforced soil wall is proposed. A user-friendly cost estimation tool and a reliability analysis method along with its validation are also introduced. As a practical application of the geo-risk-based design, the selection of the best solution from the candidate solutions and the determination of the optimum reinforcing condition are discussed.

12.1 Introduction

The concept of soil reinforcement is very ancient. The oldest case history is the "ziggurat," which was built by Babylonians 3,000 years ago. The Great Wall of China, which was built more than 2,000 years ago, is also known as an ancient reinforced soil structure. In the ancient method of soil reinforcement, natural

Y. Miyata (✉)
Department of Civil Engineering, National Defense Academy, Yokosuka, Japan
e-mail: miyamiya@nda.ac.jp

M. Maugeri and C. Soccodato (eds.), *Earthquake Geotechnical Engineering Design*, Geotechnical, Geological and Earthquake Engineering 28,
DOI 10.1007/978-3-319-03182-8_12,

materials such as tree branches were used as the reinforcing material. Terre Armee led to the birth of modern soil reinforcement that used steel strips as the reinforcing material. This method was developed in the 1960s by Henri Vidal in France. Then, geogrid, which is a polymeric reinforcing material having a grid shape, was developed in the 1970s in the United Kingdom. This accelerated the use of soil reinforcement across the world. Now, it is a popular material used in various geotechnical solutions (Jones 1985).

As a country affected by frequent earthquakes, Japan has accumulated experience on using reinforced soil wall technology as a seismic countermeasure. This paper briefly reviews the recent development of the reinforced soil wall technology in Japan. Approximately 1,600 case histories on the seismic performance of these walls during the 2011 Great East Japan Earthquake were collected and analyzed. Statistical data on the seismic damage revealed that all types of reinforced soil walls performed well during the 2011 earthquake. The case study where the damage was closely investigated was also particularly focused upon and discussed. In this paper, a new design concept that considers the lifecycle cost including the geo-risk of a reinforced soil wall. A user-friendly cost estimation tool and a reliability analysis method along with its validation are also introduced. As a practical application of the geo-risk-based design, the selection of the best solutions from several candidate solutions and the determination of the optimum reinforcing condition are discussed.

12.2 Recent Development of Reinforced Soil Wall Technology in Japan

Soil reinforcement can be applied to various types of geo-structures such as embankments, retaining walls, and foundations. In this paper, reinforced soil wall technology is focused on and a geo-risk-based design method is discussed. The reinforced soil wall technology in Japan can be classified into four types by the reinforcement type: metal or geosynthetic and development situation; primitive or innovative, as shown in Fig. 12.1. A steel strip wall, a geosynthetic wall, a multi-anchor wall (Fukuoka et al. 1980), and a geosynthetic wall with rigid facing (Tatsuoka et al. 1989) are applied to road, rail, and site preparation.

Figure 12.2 shows an increase in the cumulative number of applications with time (Ochiai 2007). The application of reinforced soil walls begun in the 1970s in Japan. The left-hand-side picture in Fig. 12.2 shows the first geogrid reinforced wall in Japan. This wall was measured carefully to understand the actual reinforcing effect (Yamanouchi et al. 1993). Such a persistent effort promoted the development of reinforced soil walls in Japan. In 1995, the Hanshin-earthquake hit Kobe. As shown in the right-hand-side picture in Fig. 12.2, the reinforced wall survived, but the other structure was seriously damaged (Tatsuoka et al. 1996). This seismic event proved that a reinforced soil wall exhibited higher seismic performance than other classical geo-structures such as a gravity-type retaining wall. Then, the use of reinforced walls became widespread.

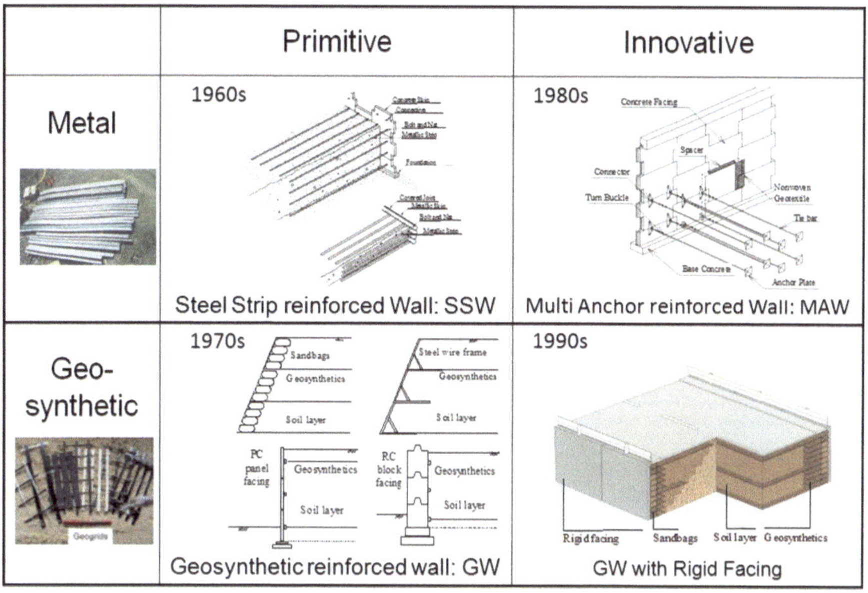

Fig. 12.1 Reinforced soil wall technology in Japan

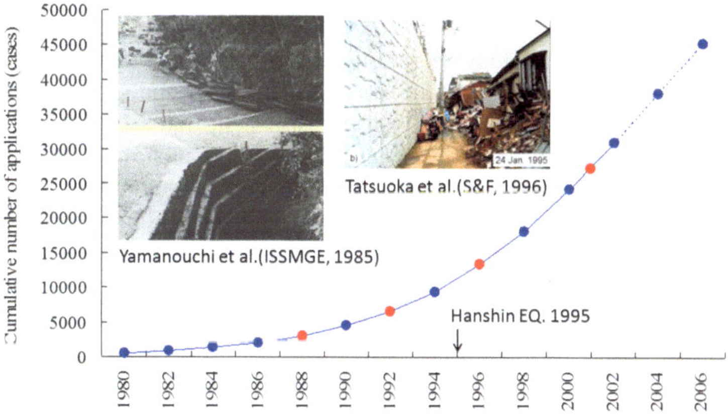

Fig. 12.2 Diffusion of reinforced soil wall technology in Japan (Ochiai 2007)

Even today, researchers are focusing on the advancement of the reinforced soil wall technology. One such advancement is a combination technology. Some examples of such technologies are soil-cement mixing, soil-recycled material mixing, ground improvement, and EPS. Some of the typical examples of such technology are also shown in Fig. 12.3. These methods were developed to achieve a higher performance, low costs, and ecological advantages. Miyata et al. (2003), Lai et al. (2006), Rowe and Taechakumthorn (2008), and Izawa et al. (2009) reported excellent case histories on the combination technology.

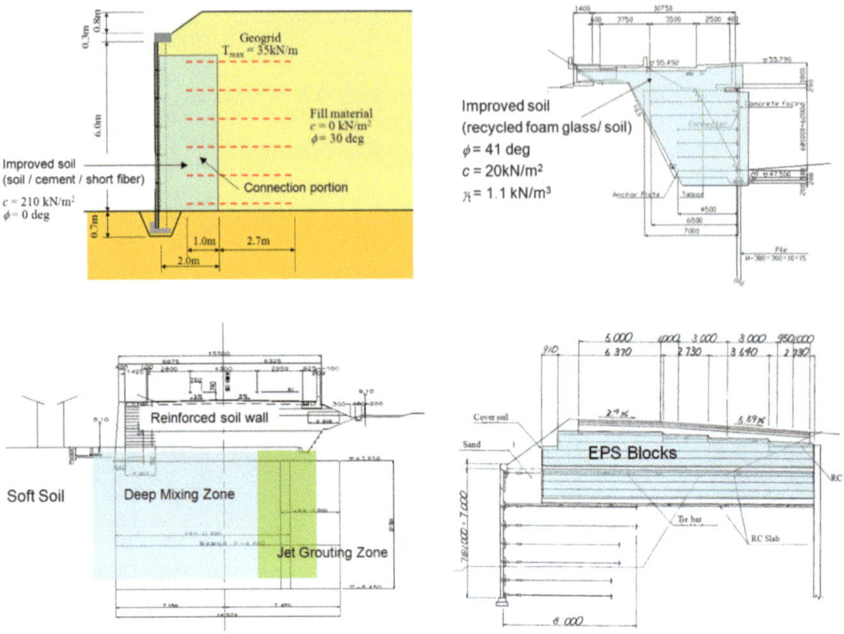

Fig. 12.3 Combination technology of reinforced soil wall

For the applied technology to restore the damaged geo-structure due to an earthquake, it is required that (1) the construction is easy and speedy, (2) it has a high seismic performance, and (3) the cost-benefit ratio is relatively high. Reinforced soil wall technology is the best solution that satisfies the above requirements. Figure 12.4 shows a case history of the restoration of a collapsed geo-structure using the reinforced soil wall technology. Reinforced soil walls have thus become a popular earthquake disaster mitigation technology (Koseki et al. 2006; Koseki 2012).

12.3 Seismic Performance of Reinforced Soil Wall During Great East Japan Earthquake

12.3.1 Great East Japan Earthquake

Most of the past researches have reported that reinforced soil walls have a high seismic performance. How did these walls perform during the great east Japan earthquake? This earthquake was the greatest earthquake ever recorded in Japan and one of the five most powerful earthquakes in the world since 1900. Figure 12.5 shows a comparison of the observed seismic waves during the 1995 Hyogoken-Nanbu earthquake, the 2004 Chuetsu earthquake, and the 2011 Great Tohoku

Fig. 12.4 Earthquake disaster rehabilitation by reinforced soil wall

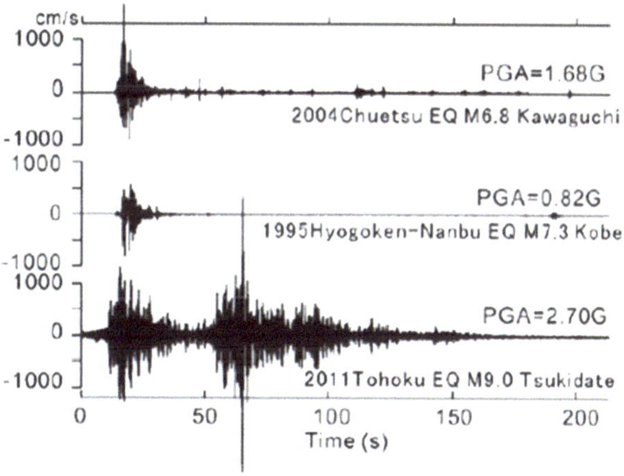

Fig. 12.5 Observed seismic motion during recent earthquakes in Japan (Yamanaka 2011)

earthquake (Yamanaka 2011). Figure 12.6 shows the distribution of the Japanese seismic intensity observed on March 11, 2011 (NIED 2011). The Japanese seismic intensity can be translated into equivalent peak accelerations by assuming the seismic wave period. The peak accelerations shown in Fig. 12.6 are values transmitted by assuming that the wave period equals 0.5 s. In general, when the seismic intensity is more than 6 upper, there is severe structural damage. In the 2011 earthquake, a strong seismic wave was observed over a wide area. The 2011 earthquake can be summarized as a strong motion observed for more than 100 s over a wide area.

Fig. 12.6 Observed seismic intensity during great East Japan earthquake (NIED 2011)

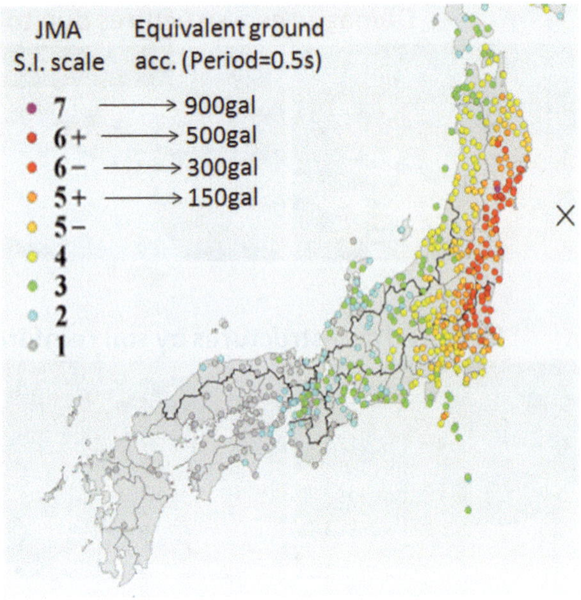

12.3.2 *Seismic Damage of Non-reinforced Geo-structure*

Figure 12.7 shows the typical damage of a highway structure, local road structure, railway structure, and geo-structures at a housing site. Excess subsidence, sliding, and slippage failure of non-reinforced geo-structures were observed. Such damage led to huge financial losses and delayed recovery.

12.3.3 *Seismic Performance of Reinforced Soil Wall During Great East Japan Earthquake*

With respect to the performance of reinforced soil walls, various technical committees for reinforced soil walls carried out an earthquake damage investigation; this investigation revealed that the estimated seismic intensity was greater than 5 upper. Figure 12.8 shows the details of this investigation for three types of reinforced soil walls. To the best of my knowledge, such a large-scale investigation has been unprecedented. A detailed analysis is still under work. In this paper, primary analysis results are explained.

The damage of reinforced soil walls can be classified into four levels, namely, ultimate limit state, restorability limit state, serviceability limit state, and no damage. The following numbers show the calculated probability for each damage level. The probability for the ultimate limit state was less than 1 % and that for the no damage state was more than 90 %. The Great East Japan Earthquake was huge, and the aftershock

Fig. 12.7 Damage to non-reinforced soil structure during great East Japan earthquake. (**a**) Subsidence of highway embankment (East Nippon Expressway Company, 2011). (**b**) Collapse of retaining wall on local road (JC-IGS Technical Committee, 2011). (**c**) Collapse of retaining wall on railway track (East Japan Railway Company, 2011). (**d**) Collapse of retaining wall at housing site (Koseki et al. 2011)

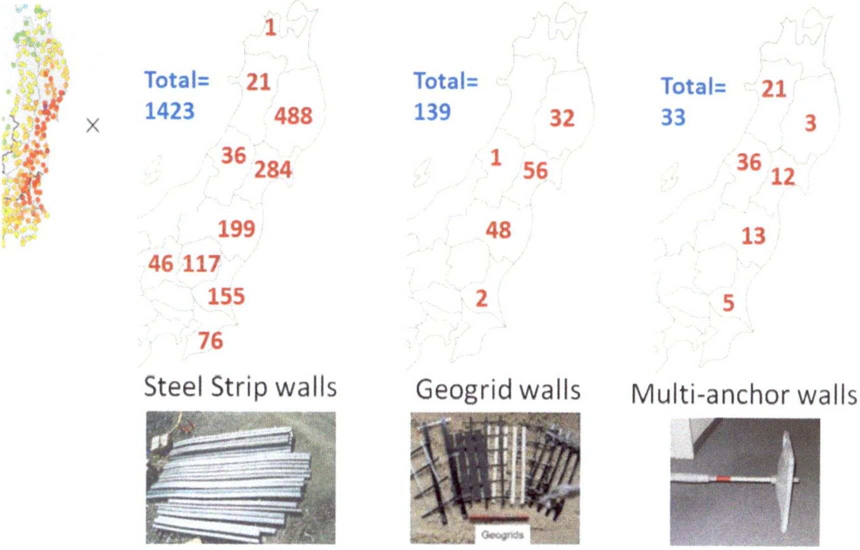

Fig. 12.8 Site investigation of earthquake damage to reinforced soil wall during 2011 Great East Japan Earthquake

Table 12.1 Damage statistics of reinforced soil wall during 2011 Great East Japan Earthquake

	Steel strip walls (%)	Geogrid walls (%)	Multi-anchor walls (%)
Ultimate limit state	0.3	0.7	0
Restorability limit state	1.0	4.3	0
Serviceability limit state	7.0	0.7	3.0
No damage	91.7	94.3	97.0

Fig. 12.9 Performance of reinforced soil wall against seismic motion. (**a**) Case 1: ultimate limit state. (**b**) Case 2: ultimate limit state. (**c**) Case 3: restorability limit state. (**d**) Case 4: serviceability limit state. (**e**) Case 5: serviceability limit state. (**f**) Case 6: no damage (*left*) and collapsed embankment (*right*). (**g**) Case 7: no damage (**a**: Kumagai and Kaneko 2011; **b**: courtesy of Dr. Otani: Hirose Co.; **c–f** and **g**: JC-IGS Technical Committee (Dr. Hironaka, Mr. Konami, and Dr. Tatta))

was also very considerable. Such a strong seismic load was not considered when designing reinforced soil walls. However, these walls exhibited a high seismic performance during the 2011 earthquake. This implies that the actual seismic performance of reinforced soil walls is considerably higher than the design target (Table 12.1).

The case study where the seismic motion damage was closely investigated is shown in Fig. 12.9. In this section, each case history is focused upon and discussed individually.

Case 1. A geogrid reinforced wall (height: 5.0 m) with battered soft facing collapsed. Not only internal slippage but also connection failure was observed.

This damage state can be judged as the "ultimate limit state." The estimated seismic intensity was 6 lower at this location. The equivalent ground acceleration was estimated to be 300 gal. The site investigation revealed that this collapse was caused by the fact that the ground water level in this wall was very high because of the lack of a drainage system.

Case 2. A steel strip wall (height: 10.0 m) on a soft soil foundation collapsed. Due to seismic motion, a 7-m horizontal sliding was observed. This damage can be also judges as the "ultimate limit state." The estimated seismic intensity was 6 lower at this location. The equivalent ground acceleration was estimated to be 300 gal. The site investigation revealed that this collapse was mainly caused by the fact that the soil foundation was not sufficiently treated. Adequate ground improvement should have been carried out.

Case 3. A geogrid reinforced wall (height: 5.4 m) with battered soft facing was damaged. An excess horizontal deformation of the facing and a difference in the deformation between the reinforced soil wall and the rigid reinforced concrete structure were observed. This damage can be judged as the "restorability limit state" damage because the horizontal deformation was more than the allowed limit. The estimated seismic intensity was 6 lower at this location. The equivalent ground acceleration was estimated to be 300 gal. Further, the site investigation revealed that this damage could be attributed to the fact that the fill material had a considerable amount of fine contents and the ground water level in the wall was very high.

Case 4. In this case, a difference in the deformation between the reinforced soil wall and the rigid reinforced concrete structure was observed as in Case 3. However, the horizontal facing deformation was within the allowed limits. Such damage can be classified into the "serviceability limit state." This differential deformation was repaired using soil nailing, soil cement, and a vegetation technique, as shown in Fig. 12.10a.

Case 5. A steel strip wall and a multi-anchor wall have concrete panel facing. After the earthquake, the damage caused to the concrete panel was observed at some points. This damage can be categorized to be in the "serviceability limit state." A repairing method for the damaged concrete panel has already been developed. Figure 12.10b shows the partial repairing of the damaged panel. Only the damaged part in the panel was cut, and concrete milk was injected into the spot. The full panel can also be replaced using a similar method.

Case 6. In this case, a geogrid reinforced wall (5.4 m) with battered soft facing was observed to have no damage. However, a non-reinforced embankment beside this reinforced soil wall collapsed. Here, a strong seismic motion having a seismic intensity of 6 lower seemed to act on these two structures. This is an important case history that exemplifies the high seismic performance of a reinforced soil wall.

Case 7. A geogrid reinforced wall (height: 6.2 m) with battered soft facing experienced no damage by a strong seismic motion having a seismic intensity of 7, which is the maximum intensity. In 2008, the earthquake whose seismic intensity

Fig. 12.10 Repairing of damaged reinforced soil wall. (**a**) Repairing of differential deformation between geogrid reinforced wall and RC structure. (**b**) Repairing of concrete panel facing of steel-reinforced soil wall

was 6 lower had hit this wall. This wall also experienced no damage during the 2008 earthquake. This case history shows that a reinforced soil wall can resist multiple earthquakes.

A case study, in which the damage due to a seismic motion and a tsunami action was closely investigated, is shown in Fig. 12.11. In this section, each case history is focused upon and discussed individually.

Case 8. A steel strip wall collapsed because of a combination of a 6-upper seismic motion and tsunami action. This damage can be categorized into the "ultimate limit state" category. In this case, the top surface of the wall was not covered with a pavement or an erosion mat. This case history shows that erosion control should be carried out at a wall constructed along a river or seashore.

Case 9. A geogrid reinforced wall (height: 5.0 m) was damaged at the soft facing by tsunami action. However, this damage was partial and could be repaired easily. Such damage can be categorized as the "serviceability limit state."

Case 10. An approximately15-m-high tsunami hit a steel strip wall. However, the wall experienced no damage. The top surface of this wall was covered with trees. Hence, it can be concluded that trees perform an erosion control function.

Case 11. A 10.6-m-high geogrid reinforced wall with battered soft facing experienced no damage by tsunami action and seismic motion having a seismic intensity of 6 upper. In this case, a differential deformation between the reinforced soil wall and the reinforced concrete structure was not observed. The differential settlement seemed to depend on the condition of the base foundation and the type of fill material. Now, this mechanism is being investigated by a numerical dynamic analysis.

Fig. 12.11 Performance of reinforced soil wall against seismic motion and tsunami action. (**a**) Case 8: ultimate limit state. (**b**) Case 9: serviceability limit state. (**c**) Case 10: no damage. (**d**) Case 11: no damage (**a** and **c**: by courtesy of Dr. Otani: Hirose Co.; **b** and **d**: JC-IGS Technical Committee (Dr. Hironaka))

12.4 Geo-Risk-Based Design for Reinforced Soil Wall

12.4.1 General Concept

All geo-structures can collapse or get damaged because of external loads despite having a structural system. Hence, while designing geo-structures, it is important to consider the possibility of failure. If a structural failure takes place, there would be considerable loss of money and/or human lives. Figure 12.12 shows the initial and recovery costs for a reinforced soil wall, an L-shaped wall, and a non-reinforced embankment.

These costs were estimated for three different wall heights based on the Japanese construction standard. On the basis of this estimation, it was understood that the recovery cost depended on the solution and that the most expensive wall was an L-shaped wall and the most inexpensive wall was the non-reinforced embankment. Further, the financial loss due to interruption increased with an increase in the importance of the structures. In 2009, an important highway had to stop service because of seismic damage of a non-reinforced embankment. The financial loss was

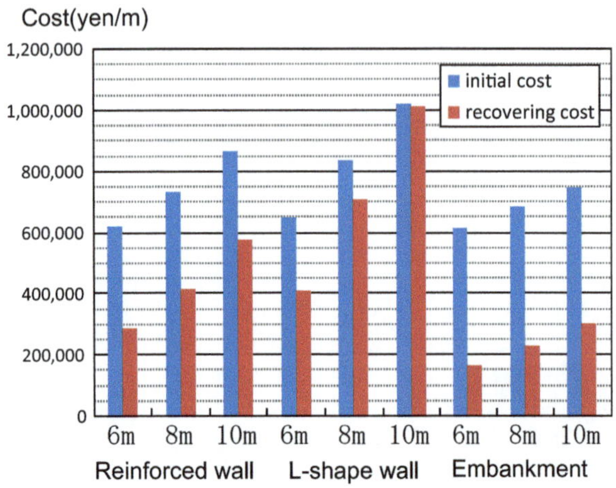

Fig. 12.12 Initial and recovery cost for reinforced wall, L-shaped wall, and embankment

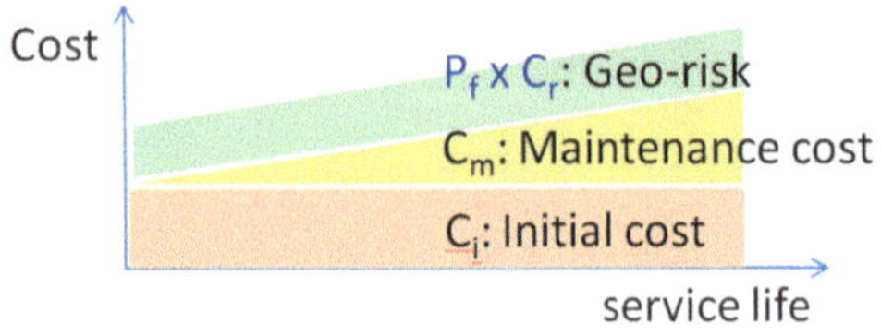

Fig. 12.13 Lifecycle cost including geo-risk

estimated to be approximately 200 million euros. In this paper, such a risk induced by the failure of a geo-structure is called "geo-risk." When we design a structure, we should consider the structure's lifecycle cost including the geo-risk.

In general, risk is defined as a product of the failure probability and the financial loss. A geo-risk-based design can be defined as one that minimizes the structure's lifecycle cost including the geo-risk. The lifecycle cost consists of initial, maintenance, and repairing costs, as shown in Fig. 12.13. To complete this design method, a cost estimation method with accountability should be developed. Further, a reliability analysis method along with its validation should be developed. In the next section, research achievements based on the above two topics are introduced.

12.4.2 Cost Estimation Tool

With respect to cost estimation, standardization is a challenging subject. To solve this problem, a technical committee established in the Japan chapter of International Geosynthetics Society corrected the case histories on a rehabilitation project of geo-structures damaged in earthquakes and discussed a standard to estimate the initial, maintenance, and recovery costs. Finally, a user-friendly cost estimation tool shown in Fig. 12.14 was developed.

初期建設費概略単価

工種	作業内容		単価	単位	数量	金額	比率
盛土	ブルドーザの敷均し 路体(普通15級)		107	円/m³	31.8	3,403	0.5%
	タイヤローラの締固め		26	円/m³	31.8	827	0.1%
	法面整形工		351	円/m²	0.0	0	0.0%
	購入土		2,000	円/m³	0.0	0	0.0%
道路舗装	不陸整正		93	円/m²	7.5	698	0.1%
	路盤工 (下層20cm, 上層15cm)		923	円/m²	7.5	6,923	1.1%
	アスファルト舗装工 (車道b>3m)		1,420	円/m²	7.5	10,650	1.7%
道路付属施設	防護柵設置 (ガードケーブル支柱コンクリート立込)		2,000	円/m	1.0	2,000	0.3%
	防護柵設置 (Gr-A-2B)		8,000	円/m	1.0	8,000	1.3%
ジオテキスタイル工	ジオテキスタイル工鋼製ユニット		10,000	円/m²	6.0	60,000	9.6%
	ジオテキスタイル工敷設, まき出し, 敷均し, 締固め		1,718	円/m²	28.0	48,104	7.7%
	ジオテキスタイル材料費	安定補助材	670	円/m²	7.5	5,025	0.8%
		Ta=21.6kN/m (32.4kN/m)	1,200	円/m²	11.2	13,440	2.2%
		Ta=30.0kN/m (45.0kN/m)	1,350	円/m²	5.6	7,560	1.2%
		Ta=36.0kN/m (54.0kN/m)	1,500	円/m²	0.0	0	0.0%
		Ta=54.0kN/m (81.0kN/m)	1,870	円/m²	11.2	20,944	3.4%
		Ta=60.0kN/m (90.0kN/m)	2,100	円/m²	0.0	0	0.0%
		Ta=75.0kN/m (112.5kN/m)	2,250	円/m²	0.0	0	0.0%
	ジオテキスタイル水平排水材		490	円/m²	11.2	5,488	0.9%
擁壁工	L型擁壁(H=6m)		215,323	円/m	0.0	0	0.0%
	L型擁壁(H=8m)		403,588	円/m	0.0	0	0.0%
	L型擁壁(H=10m)		588,420	円/m	0.0	0	0.0%
一般国道(補助)	「道路投資の評価に関する指針(案)」参考		240,000	円/m	1.0	240,000	38.6%
用地費(山林)	(財)資産評価システム研究センター, 平成20年度掲示平均価額		16,000	円/m²	11.8	188,800	30.4%
				初期コスト: C_0	合計	621,860	

Fig. 12.14 Tool for lifecycle cost estimation

The abovementioned cost estimation tool is created in a spreadsheet of Microsoft Excel. In this spreadsheet, the required work and materials are entered in an initial column for all the construction work. For each work and material entry, the unit cost has also been decided. Therefore, the user just inputs the volume based on the design conditions, and the initial, maintenance, and repairing costs are automatically estimated. This tool can estimate the lifecycle cost for not only reinforced soil walls but also for RC-walls and non-reinforced embankments. Now, the technical committee is trying to develop integrated software combining this tool with design and construction management software.

12.4.3 Reliability Analysis with Validation

The other challenge is to develop a reliability analysis method along with its validation. The basic flow of the classical approach is as follows: First, the statistical structure of the failure is analyzed as shown in Fig. 12.15. In the next step, input parameters such as the most frequent load or resistance and its uncertainty are prepared. Finally, the probability of failure is calculated using a rigorous closed formulation of the reliability index, β, or a Monte-Carlo simulation.

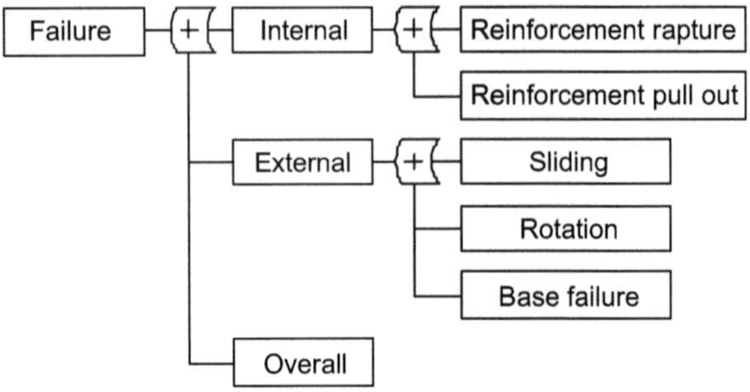

Fig. 12.15 Statistical structure of failure of reinforced soil wall

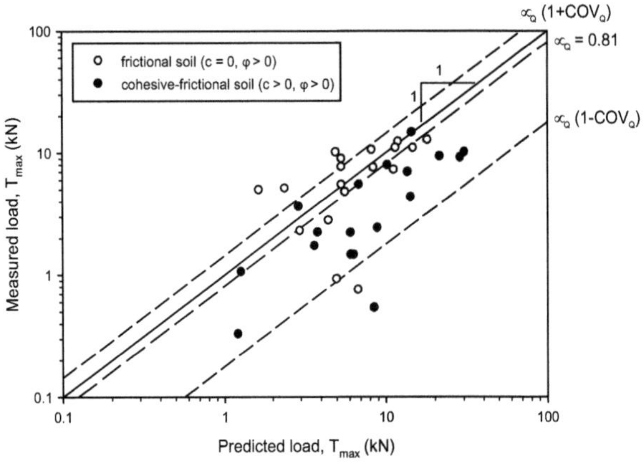

Fig. 12.16 Measured and predicted reinforcement loads on multi-anchor wall

Assuming that the road and the resistance have a normal distribution, we can estimate the probability of failure using the reliability index, β.

$$\beta = \frac{\mu_R - \mu_Q}{\sqrt{\sigma_R^2 + \sigma_Q^2}} \tag{12.1}$$

where μ_Q and μ_R are the average load and resistance, respectively, which is almost the same as the most frequent load and resistance. σ_R and σ_Q are the standard deviation of the load and the resistance, respectively.

The simplest way for evaluating μ_Q and μ_R is to use the current design model. However, the current design model was developed for an allowable stress design. It captures the lower (safer) bound of the actual value but cannot predict the most frequent value. When the current design model is applied to the reliability analysis, this point should be carefully considered. Figure 12.16 shows the measured and predicted reinforcement load relation in a multi-anchor wall system. The average of the bias defined as the ratio of the measured to the predicted value is 0.80. This implies that the current design model predicts on an average a 20 % safer load. This bias property should be considered in the reliability analysis. Further, there is a

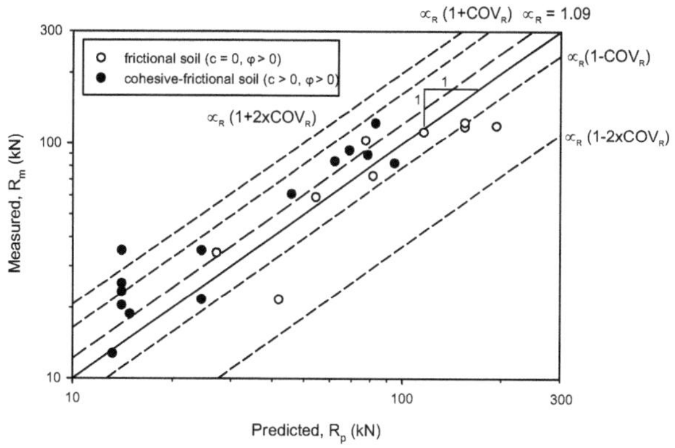

Fig. 12.17 Measured and predicted reinforcement pull-out capacity of multi-anchor wall

model bias problem on the resistance side. Figure 12.17 shows the measured and the predicted pull-out capacity of the reinforcing material used in a multi-anchor wall system in soil. The average of the bias is 1.20. This implies that the current design model predicts on an average a 20 % safer pull-out capacity. This bias statistic also should be considered in the reliability analysis.

The simplest way to consider the model bias in the reliability analysis is to estimate the average value of the load or the resistance by using the bias that is defined as the ratio of the measured to the predicted value. Assuming the simple bias statistics property, we find that the COV of the load or resistance is the same as the COV of their bias. Then, the reliability index, β, shown in Eq. (12.1) can be rewritten as follows (Allen et al. 2005):

$$\beta = \frac{(\gamma/\phi)\mu_{XR} - \mu_{XQ}}{\sqrt{[\mathrm{COV}_{XR}(\gamma/\phi)\mu_{XR}]^2 + [\mathrm{COV}_{XQ}\mu_{XQ}]^2}} \tag{12.2}$$

where μ_{XQ} and μ_{XR} are the average of the bias of the load and the resistance model, which implies the most frequent load and resistance, respectively. COV_{XR} and COV_{XQ} denote the standard deviation of the bias of the load and the resistance model, respectively. γ represents the partial safety factor for the load and ϕ represents that for the resistance. γ/ϕ is equivalent to the safety factor considered in the allowable stress design.

Figure 12.18 shows the relation between the failure probability and the uncertainty of the load and the resistance. The averages of the bias are assumed to be 0.80 for the load and 1.20 for the resistance based on the previous analysis results shown in Figs. 12.16 and 12.17. For γ/ϕ, a value of 3.0 is assumed because the safety factor in the current design is 3.0. The general approach can estimate the failure probability by considering the design model error, scattering of the load or resistance value, and safety level considered in the current design.

Now, my team is attempting to develop a reliability analysis for these three types of reinforced soil walls by using a general approach. First, the bias statistic of the

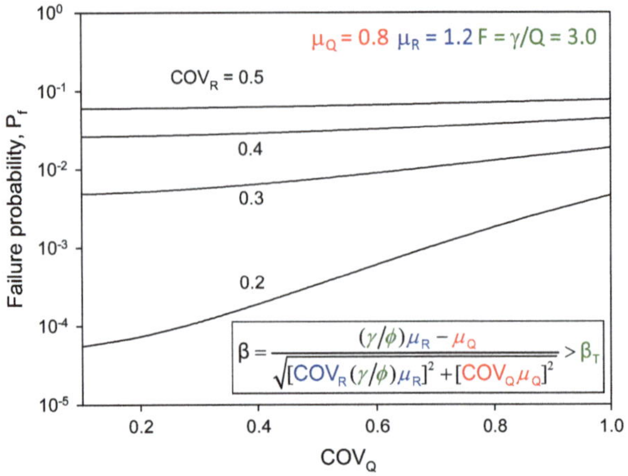

Fig. 12.18 Failure probability and uncertainty of load and resistance

Table 12.2 Miyata and Bathurst's recent work on reliability analysis method for reinforced soil wall

	Steel strip walls	Geogrid walls	Multi-anchor walls
Load model	Miyata and Bathurst (2012a)	Miyata and Bathurst (2007a)	Miyata et al. (2009)
		Miyata and Bathurst (2007b)	Miyata et al. (2010)
		Bathurst et al. (2008)	
Resistance model	Miyata and Bathurst (2012b)	Miyata and Bathurst (2012c)	Miyata et al. (2011)
LRFD calibration	Under work	Under work	Bathurst et al. (2011)

load model is investigated. When the bias is too high, a new model is developed as that the bias is equal to one. For the resistance side, a similar analysis is performed. Finally, the load and the resistance factors are calibrated as that the estimated failure probability is smaller than the acceptable level. Our activity is summarized in Table 12.2. Currently, our achievement is limited to a statics problem. The next challenge is a dynamics problem.

12.4.4 Practical Application

Finally, we move to the application of a geo-risk-based design in practice. The first step of the design is to select the best solution from some candidate solutions. In this section, an example to select the best from a non-reinforced embankment, an RC-retaining wall, and a reinforced wall will be shown. On the basis of the current design, the primary condition is assumed and the lifecycle cost is estimated under

Table 12.3 Analysis condition of lifecycle estimation

Fill material (uncertainty)	Unit weight $\gamma_t = 19.0\ kN/m^3$ (5 %), Cohesion c = 0 kN/m^2 (10 %), Friction angle $\phi = 30°$ (10 %)
Required performance	Safety factor in static $F_s = 1.2$, in seismic $F_s = 1.0$
Seismic load	kh = 0.2, COV = 50 %, Dan and Kanda (1998)
Design life	40 years

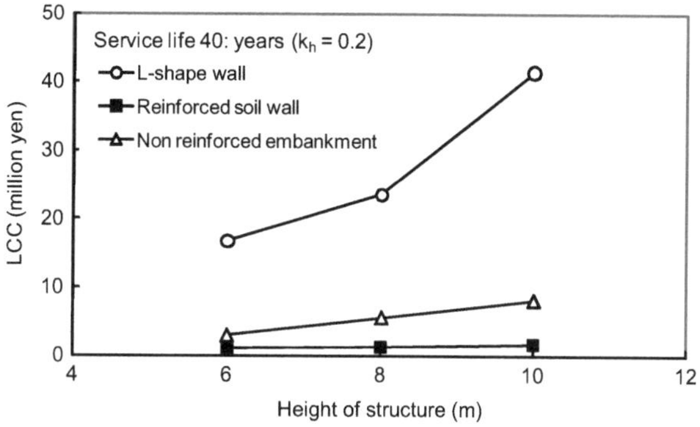

Fig. 12.19 Application of lifecycle analysis for selecting best solution

the analysis condition shown in Table 12.3. Figure 12.19 shows the analysis result; it shows the relation between the estimated lifecycle cost and the structure height. For all the considered structure heights, the lifecycle cost of the reinforced soil wall is the lowest. On the other hand, that for the L-shaped wall is the highest. That for the non-reinforced soil is medium. The failure probability of a reinforced soil wall is smaller than that of a classical geo-structure because it has a ductile structural system. Then, the lifecycle cost of the reinforced soil structure is the lowest, and it is selected as the best solution. This analysis result shows that the reinforced soil wall generally has an advantage over the other classical geotechnical solutions in terms of the lifecycle.

When the best solution is selected, the detailed design conditions such as the reinforcing conditions can be determined using the geo-risk-based design. Figure 12.20 shows a schematic representation of this analysis. The initial cost and the maintenance cost are higher when the reinforcement length is longer or when the strength of the reinforcing material is higher. On the other hand, the failure probability is smaller when the length or strength of the reinforcing material is longer or higher, respectively. Therefore, the calculated lifecycle cost-length of the strength relation is as given below. This implies that there is an optimum length and strength for the reinforcement based on the geo-risk evaluation. To apply this method practically, further discussion should be carried out. However, research on the optimum design is a challenging subject for the future. By using this approach, we can easily design a structure by considering the importance of the structure and the uncertainty of the project conditions.

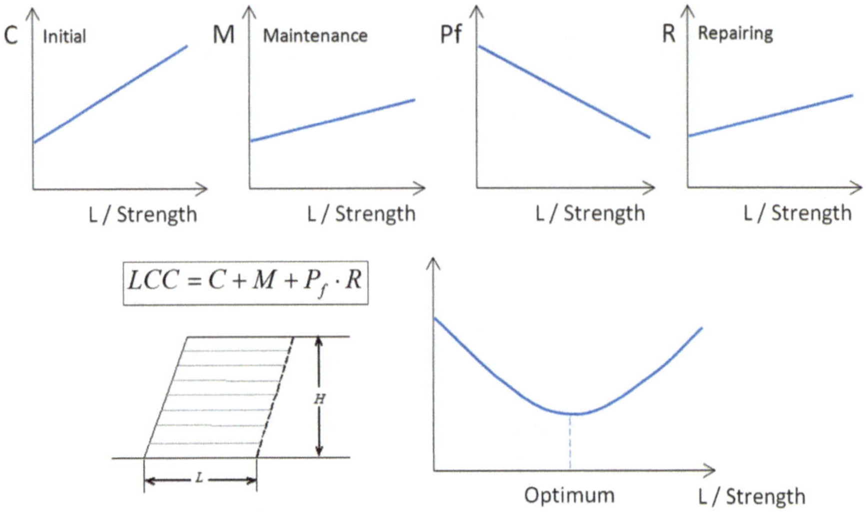

Fig. 12.20 Application of lifecycle analysis for determining optimum reinforcing conditions

12.5 Concluding Remarks

This paper briefly reviews the recent development of the reinforced soil wall technology in Japan and proposes a new design concept for reinforced soil walls. The main conclusions are summarized as follows:

1. Recent development of the reinforced soil wall technology is remarkable. The validity of the reinforced soil wall for earthquake damage mitigation has been proved through many seismic events.
2. Approximately 1,600 site investigations show that reinforced soil walls performed well during the Great East Japan earthquake. The repairs of the damaged reinforced soil walls are progressing smoothly.
3. Geo-risk-based design to minimize the lifecycle cost is useful for selecting the best solution for seismic damage mitigation. It can be applied for designing the optimum reinforcing conditions. To apply this method, a user-friendly cost estimation tool and a reliability analysis method along with its validation are important. Hence, a simple tool and analysis method were proposed.

Acknowledgements The author is grateful for the funding awarded by the Japan Ministry of Education, Culture, Sports, Science and Technology (Grant-in-Aid for Scientific Research (B) No. 24360195) and the Japan Ministry of Defense. The analysis of many case histories and the development of the lifecycle cost estimation were performed under the support of the technical committee established by the Japan chapter of International Geosynthetics Society. The achievement of the reliability analysis is a result of a joint research with Dr. Richard J. Bathurst, who is a professor at the Royal Military College of Canada. The author would like to express his sincere gratitude to all their support.

References

Allen TM, Nowak AS, Bathurst RJ (2005) Calibration to determine load and resistance factors for geotechnical and structural design, Transportation Research Board circular E-C079. Transportation Research Board, Washington, DC, p 93

Bathurst RJ, Miyata Y, Nernheim A, Allen MT (2008) Refinement of K-stiffness method for geosynthetic reinforced soil walls. Geosynth Int 15(4):269–295

Bathurst RJ, Miyata Y, Konami T (2011) Limit states design calibration for internal stability of multi-anchor walls. Soils Found 51(6):1051–1064

Dan K, Kanda J (1998) Theoretical approach to extreme distributions with lower and upper limits for modeling annual maxima of earthquake ground motions in seismic risk analysis. J Struct Constr Eng 506:57–65 (in Japanese)

Fukuoka M, Imamura Y, Kudoh K, Naitoh S, Fukahori T (1980) Research on multi-anchor wall. Proceedings of 15th Japanese Geotechnical Society annual meeting, Hiroshima, Japan, pp 1525–1528 (in Japanese)

Izawa J, Ito H, Saito T, Ueno M, Kuwano J (2009) Development of rational seismic design method for geogrid-reinforced soil wall combined with fibre-mixed soil-cement and its applications. Geosynth Int 16(4):286–300

Jones CJFP (1985) Earth reinforcement and soil structures. Butterworth, London (Thomas Telford Edition, London, 1996)

Koseki J (2012) Use of geosynthetics to improve seismic performance of earth structures. Geotext Geomembr 34:51–68

Koseki J, Bathurst RJ, Guler E, Kuwano J, Maugeri M (2006) Seismic stability of reinforced soil walls. In: Proceedings of 8th international conference on geosynthetics, vol 1, Yokohama, pp 51–77

Koseki J, Wakai A, Mitsuji K (2011) Investigation report on damage inflicted by the 2011 off the Pacific Coast of Tohoku Earthquake – Damage in inland of Miyagi Prefecture. Geotechnical Engineering Magazine 59(6) Ser.No.641: 40–73, Japanese Geotechnical Society (in Japanese)

Kumagai K, Kaneko K (2011) Damage of reinforced earth walls at Iwate area in the 2011 off the Pacific coast of Tohoku Earthquake and Tsunami. Geosynthetics Technical Information, JC-IGS, 27–2, pp 16–23 (in Japanese)

Lai YP, Bergado DT, Lorenzo GA, Duangchan T (2006) Full-scale reinforced embankment on deep jet mixing improved ground. Ground Improv 10(4):153–164

Miyata Y, Bathurst RJ (2007a) Evaluation of K-stiffness method for vertical geosynthetic reinforced granular soil walls in Japan. Soils Found 47(2):319–335

Miyata Y, Bathurst RJ (2007b) Development of K-stiffness method for geosynthetic reinforced soil walls constructed with c-ϕ soils. Can Geotech J 44:1391–1416

Miyata Y, Bathurst RJ (2012a) Measured and predicted loads in steel strip c-ϕ reinforced soil walls in Japan. Soils Found 52(1):1–17

Miyata Y, Bathurst RJ (2012b) Analysis and calibration of default steel strip pullout models used in Japan. Soils Found 52(3):481–497

Miyata Y, Bathurst RJ (2012c) Reliability analysis of soil-geogrid models in Japan. Soils Found 52 (4):620–633

Miyata Y, Ochiai H, Otani J (2003) Recent case histories on combined technology with earth reinforcement. In: Proceedings of Asia regional conference, vol 2, Singapore, pp 1323–1328

Miyata Y, Bathurst RJ, Konami T (2009) Measured and predicted loads in multi-anchor reinforced soil walls in Japan. Soils Found 49(1):1–10

Miyata Y, Bathurst RJ, Konami T, Dobashi K (2010) Influence of transient flooding on multi-anchor walls. Soils Found 50(3):371–382

Miyata Y, Bathurst RJ, Konami T (2011) Evaluation of two anchor plate capacity models for MAW systems. Soils Found 51(5):885–896

NIED (National Research Institute for Earth Science and Disaster Prevention) (2011) Off the Pacific Coast of Tohoku earthquake. Strong Ground Motion (in Japanese)

Ochiai H (2007) Earth reinforcement technique a role of new geotechnical solutions – memory of IS Kyushu. In: Proceedings of the international symposium on earth reinforcement, IS-Kyushu 2007, Fukuoka, Japan, pp 1–23

Rowe RK, Taechakumthorn C (2008) Combined effect of PVDs and reinforcement on embankments over rate-sensitive soils. Geotext Geomembr 26:239–249

Tatsuoka F, Tateyama M, Murata O (1989) Earth retaining wall with a short geotextile and a rigid facing. In: Proceedings of the 12th ICSMFE, vol 2, Rio de Janeiro, pp 1311–1314

Tatsuoka F, Tateyama M, Koseki J (1996) Performance of soil retaining walls for railway embankments. Soils and Foundations, Special issue for the 1995 Hyogoken–Nambu Earthquake, pp 311–324

Yamanaka H (2011) Strong motion characteristics and aftershock observations. Center for Urban Earthquake Engineering (CUEE), Tokyo Institute of Technology Newsletter No.11—2011 Tohoku Pacific Earthquake, Tokyo

Yamanouchi T, Fukuda N, Ikegami M (1993) Geogrid reinforcement for a steel road embankment slope Kyushu island, Japan. In: Raymond GP, Giroud JP (eds) Geosynthetics case histories (ISSMGE TC9). BiTech Publishers, Richmond, pp 242–243

Chapter 13
Performance Based Design for Seismic Design of Geosynthetics-Lined Waste Containment Systems

Edward Kavazanjian Jr., Mohamed Arab, and Neven Matasovic

Abstract A performance-based methodology for seismic analysis and design of the geosynthetic elements of waste containment systems, including landfills and heap leach pads, has been developed. The methodology offers a rational alternative to the current state of practice for seismic design of geosynthetic containment system elements in which a decoupled Newmark-type displacement analysis is used to calculate a permanent seismic displacement. This calculated displacement is generally considered to be an index of the performance of the containment system in an earthquake. In the Newmark-type design methodology, no explicit evaluation is made of the stresses and strains in the geosynthetic elements of the containment system. In order to explicitly assess the ability of the geosynthetic elements of a containment system to maintain their integrity in a design earthquake, a finite difference model of waste-liner system interaction has been developed using the computer code FLAC^{TM}. A beam element with zero moment of inertia and with interface elements on both sides is employed in the model to represent a geosynthetic element in the liner system. This enables explicit calculation of the axial forces and strains within the liner system element. The beam element model was calibrated using available experimental data from shaking table tests of rigid and compliant blocks sliding on geomembranes. The model was then used to analyze the behavior of the Chiquita Canyon landfill in the Northridge earthquake. Results of the analysis provide insight into the reasons for the tears in the liner system at Chiquita Canyon observed after the Northridge event. This model provides a basis for direct performance based seismic design of geosynthetic elements

E. Kavazanjian Jr. (✉)
Department of Civil Engineering, Arizona State University, Tempe, AZ 85287-5306, USA
e-mail: edkavy@asu.edu

M. Arab
Department of Structural Engineering, Mansoura University, Mansoura, Egypt

N. Matasovic
Landfill and Geotechnical Engineering Group, Geosyntec Consultants,
Huntington Beach, CA, USA

M. Maugeri and C. Soccodato (eds.), *Earthquake Geotechnical Engineering Design*, Geotechnical, Geological and Earthquake Engineering 28,
DOI 10.1007/978-3-319-03182-8_13, © Springer International Publishing Switzerland 2014

not only in waste containment systems but in a variety of other civil structures that employ geosynthetic elements wherein earthquake ground motions cause relative displacement between the geosynthetic element and the surrounding soil.

13.1 Introduction

The state of practice for seismic design of geosynthetics-lined waste containment facilities has changed little in the last two decades. While great strides have been made in the understanding of the dynamic properties of solid waste, seismic analyses of the performance geosynthetic liner system elements still generally follows the methodology reported on by Seed and Bonaparte (1992). In this type of analysis, the ability of the geosynthetic elements of the liner system to resist earthquake strong ground motions is based upon the displacement calculated in a decoupled Newmark-type analysis (Newmark 1965). The seismic response of the waste mass is calculated without consideration of the influence of the relative displacement (slip) at liner system interfaces on the response. The response is then used to calculate the relative displacement (slip) at the liner interface (hence the calculation of seismic response is decoupled from the calculation of relative displacement). This calculated relative displacement is then used as an index of seismic performance of the liner system, with calculated values of less than 150 mm generally accepted as being indicative of satisfactory performance, i.e. of the liner system maintaining its integrity in the earthquake. In this type of analysis, the loads on the liner system elements, e.g. tensile strains and axial forces, are never explicitly evaluated. Furthermore, no consideration is given to transient strains and forces induced in the liner system by seismic loading and the method fails to predict liner system performance in the one case history of damage to a geosynthetic liner system in an earthquake, the tearing of the geomembrane at the Chiquita Canyon landfill in the 1994 Northridge, California (USA) earthquake (EMCON 1994).

This paper presents a rational methodology for performance-based analysis and design of the geosynthetic elements of waste containment systems at landfills and mine sites. In this methodology, the strains and axial forces induced in the geosynthetic elements of the containment system in a design earthquake can be explicitly evaluated. A finite difference model of waste-liner system interaction is employed in which a beam element with zero moment of inertia and with interface elements on both sides is used to represent a geosynthetic element in the liner system. The interface model was calibrated using available experimental data from shaking table tests of rigid and compliant blocks sliding on geomembranes and large scale cyclic direct shear testing of high density polyethylene (HDPE) geomembrane/geosynthetic clay liner (GCLs) combinations. The beam model was based upon available information on the stress-strain behavior of HDPE geomembranes subject to tensile loading and the internal shear behavior of geosynthetic clay liners (GCLs). Analysis of the behavior of the Chiquita Canyon landfill in the Northridge earthquake demonstrates the capabilities of the model and provides insight into the reasons for the tears in the liner system at the landfill in the Northridge event. This model provides a basis for direct performance based seismic design not only of the geosynthetic elements of waste containment systems and can also be used in a variety of other situations wherein earthquake ground motions cause relative displacement between soil or waste and other elements of the geotechnical system.

13.2 Current State of Practice

The current state of the practice for seismic analysis and design of geomembranes in waste containment systems is described by Bray and Rathje (1998), Bray et al. (1998), Kavazanjian (1999), and Matasovic and Kavazanjian (2006). In current practice, the adequacy of a geomembrane to withstand seismic loading is based upon the permanent seismic displacement calculated in a decoupled Newmark-type analysis. In this type of analysis seismic response is calculated assuming there is no relative displacement at the liner system interfaces and then the calculated response is used as input to calculate permanent seismic displacement in a Newmark analysis (i.e. the seismic displacement is decoupled from the seismic response). As reported by Seed and Bonaparte (1992), a calculated permanent seismic displacement no greater than 150–300 mm is generally used as the seismic performance criterion in this type of analysis (Kavazanjian 1999; Kavazanjian et al. 1998), with the more stringent criterion of 150 mm typically employed in California. However, as noted by Kavazanjian (1998, 1999), this calculated displacement should not be considered the actual seismic displacement, but rather serves as an index of seismic performance. Bray and Rathje (1998) showed that for typical landfill problems, the decoupled approach significantly overestimates the actual seismic displacement at the liner interface. Matasovic et al. (1998) showed that if the large displacement shear strength is used to calculate the yield acceleration in the Newmark analysis, the actual permanent seismic displacement may be negligibly small for calculated Newmark displacements of 300 mm or more (depending on the ratio of the peak to large displacement shear strength and the displacement necessary to mobilize the large displacement strength). Furthermore, regardless of the magnitude of the permanent seismic displacement, this type of analysis does not reflect the actual strains and displacements induced in the liner system and thus can never be more than an index of seismic performance.

Augello et al (1995) and Matasovic et al. (1995) summarize the performance of landfills in the 1994 Northridge earthquake. This event remains the only major earthquake in which landfills with modern geomembrane liner systems were subjected to high levels of seismic shaking. Most notable of the case histories summarized by these investigators is the performance of the Chiquita Canyon landfill, where tears were observed in the side slope geomembrane liner system at or near the top of the slope after the earthquake at two different locations, Canyon C

Fig. 13.1 Tear observed in the Canyon D side slope geomembrane at Chiquita Canyon following the Northridge Earthquake (Photo courtesy of California Integrated Waste Management Board)

and Canyon D. Figure 13.1 shows the tear observed in the Canyon D liner system after the earthquake.

Augello (1997) reported that the displacement predicted in a conventional seismic performance assessment (i.e. a decoupled Newmark-type analysis) at both of these locations was less than 150 mm, bringing into question the adequacy of the conventional seismic performance criteria for geomembrane liner systems. EMCON (1994) noted that the tears in the Chiquita Canyon liner systems at both locations appear to have emanated from locations where a sample had been cut from the geomembrane for construction quality assurance (CQA) testing and a patch had been extrusion welded over the area where the sample had been taken. While not explicitly stated, this observation suggests that stress and/or strain concentrations associated with removal of samples and placement of patches for CQA provide a possible explanation as to why conventional seismic performance analyses fail to predict the tears in the geomembrane at the Chiquita Canyon landfill in the Northridge earthquake. Giroud (2005) presents strain concentration factors that account for the presence of geomembrane seams perpendicular to the direction of loading of a geomembrane (i.e. due to the presence of a patch) as well as for the presence of scratches and defects in the geomembrane. These factors indicate that the strain in a geomembrane can more than double in the vicinity of an extrusion-welded patch and that additional strain amplification can occur around scratches or gouges in the geomembrane. However, as geomembrane strains are not explicitly evaluated, the Giroud strain concentration factors do not influence current practice for the seismic performance of the geosynthetic elements of waste containment systems.

13.3 Performance-Based Analysis of Geosynthetic Liner System Elements

13.3.1 Methodology

Fowmes et al. (2006) and Fowmes (2007) developed a methodology for calculating actual axial forces and strains in waste containment geomembranes subject to relative displacement at the interface between the geomembrane and an adjacent material due to waste settlement. In their analyses, conducted using the large displacement finite difference formulation in the computer program FLAC™ (Itasca 2008), these investigators modeled the geomembrane as a linear elastic beam with zero moment of inertia (to allow for buckling) and modeled the interface using a linear elastic-perfectly plastic stress-stain model with a high shear stiffness and a yield strength described by the Mohr-Coulomb criterion (with parameters based upon direct shear interface shear strength test results). The tensile stiffness of the geomembrane was based upon manufacturer-cited stiffness for high density polyethylene (HDPE), the typical material used for waste containment geomembrane liner elements. The compressive stiffness of the geomembrane was assumed equal to tensile stiffness in the analyses conducted by these investigators.

Arab and Kavazanjian (2010) and Arab et al. (2010) describe extension of the elastic-perfectly plastic interface model of Fowmes et al. (2006) and Fowmes (2007) to include cyclic loading. Arab and Kavazanjian (2010) and Arab et al. (2010) demonstrated the ability of this interface model to describe relative displacement across an interface subject to cyclic loading by comparison of numerical analysis to the results of shaking table model tests of a sliding block on a plane. Figure 13.2 illustrates the numerical models used to analyze a block on a horizontal plane and a block on an inclined plane.

13.3.2 Slip at the Interface

Arab and Kavazanjian (2010) report on comparison of results from numerical analyses compared to shaking table tests of a geomembrane lined block on a horizontal plane lined with a geomembrane, tests conducted to evaluate the potential use of geomembranes for frictional base isolation. Values of bulk and shear modulus representative of structural steel were used to model both the rigid block and shaking table. The interface was assigned an elastic shear and normal stiffness

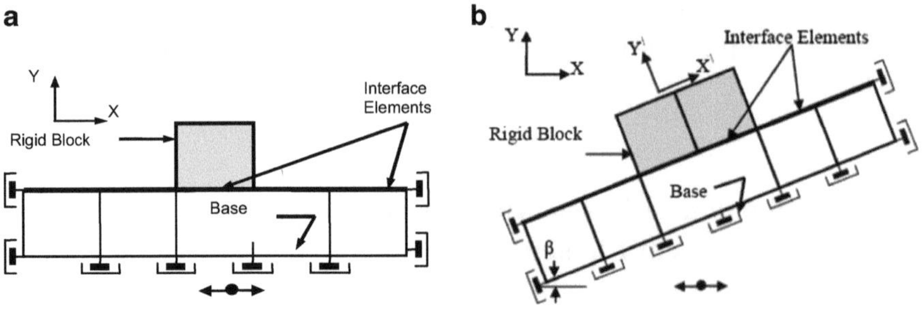

Fig. 13.2 Finite difference models of a block on a sliding plane. (**a**) Block on a horizontal plane. (**b**) Block on an inclined plane

approximately equal to ten times the bulk modulus of the mesh elements and a shear strength based upon tilt-table testing. Figure 13.3 presents the elasto-plastic interface stress-displacement model along with the results of the comparison for uniform sinusoidal loading from Kavazanjian et al. (1991). Arab and Kavazanjian (2010) also show a favorable comparison between numerical analysis and experimental results for a geomembrane lined (base isolated) block on a horizontal plane subject to earthquake loading from Yegian and Kadakal (2004).

Arab et al. (2012) report on results from numerical analyses for the block on an inclined plane model compared to shaking table test results from Elgamal et al. (1990) and from Wartman (1999) and Wartman et al. (2003). In the Elgamal et al. (1990) shaking table tests, the plane was coated with sandpaper, sand was glued to the base of the rigid block and the plane was inclined at an angle of 10° and subject to horizontal sinusoidal loading. Comparison between the block acceleration from the numerical analysis and the block acceleration measured in the Elgamal et al. (1990) tests for sinusoidal loading using a friction angle based upon the static friction coefficient of 0.36 reported by Elgamal et al. (1990) is presented in Fig. 13.4. Similar good agreement was shown between predicted and observed block displacement by Arab et al. (2010).

Wartman (1999) and Wartman et al. (2003) reported the results of shaking table tests of a rigid block on a plane inclined at 13.37° and subject to horizontal shaking. The interface between the block and the plane was a smooth high-density polyethylene (HDPE)/non-woven geotextile interface similar to one that might be found in a side-slope liner system for a landfill. Wartman (1999) and Wartman et al. (2003) used a suite of 22 uniform sinusoidal motions, three sinusoidal frequency sweep motions, and one earthquake-like input motion. For each test, Wartman et al. (2003) varied the interface friction angle in a Newmark-type displacement analysis until a calculated cumulative relative displacement approximately equal to the one observed in the shaking table test was achieved. Results of this analysis indicated that the interface friction angle depended upon the frequency of the input motion. This frequency dependence was interpreted by Wartman et al. (2003) as a dependence of interface friction angle on sliding velocity. Arab et al. (2010) repeated the Wartman et al. (2003) analysis using the numerical model. Figure 13.5 presents a

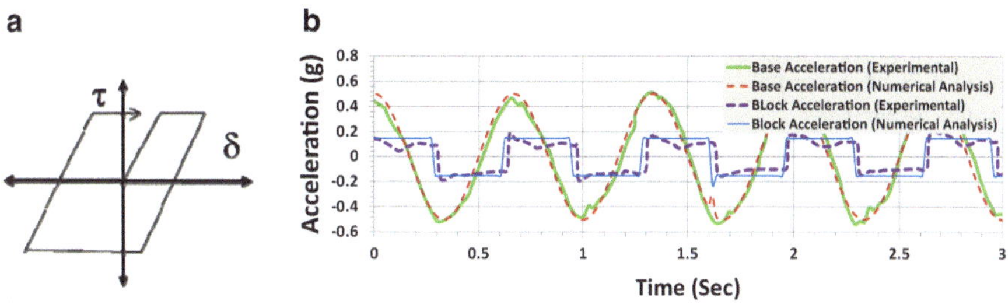

Fig. 13.3 Measured and calculated accelerations of a geomembrane lined black subject to uniform sinusoidal loading (Arab et al. 2012). (**a**) Interface model. (**b**) Comparison of numerical analysis with test results

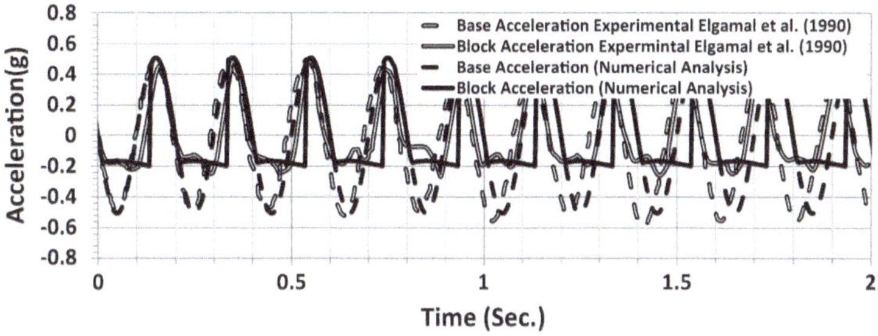

Fig. 13.4 Comparison of numerical analyses to the experimental results of Elgamal et al. (1990) for a block on an inclined plane (Arab et al. 2010)

Fig. 13.5 Comparison of back-calculated friction angles for a sliding block on a plane from Wartman et al. (2003) and the numerical model (Arab et al. 2010)

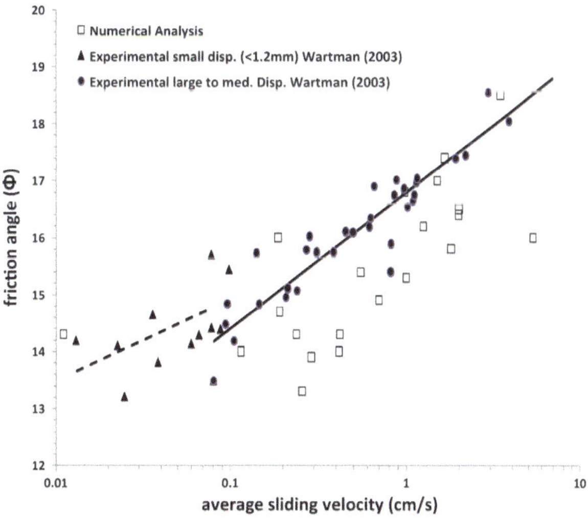

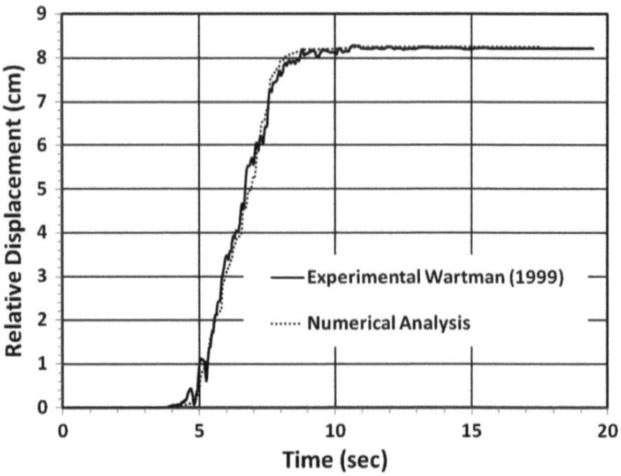

Fig. 13.6 Comparison of results from the numerical analysis to experimental results for the Kobe earthquake motion from Wartman (1999) (Arab et al. 2010)

comparison the results of the Arab et al. (2010) analysis to the Wartman et al. (2003) analysis.

Figure 13.6 presents a comparison of the relative displacement of the rigid block as measured in the shaking table test reported by Wartman (1999) for the Kobe earthquake acceleration input and as calculated by Arab et al. (2010) using a best-fit interface friction angle. The relative displacement time history from the numerical analysis closely matches the relative displacement time history observed in the laboratory test, including episodes of upslope relative displacement at around 5 s. The results shown in Figs. 13.2, 13.3, 13.4, 13.5 and 13.6 indicate that the simple elasto-plastic interface model can accurately predict the impact of slip at the interface on the response and cumulative deformation of a sliding mass above the interface provided the interface strength is accurately characterized.

13.3.3 Enhanced Constitutive Models

To facilitate accurate prediction of the axial strains and forces in geosynthetic liner system elements as well as the impact of these elements on seismic response, a series of enhanced constitutive models were developed for liner system elements and interfaces. These advanced models included a non-linear elastic stress-strain model for the geomembrane beam element and models for interface behavior and for the internal in-plane shear behavior of a hydrated geosynthetic clay liner (GCL). The interface and GCL internal shear behavior models both account for potential post-peak degradation of shear strength.

The non-linear elastic constitutive model for the stress-strain behavior of geomembranes was developed based upon the work of Giroud (1994, 2005)). Giroud (2005) showed that an N-order parabola, with $N = 4$, provided a good approximation of a unique normalized stress–strain curve for HDPE geomembranes. Figure 13.7 compares the curve described by Eq. 13.1 to the normalized curve for HDPE

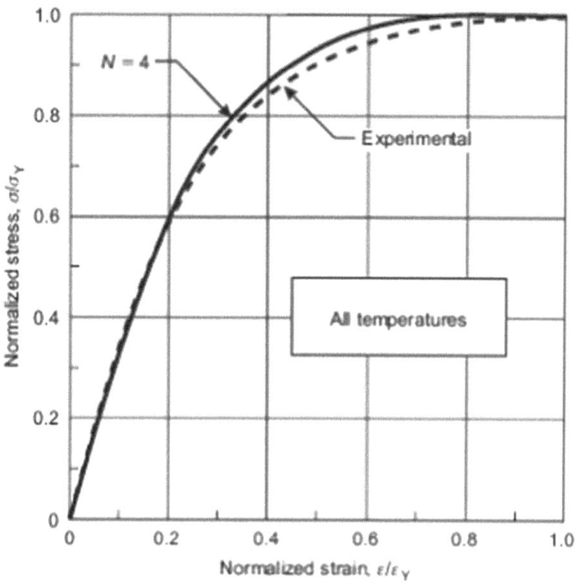

Fig. 13.7 Normalized uniaxial stress– strain curve for HDPE geomembrane (Giroud 2005)

geomembranes developed by Giroud (1994) from analysis of over 500 uniaxial tension tests on HDPE.

The equation for the tangent moduli, E_{tan}, of the geomembrane for any strain below the yield strain is given by Giroud (2005) as:

$$E_{\tan} = \frac{N\sigma_y}{\varepsilon_y}\left(1 - \frac{\varepsilon}{\varepsilon_y}\right)^{n-1} \tag{13.1}$$

Using Eq. 13.1, only the stress and strain at the yield peak need be known to develop the uniaxial stress–strain curve of the geomembrane between the origin and the yield peak. However, Eqs. 13.1 and 13.2 are based upon uniaxial tests. In the field, geomembranes are generally subjected to biaxial or triaxial stress states. For the case of the axial strain in a plane strain biaxial stress states, Giroud (2005) conducted an analysis based on energy conservation that led to the following relationships between the uniaxial yield strain, ε_y, and the plane strain biaxial yield strain, ε_{yps}, in the geomembrane as a function of the Poisson's ratio, ν, as follows:

$$\varepsilon_{yps} = \varepsilon_y \frac{(1 - \nu^2)}{\sqrt{1 - \upsilon + \nu^2}} \tag{13.2}$$

Equations 13.1 and 13.2 for the tangent moduli of an HDPE geomembrane subject to plane strain loading were coded into a FLAC^{TM} subroutine to describe the non-linear elastic stress-strain behavior of a geomembrane beam element subject to tensile loading. For compressive loading, the beam element was assumed

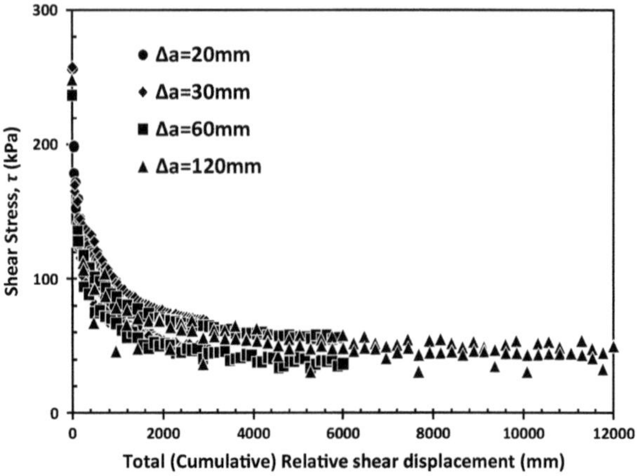

Fig. 13.8 Mobilized shear strength vs. cumulative relative shear displacement for uniform amplitude, Δa, cyclic shear tests at 692 kPa normal stress (Arab et al. 2011a, b)

to behave in an elastic manner with modulus equal to the initial tangent plane strain modulus of the HPDE geomembrane in tension. As recommended by Fowmes et al. (2006) and Fowmes (2007), geomembrane beam elements were always assigned a zero moment of inertia to account for the potential for buckling.

Arab et al. (2011a, 2012) describe the development of a displacement-softening model for geosynthetic interfaces. The model assumes that a unique relationship exists between the interface friction angle and cumulative relative shear displacement, as suggested by Esterhuizen et al. (2001) for static deformations. This assumption is consistent with the data in Fig. 13.8 for uniform cyclic loading of a textured geomembrane/ nonwoven geotextile interface from test conducted by Ross et al. (2010).

The generalized shear strength-displacement relationship for the mobilized interface friction angle, ϕ, assumed in the model is illustrated in Fig. 13.9a. Under cyclic shear loading, the model initially behaves elastically until the cumulative displacement exceeds the displacement at the peak friction angle ($\delta^{e}{}_{p}$), at which point cumulative plastic shear displacements will start to accumulate. Once plastic shear displacements begin to accumulate, the mobilized friction angle (or shear strength) will follow Fig. 13.9a until unloading begins. Plastic shear displacements begin to accumulate in the reverse direction when the shear stress exceeds the mobilized shear strength from any prior loading cycle. The hysteretic cyclic shear stress-displacement behavior that develops based these assumptions are illustrated in Fig. 13.9b. Key parameters in this relationship include the initial elastic stiffness, E_i, plastic cumulative relative shear displacement (δ_p), the peak friction angle (ϕ_p), the large displacement friction angle (ϕ_r), the cumulative displacement at the peak friction angle ($\delta^{e}{}_{p}$) and cumulative displacement at the large displacement friction angle ($\delta^{p}{}_{r}$).

Figure 13.10 compares the hysteresis loops generated by this model to the results of a uniform cyclic direct shear test on a textured geomembrane/nonwoven geotextile interface reported by Ross et al. (2010). In Fig. 13.10a, the hysteretic cyclic shear behavior calculated using the model is compared to the results of a

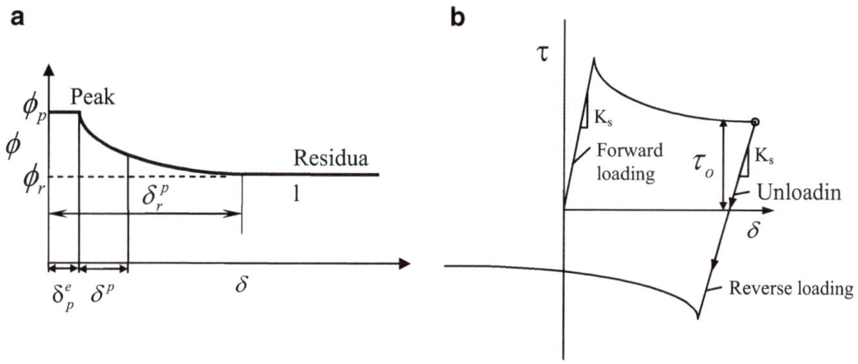

Fig. 13.9 Displacement softening model for geosynthetic interfaces (Arab et al. 2011a, b). (**a**) Mobilized shear strength relationship. (**b**) Hysteretic stress-displacement relationship

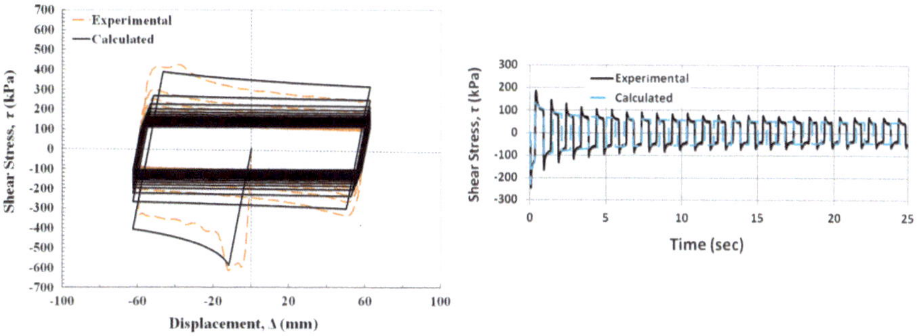

Fig. 13.10 Comparison of numerical results to uniform cyclic direct shear tests by Ross et al. (2010) for ±60 mm displacement amplitude and 2,071 kPa normal stress (Arab et al. 2011a, b). (**a**) Cyclic stress-displacement behavior. (**b**) Shear stress time history

displacement controlled uniform cyclic direct shear test with an amplitude of 60 mm at a normal stress of 2,071 kPa. The general trend of the cyclic shear stress time history is captured well by the numerical model results presented in Fig. 13.10a. Figure 13.10b shows the excellent agreement between the model prediction and the observed shear stress time history for this displacement-controlled test.

Arab et al. (2012) describe the development of a constitutive model for the internal in-plane stress-strain behavior of a hydrated GCL subject to cyclic loading. To capture the unusual shape of the back bone curve and the pre-peak hysteresis loops and the post-peak softening observed in uniform cyclic direct shear testing of a hydrated needlepunched GCL with two non-woven carrier geotextiles reported by Nye (2006) and Sura (2009), the multiple yield surface kinematic hardening isotropic softening plasticity model developed by Salah-Mars and Kavazanjian (1992) was employed. Multiple yield surface (MYS) kinematic hardening plasticity was first developed by Iwan (1967) and Mroz (1967). MYS constitutive models use

Fig. 13.11 Nested circular yield surfaces (Prevost 1978)

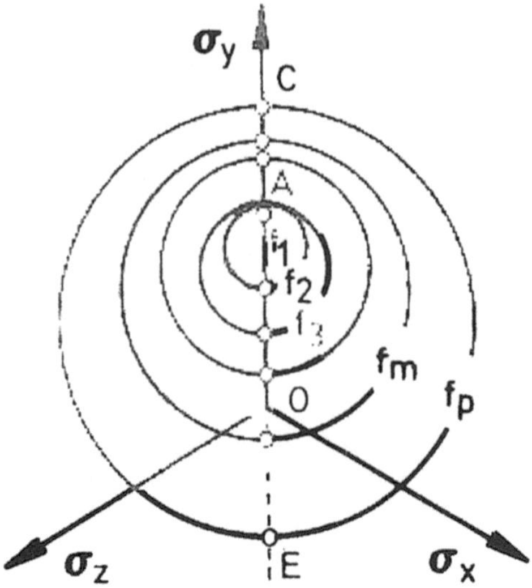

nested circular yield surfaces combined with a non-linear backbone curve to produce hysteresis loops that obey the Masing (1926) criterion for cyclic loading. Prevost (1978, 1985) popularized this type of model for use in soil mechanics. Figure 13.11 illustrates conceptually the MYS model with nested circular yield surfaces developed by Prevost (1978) for soils.

Salah-Mars and Kavazanjian (1992) extended MYS plasticity theory to include post-peak softening behavior by adding an outermost isotropic softening yield surface and an inner final perfectly plastic yield surface. The Salah-Mars and Kavazanjian (1992) constitutive model has two main characteristics that make it suitable to reproduce the GCL hysteresis loops observed in the laboratory testing program. First, because the model belongs to the MYS family, it can employ a piecewise linear model to approximate a backbone curve of any arbitrary shape. Second, the model has the capability to exhibit softening after the material reaches its peak strength. These two characteristics make the model well-suited for modeling the observed GCL in-plane cyclic shear behavior.

The Drucker-Prager yield function was employed by Salah-Mars and Kavazanjian (1992) in their MYS model. This yield function can be represented by a straight line in principal stress space and the outermost surface can be established using the Mohr-Coulomb peak shear strength parameters. The final inner yield surface, representing the large displacement strength of the material, is established using the large displacement shear strength. The associative flow rule is employed with this MYS formulation. Thus the plastic potential surfaces coincide with the yield surfaces. The MYS model behaves in a linear elastic manner until the innermost yield surface is engaged. As the material is assumed to be an isotropic material, two parameters are required to describe the elastic behavior, e.g. shear modulus, G, and the bulk modulus, B, or Young's modulus, E, and Poisson ratio, ν. Soil nonlinearity is represented by a nonlinear stress-strain backbone curve.

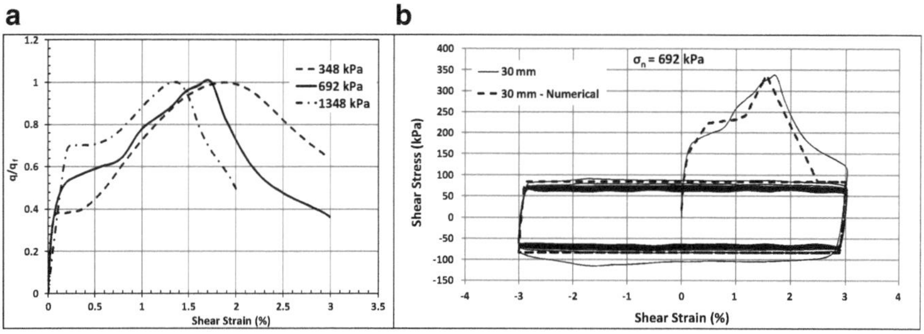

Fig. 13.12 Comparison of behavior of hydrated needlepunched GCL tested by Nye (2006) and Ross (2009) to model predictions (Arab et al. 2012). (**a**) Hydrated GCL backbone curves. (**b**) Hydrated GCL hysteretic stress-strain behavior

In the MYS model, this backbone curve is modeled by a piecewise linear approximation. The end point of each segment of this piecewise linear approximation represents a yield surface, f_i, which is characterized by an elasto-plastic modulus, H_i, and size, l_i, where $i = 1, 2\ldots$ *nys* and *nys* is the number of yield surfaces employed in the model. The maximum number of yield surfaces is only limited by the numerical implementation of the model. An increase of the number of surfaces may lead to more accuracy but will also lead to an increase in computing time.

Figure 13.12a presents the normalized backbone curves developed from the tests conducted by Nye (2006) and Sura (2009) that were used to develop the parameters for the MYS model. Figure 13.12b shows the excellent agreement between the observed hysteretic behavior of the hydrated GCL tested by Nye (2006) and Sura (2009) and the behavior predicted by the MYS model.

13.3.4 Axial Strains and Forces in Geosynthetic Elements

To evaluate the ability of the numerical model for geosynthetic liner system elements to calculate the axial strains and forces in the geosynthetic elements, a series of parametric studies were conducted of Arab (2011) and Wu (2012) of landfill side slope liner systems. Figure 13.13a presents the tensile strains induced in the geomembrane liner on a 35 m-tall landfill side slope with a 12 m-tall 1H:1V (horizontal:vertical) side slopes separated by 3.6 m-wide benches subject to settlement equal to approximately 20 % of the total waste thickness. In the analyses, the waste was placed in 6 m lifts, the geomembrane was anchored at crest of each slope segment, and the interface friction angle on top of the geomembrane was 5° greater than the interface friction angle on the bottom of the geomembrane. The parametric study showed that the axial strains and forces induced in the side slope liner were proportional to the difference between the upper and lower interface friction angle, provided the interface friction angle on top of the geomembrane was less than the

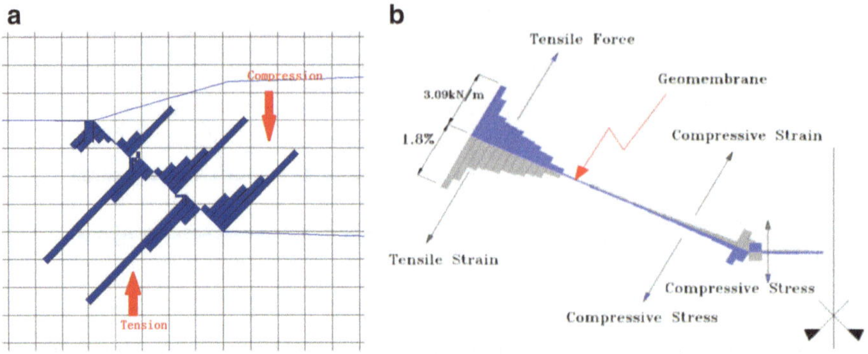

Fig. 13.13 Results of numerical analyses for axial strains induced in side slope geomembranes. (**a**) Strains induced by waste settlement (Wu 2012). (**b**) Strains induced by seismic loading (Arab 2011)

interface friction angle on the bottom of the geomembrane. Furthermore, the axial strains and forces in the analyses decreased as the side slope inclination decreased from 1H:1V to 3H:1V, the tensile strain and force at the top of each slope segment, where the tensile strains and forces were the greatest, depended upon the anchorage of the geomembrane at that point, and there was no tension induced in the geomembrane when the upper interface friction angle was less than the lower interface friction angle. The pattern and magnitude of tensile strains induced in the geomembrane in these analyses are consistent with the numerical results of Fowmes et al. (2006) and Fowmes (2007) and field data from Yazdani et al. (1995) and Gourc et al. (2004). Figure 13.13b presents the axial forces and strains induced in a side slope liner with an upper interface shear strength greater than the lower interface shear strength by seismic loading. Again, the pattern of axial strain and forces is consistent with what was expected, with tension at the top of the slope, compression at the bottom of the slope, and minimal strain in the mid-section of the slope.

13.4 Heap Leach Pad Seismic Response

Following demonstration of the ability of the interface model capture geomembrane behavior, Kavazanjian et al. (2011) used the numerical model to analyze the seismic response of a geomembrane in a heap leach pad. In this analysis, the geomembrane was modeled with a constant elastic modulus. The heap leach pad model, illustrated in Fig. 13.14, had a 3 % geomembrane-lined base slope and 2H:1V side slopes for the ore pile. The heap ore was modeled using material properties for coarse gravel. The three cases illustrated in Fig. 13.14 were analyzed using the numerical model. In Case I, the impact of slip at the liner interface was modeled using a single interface between the foundation and the ore pile but with

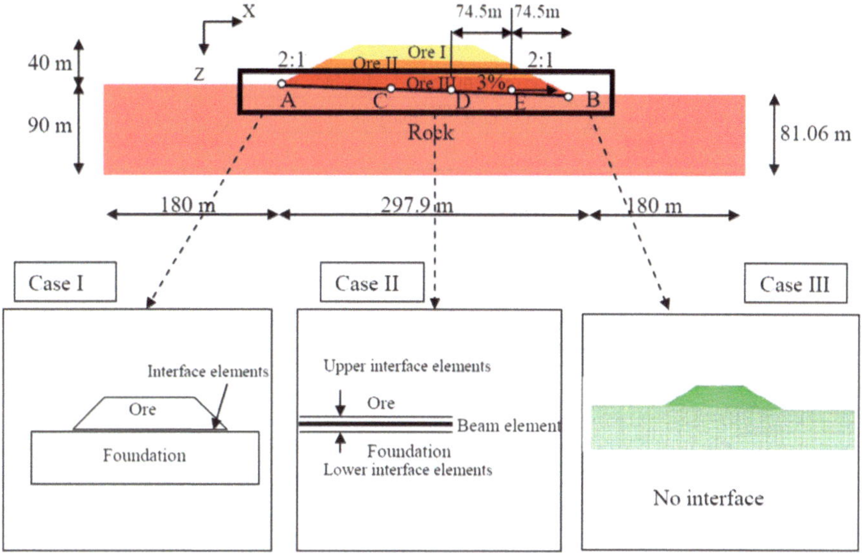

Fig. 13.14 Heap leach pad numerical model (Kavazanjian et al. 2011)

no beam element representing the liner itself. In Case II, the liner was modeled as a beam element with two interfaces (one on each side) connecting the beam (liner) to the underlying foundation and the overlying ore pile. In Case III, neither interfaces nor a beam element were employed in order to provide a basis for evaluating the impact of the liner system models used in Cases I and II on seismic response. The geomembrane was assigned the elastic modulus at 50 % yield stress reported by a manufacturer for a typical HDPE geomembrane (based upon unconfined wide-width tensile loading) and was assumed to be the same in compression as in tension. Results for Cases I and II were essentially identical when the upper interface in Case II had the same properties as the interface in Case I and was weaker than the lower interface. Results from Cases I and II were then compared to Case III results to establish the impact of slip at the liner interface on seismic response and to calculate the forces and strains induced by seismic loading in the geomembrane.

Results from the numerical analysis in terms of strains in the geomembrane and permanent seismic displacement from Cases I and II were also compared to the results from a state-of-the practice analysis of seismic response and geomembrane performance, i.e. to the results of a decoupled analysis using one dimensional equivalent linear seismic response analysis, Newmark displacement analysis, and the yield acceleration from a limit equilibrium analysis.

Permanent deformations calculated in numerical analysis at four different locations (locations labeled A, B, C and D in Fig. 13.14) are presented in Table 13.1 along with the results of the Newmark displacement analysis and the maximum strain in the geomembrane from the numerical analysis for six different earthquake records. The Newmark displacement analyses indicated unacceptable seismic

Table 13.1 Calculated permanent deformation and maximum tensile strains in the heap leach pad geomembrane liner (Kavazanjian et al. 2011)

		Permanent seismic displacement (mm)					
		2-D FLAC^{TM} analysis					
Earthquake record	PGA (g)	A	B	D	C	Newmark analysis	Tensile strain (%)
Coalinga (1983)	0.60	−1,050	1,670	160	15	640.0	2.2
Loma Prieta (1989)	0.60	−1,089	1,540	193	47	1,253.2	3.0
Imperial Valley (1979)	0.45	−976	1,200	135	32	346.0	2.4
Kobe (1995)	0.45	−775	1,210	40	2	255.3	2.2
Landers (1992)	0.25	−1,300	2,070	37	0	464.0	12.8
Tabas (1978)	0.25	−137	230	0	0	9.4	0.7

performance in 4 or 5 of the cases (depending on whether the limiting displacement is 150 or 300 mm) and the displacements calculated by numerical analysis at points A to B also exceeded the allowable displacement limit for all records except the Tabas record. However, except for the Landers record, the maximum tensile strain in the geomembrane (typically at a point in between points D and B) was well below the value at yield, indicating acceptable performance.

Arab et al. (2011b) also applied the numerical model to analysis of the performance of typical canyon and gravel pit landfill configurations. These analyses yielded similar results to those reported by Kavazanjian et al. (2011) for the heap leach pad liner system. In several cases where the conventional decoupled approach to seismic design gave Newmark displacements significantly in excess of 300 mm, the numerical model suggested that geomembrane strains and axial forces remained well within acceptable limits. Arab et al. (2011b) also showed that the interface model gave results comparable to those of Rathje and Bray (2000) with respect to the effect of slip at the interface on seismic response at the top of the landfill (with the added benefit of predicting strains and forces in the geomembrane liner).

13.5 The Chiquita Canyon Landfill Case History

Arab (2011) used the numerical model to analyze the behavior of the Chiquita Canyon landfill in the 1994 Northridge earthquake. As noted previously, despite the fact that the geomembrane liner for the landfill was observed to be torn in two places following the earthquake, conventional decoupled liner performance analyses using time histories recorded near the landfill during the earthquake as input suggested that liner performance should have been satisfactory, i.e. calculated permanent seismic displacements were less than 150 mm at both locations (Augello 1997).

The two finite difference meshes employed by Arab (2011) to back analyze the seismic performance of the Chiquita Canyon landfill liner system are shown in

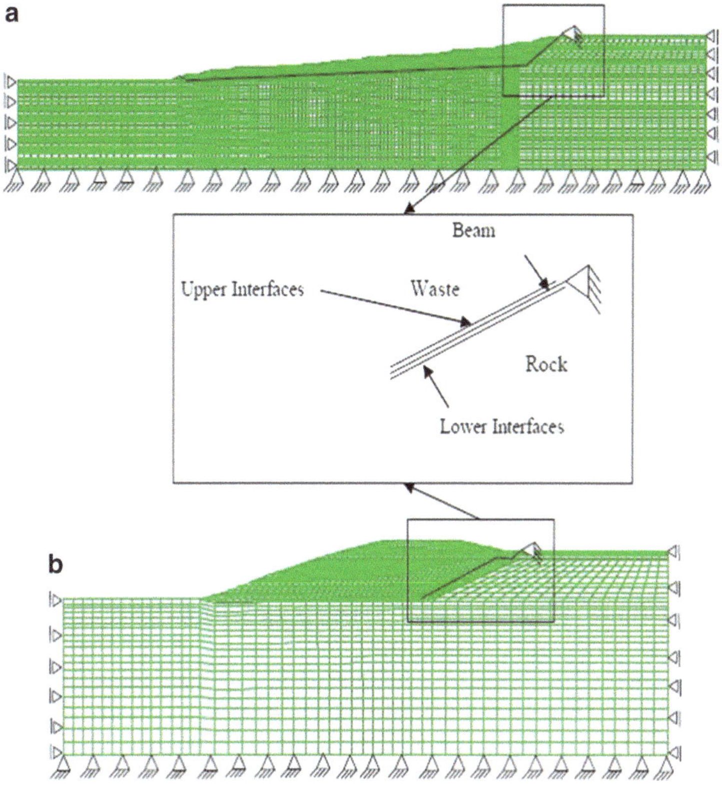

Fig. 13.15 Finite difference meshes for Chiquita canyon landfill cross sections: (**a**) cross section C1-C1 (**b**) cross section D1-D1 (Arab 2011)

Fig. 13.15. In both cases the geomembrane was modeled as a non-linear elastic beam element with interface elements on both sides. Geomembrane properties and interface strength properties were based upon results of the forensic analysis conducted by EMCON (1994) after the earthquake. Table 13.2 presents the interface strength and stiffness for the Canyon B and Canyon C liner systems (the two locations where tears were observed) at the Chiquita Canyon landfill from interface direct shear tests reported by EMCON (1994).

Two strong motion records captured in the vicinity of the landfill, scaled to the peak ground acceleration predicted for the site using state-of-the-art Next Generation Attenuation (NGA) relationships, were employed as input to the analysis. In the numerical analyses, the waste was placed in 6 m lifts and then, after reaching the waste elevation at the time of the Northridge earthquake, subjected to each of the strong ground motions records. During placement of the waste lifts waste behavior was modeled using the modified Cam-Clay constitutive model using properties based upon back analysis of the static behavior of the Operating Industries, Inc. (OII) landfill conducted by Geosyntec Consultants (1996). During seismic loading

Table 13.2 Chiquita canyon interface shear test results from EMCON Associates (1994)

Sample ID	Sample location	Peak friction angle (degree)	Large displacement friction angle (degree)	Initial interface shear stiffness, E_i (Pa/m)
CA	Above Canyon C geomembrane	27	24	8e7
CB	Below Canyon C geomembrane	24	22	5.5e7
DA	Above Canyon D geomembrane	26	24	4.7e7
DB	Below Canyon D geomembrane	29	28	4.5e7

waste behavior was modeled using equivalent linear properties reported by Matasovic and Kavazanjian (1998) based upon back analysis of the seismic response of the OII landfill. Complete details of the analysis are provided by Arab (2011).

The maximum tensile strains in the geomembrane from the numerical analysis were 4.3 and 2.2 % for cross section C1-C1 and cross section D1-D1, respectively. In both cases, the maximum tensile strain was at the geomembrane anchor point at the top of the slope. In both cases the calculated maximum tensile strain was well below the yield strain of 11 % calculated for the geomembrane after adjusting the values reported by EMCON (1994) for strain rate and plane strain conditions.

Despite varying the input parameters for the analysis over a wide range of values, geomembrane strains for the two Chiquita Canyon landfill cross sections never approached the reported yield strain of the geomembrane. The only way the tears in the geomembrane liner system could be accounted for in the analysis was by invoking the strain concentration factors developed by Giroud (2005) for geomembrane irregularities (e.g. seams, patches, scratches). Strain concentrations factors for both the increase in thickness of a geomembrane at a seam perpendicular to the direction of loading and for a scratch in the geomembrane had to be invoked in order to account for the observed tears in the geomembrane, a result consistent with the observation that the tears at both locations appeared to emanate from locations patches had been welded over areas where samples of the seam had been removed for testing during construction quality assurance testing (EMCON 1994). Figure 13.16 shows the strain concentration factor for a scratch in a geomembrane (the more significant of the two sources of strain concentration) from Giroud (2005).

The strain concentration factor is Fig. 13.16a is a reduction factor applied to the calculated geomembrane strain as a function of the depth of the scratch in the geomembrane. Even a modest scratch with a depth on the order of 4 % of the thickness of the geomembrane (i.e. on the order of 0.08 mm for a 2 mm geomembrane) results in a 40 % reduction in the yield strain of the geomembrane. The strain concentration factor in Fig. 13.16b is an incremental strain due to bending induced by the overlap of geomembrane sections at a seam that gets added to the calculated geomembrane strain

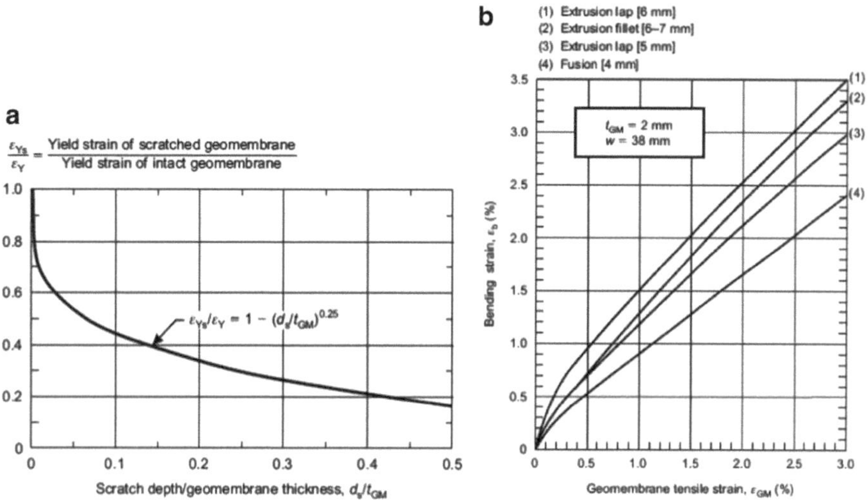

Fig. 13.16 Strain concentration factors for geomembranes (Giroud 2005). (**a**) Strain concentration factors for scratches. (**b**) Strain concentration factors for seams

Fig. 13.17 Crack through a doubly seamed region consisting of a fillet extrusion seam on top of a flat extrusion seam (EMCON 1994)

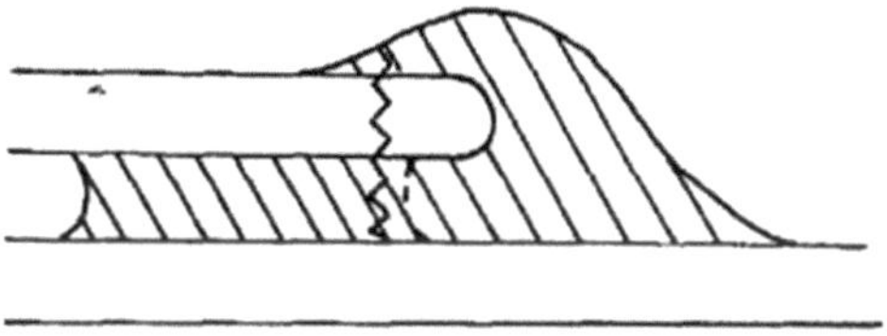

perpendicular to the seam. This strain concentration factor more than doubles the strain in a geomembrane with a 6–7 mm extrusion fillet weld.

EMCON (1994) reported that the geomembrane tears at the Chiquita Canyon landfill appear to have emanated from extrusion welded patches near the crest of the slope (where the strain induced in the geomembrane by the earthquake was the greatest). The oxidized surface layer on the geomembrane was removed by grinding before applying these extrusion welded patches, a process likely to have caused at least modest scratches in the geomembrane, reducing the yield strain by at least 40 %. Furthermore, EMCON (1994) reported that the geomembrane fractured in the double seamed region of a patch extrusion weld, as illustrated in Fig. 13.17. The Giroud (2005) seam strain concentration factors could result in tripling of the geomembrane strain in such a double seamed region. Thus, considering the Giroud (2005) strain concentration factors, the strains predicted in the numerical analysis are sufficient to explain the failure of the geomembrane at Chiquita Canyon landfill when subjected to the Northridge earthquake ground motions.

13.6 Conclusion

A performance based model for seismic analysis and design of the geosynthetic elements of waste containment systems has been developed. The model allows for explicit calculation of axial strains and forces in the geosynthetic elements of the liner system. The model was developed within the framework of a large strain finite difference formulation for solving boundary value problems in geotechnical engineering. The model employs beam elements with zero moment of inertia and with interface elements on both sides of the beam to represent the geosynthetic elements of the containment system. The stress-strain behavior of HDPE geomembranes is modeled using the hyperbolic non-linear elastic stress-strain relationship developed by Giroud (2005). Interface shear stress-shear displacement relationships available for use with the model include a simple linear elastic-perfectly plastic relationship and displacement softening relationship in which the mobilized shear resistance degrades from a peak strength to a large displacement strength as a function of the cumulative plastic shear deformation. The model also includes a kinematic hardening, isotropic softening multiple yield surface model for the internal shear deformation of hydrated GCLs.

The model has been shown to accurately predict the behavior of a rigid block on horizontal and inclined planes subject to uniform cyclic loading and earthquake strong ground motion time histories. Both block acceleration and block displacement area accurately predicted provided the correct interface shear strength is employed in the model. The model is also shown to give reasonable results for the axial forces and strains in side slope liner system geomembranes subject to waste settlement and seismic loading. The model is also shown to predict the tears observed in the liner system at the Chiquita Canyon landfill following the Northridge earthquake once the strain concentration factors developed by Giroud (2005) for geomembrane seams and scratches in the geomembrane are taken into account.

Important conclusions drawn from analyses conducted using the model include:

- Calculated permanent displacements from a decoupled Newmark analysis are not a reliable index of liner system seismic performance;
- When the interface shear strength on the top of a geosynthetic liner element is less than the interface shear strength on the bottom of the element, the element can sustain very large relative displacements without inducing significant tension in the element;
- When the interface shear strength on the bottom of a geosynthetic liner element is less than the interface shear strength on the top of the element, tension proportional to the difference in the top and bottom interface shear strengths is induced in the element;
- Tensile strains induced in liner system elements are greatest at the anchorage points for the element, e.g. at the crest of a slope or the edge of a base liner; and
- Strain concentration factors due seams and scratches in the geomembrane must be considered in evaluating the allowable strain in a geomembrane.

The model presented in this paper provides a rational basis for seismic analysis and design of the geosynthetic elements of containment systems for landfills, heap leach pads, and other engineered systems subject to earthquake shaking. However, additional research is needed on the influence of a variety of model parameters on liner system performance, including the normal and shear stiffness of liner system interfaces and the tension and compression modulus of geomembranes under confinement.

Acknowledgements The work described in this paper is part of a joint Arizona State University/ University of California at San Diego/Geosyntec Consultants/CETCO, Inc. research program titled GOALI: Collaborative Research: The Integrity of Geosynthetic Elements of Waste Containment Barrier Systems Subject to Large Settlements or Seismic Loading. This project is funded by the Geomechanics and Geotechnical Systems, Geoenvironmental Engineering, and Geohazards Mitigation program of the National Science Foundation (NSF) Division of Civil, Mechanical, and Manufacturing Innovation under grant numbers CMMI-0800873 and CMMI-0800030. The authors are grateful for this support. Opinions and positions expressed in this paper are the opinions and positions of the authors only and do not reflect the opinions and positions of the NSF.

References

Arab GA (2011) The integrity of geosynthetic elements of waste containment barrier systems subject to seismic loading. PhD thesis, Arizona State University, Tempe

Arab MG, Kavazanjian E Jr (2010) Time-domain analysis of frictional base isolation using geosynthetics. In: Proceedings of 9th international conference on geosynthetics, vol 2. IGS-Brazil, pp 695–698

Arab MG, Kavazanjian E Jr, Matasovic N (2010) Non-linear time domain analysis of a sliding block on a plane. In: Proceedings of 5th international conference on recent advances in geotechnical earthquake engineering and soil dynamics, San Diego, May, Missouri University of Science and Technology Press, on CD ROM, Paper no 4.08b, 12pp

Arab MG, Kavazanjian E Jr, Fox PJ, Ross JD (2011a) Displacement-softening constitutive model for geosynthetic interfaces. In: Proceedings of 14th Pan American conference on soil mechanics and geotechnical engineering, Canadian Geotechnical Society, on CD ROM, 7 pp

Arab MG, Kavazanjian E Jr, Matasovic N (2011b) Seismic analysis of a geosynthetic liner system. ASCE Geotech Spec Publ 211:1981–1990

Arab MG, Kavazanjian E Jr, Fox PJ, Matasovic N (2012) In plane-behavior of geosynthetic barrier layers subject to cyclic loading. In: Proceedings of 2nd international conference on performance based design in earthquake geotechnical engineering, Taormina, Sicily, Paper no 3.11, May, University of Catania, 12pp (on CD ROM)

Augello AJ (1997) Seismic response of solid-waste landfills. PhD dissertation, Department of Civil and Environmental Engineering, University of California, Berkeley, p 345

Augello AJ, Matasovic N, Bray JD, Kavazanjian E Jr, Seed RB (1995) Evaluation of solid waste landfill performance during the Northridge Earthquake. Earthquake Design and Performance of Solid Waste Landfills, ASCE Geotechnical Special Publication no 54: 17-50

Bray JD, Rathje EM (1998) Earthquake-induced displacements of solid-waste landfills. J Geotech Geoenviron Eng ASCE 124(3):242–253

Bray JD, Rathje EM, Augello AJ, Merry SM (1998) Simplified seismic design procedure for geosynthetic-lined, solid-waste landfills. Geosynth Int 5(1–2):203–235

Elgamal AWM, Scott RF, Succarieh MF, Yan L (1990) La Villita dam response during five earthquakes including permanent deformation. J Geotech Eng ASCE 116(10):1443–1462

EMCON (1994) Northridge earthquake seismic evaluation, Chiquita Canyon landfill. EMCON Associates, San Jose, CA, April

Esterhuizen JJB, Filz GM, Duncan JM (2001) Constitutive behavior of geosynthetic interfaces. J Geotech Geoenviron Eng ASCE 127:834–840

Fowmes GJ (2007) Analysis of steep sided landfill lining systems. Eng.D. dissertations, Department of Civil and Building Engineering, Loughborough University, Loughborough, UK

Fowmes GJ, Dixon N, Jones DRV (2006) Modelling of lining system integrity. In: Proceedings of 8th international conference on geosynthetics, Yokohama, Japan, pp 207–212

Giroud JP (1994) Mathematical model of geomembrane stress-strain curves with a yield peak. Geotext Geomembr 13(1):1–22

Giroud JP (2005) Quantification of geosynthetic behavior. Geosynth Int, Special Issue on the Giroud Lectures 12(1): 2–27

Gourc JP, Reyes-Ramirez R, Villand P (2004) Assessment of geosynthetics interface friction for slope barriers of landfills. In: Proceedings of 3rd Asian regional conference on geosynthetics, Seoul, Korea, June, pp 116–149

Itasca Consulting Group, Inc (2008) FLAC – Fast Lagrangian Analysis of Continua, user's manual, www.itasca.com

Iwan WD (1967) On a class of models for the yielding behavior of continuous and composite systems. J Appl Mech 34:612–617

Kavazanjian E Jr (1998) Current issues in seismic design of geosynthetic cover systems. In: Proceedings of sixth international conference on geosynthetics, vol I. Atlanta, Georgia, pp 219–226

Kavazanjian E Jr (1999) Seismic design of solid waste containment facilities. In: Proceedings of eighth Canadian conference on earthquake engineering, Vancouver, BC, 13–16 June pp 51–68

Kavazanjian E Jr, Hushmand B, Martin GR (1991) Frictional base isolation using a layered soil-synthetic liner system. In: Proceedings of 3rd U.S. conference on lifeline earthquake engineering, ASCE Technical Council on Lifeline Earthquake Engineering Monograph 4: 1139–1151

Kavazanjian E Jr, Matasovic N, Caldwell J (1998) Damage criteria for solid waste landfills. In: Proceedings of 6th U.S. national conference on earthquake engineering (on CD ROM), Seattle, WA, 31 May – 4 June

Kavazanjian E Jr, Arab MG, Matasovic N (2011) Seismic analysis of heap leach pad liner systems. In: Proceedings of 5th international conference on earthquake geotechnical engineering, Santiago, Chile, Paper no SEIKA, on CD ROM, 12pp

Masing G (1926) Eigenspannungen and verfertigung beim messing. In: Proceedings of 2nd International, Congress on Applied Mechanics, Zurich

Matasovic N, Kavazanjian E Jr (1998) Cyclic characterization of OII landfill solid waste. J Geotech Geoenviron Eng ASCE 124(3):197–210

Matasovic N, Kavazanjian E Jr (2006) Seismic response of a composite landfill cover. J Geotech Geoenviron Eng ASCE 132(4):448–455

Matasovic N, Kavazanjian E Jr, Augello AJ, Bray JD, Seed RB (1995) Solid waste landfill damage caused by 17 January 1994 Northridge Earthquake. In: Woods MC, Seiple RW (eds) The Northridge, California, Earthquake of 17 January 1994, vol 116, Division of Mines and Geology Special Publication. California Department of Conservation, Sacramento, pp 221–229

Matasovic N, Kavazanjian E Jr, Giroud JP (1998) Newmark seismic deformation analysis for geosynthetic covers. Geosynth Int 5(1 – 2):237–264

Mroz Z (1967) On the description of anisotropic work hardening. J Mech Phys Solids 15:163–175

Newmark NM (1965) Effects of earthquakes on dams and embankments. Geotechnique 15:139–160

Nye CJ (2006) Research on the dynamic internal shear behavior of a needle-punched geosynthetic clay liner. MS thesis, The Ohio State University, Columbus

Prevost JH (1978) Plasticity theory for soil stress-strain behavior. J Eng Mech Div ASCE 104:1177–1194

Prevost JH (1985) A simple plasticity theory for frictional cohesionless soils. Soil Dyn Earthq Eng 4(1):9–17

Rathje EM, Bray JD (2000) Nonlinear coupled seismic sliding analysis of earth structures. J Geotech Geoenviron Eng ASCE 126(11):1002–1014

Ross JD (2009) Static and dynamic shear strength of a geomembrane/geosynthetic clay liner interface. MS thesis, The Ohio State University, Columbus

Ross JD, Fox PJ, Olsta JT (2010) Dynamic shear testing of a geomembrane/geosynthetic clay liner interface. In: Proceedings of 9th international conference on geosynthetics, vol 2. IGS-Brazil, pp 649–652

Salah-Mars YE, Kavazanjian E Jr (1992) A virutal surface concept for nested yield surface plasticity. Int J Numer Methods Geomech 16:779–796

Seed RB, Bonaparte R (1992) Seismic analysis and design of lined waste fills: current practice. In: Proceedings of conference on stability and performance of slopes and embankments – II, ASCE Special Geotechnical Publication no 31, pp 1521–1545

Sura JM (2009) Monotonic and cyclic shear response of a needle punched geosynthetic clay liner at high normal stresses. MS thesis, The Ohio State University, Columbus

Wartman J (1999) Physical model studies of seismically induced deformation in slopes. PhD dissertation, Department of Civil Engineering, University of California, Berkeley

Wartman J, Bray JD, Seed RB (2003) Inclined plane studies of the Newmark sliding block procedure. J Geotech Geoenviron Eng ASCE 129(8):673–684

Wu X (2012) Performance-based design of geosynthetic waste containment systems. MS thesis, Arizona State University, Tempe (in preparation)

Yazdani R, Campbell JL, Koerner GR, (1995) Long-term in situ strain measurements of a high density polyethylene geomembrane in a municipal solid waste landfill. In: Proceedings of geosynthetics'95, Nashville, TN, pp 893–906

Yegian MK, Kadakal U (2004) Foundation isolation for seismic protection using a smooth synthetic liner. J Geotech Geoenviron Eng 130(11):1121–1130

Subject Index

M. Maugeri and C. Soccodato (eds.), *Earthquake Geotechnical Engineering Design*, Geotechnical, Geological and Earthquake Engineering 28, DOI 10.1007/978-3-319-03182-8, © Springer International Publishing Switzerland 2014

MIX
Papier aus verantwortungsvollen Quellen
Paper from responsible sources
FSC® C105338

If you have any concerns about our products,
you can contact us on
ProductSafety@springernature.com

In case Publisher is established outside the EU,
the EU authorized representative is:
Springer Nature Customer Service Center GmbH
Europaplatz 3, 69115 Heidelberg, Germany

Printed by Libri Plureos GmbH
in Hamburg, Germany